Robotic Fabrication in Architecture, Art and Design 2024

Maria Yablonina · Paul H. Harrison ·
Brady Peters · Jonathon Anderson ·
Nicholas Hoban · Zachary Mollica
Editors

Robotic Fabrication in Architecture, Art and Design 2024

Foreword by Sigrid Brell-Cokcan and Johannes Braumann, Association for Robotics in Architecture

Springer

Editors
Maria Yablonina
John H. Daniels Faculty of Architecture, Landscape, and Design
University of Toronto
Toronto, ON, Canada

Brady Peters
John H. Daniels Faculty of Architecture, Landscape, and Design
University of Toronto
Toronto, ON, Canada

Nicholas Hoban
John H. Daniels Faculty of Architecture, Landscape, and Design
University of Toronto
Toronto, ON, Canada

Paul H. Harrison
John H. Daniels Faculty of Architecture, Landscape, and Design
University of Toronto
Toronto, ON, Canada

Jonathon Anderson
The Creative School
Toronto Metropolitan University
Toronto, ON, Canada

Zachary Mollica
John H. Daniels Faculty of Architecture, Landscape, and Design
University of Toronto
Toronto, ON, Canada

ISBN 978-3-032-22606-8 ISBN 978-3-032-22607-5 (eBook)
https://doi.org/10.1007/978-3-032-22607-5

This work was supported by Sigrid Brell-Cokcan, Johannes Braumann.

This Springer imprint is published by the registered company Springer Nature Switzerland AG
The registered company address is: Gewerbestrasse 11, 6330 Cham, Switzerland

Foreword: ROB|ARCH 2024

The Association for Robots in Architecture was founded in 2010 with the goal of making robots accessible to the creative industries, sharing ideas, and creating a global platform for robot enthusiasts from academia and industry to meet in a biannual ROB|ARCH conference series.

It has been six years since the iconic ROB|ARCH 2018 conference, hosted by ETH Zurich in Switzerland. For ROB|ARCH 2020 - for the first time ever - we wanted to bring the conference series to Asia. In collaboration with Tsinghua University, the Beijing Biennale, and many Chinese academic partners, we were looking forward to a great setup for the global community to meet in this endeavor. However, the global pandemic prevented the conference from taking place - first in China, then around the world.

An initial idea to move the conference into the digital world and host an online event was carefully evaluated, however we concluded that the core spirit of the ROB|ARCH community is to meet, share and work with real robots and in-person. The conference, especially the workshops, is what brings everyone together.

In the meantime, the Springer Journal of Construction Robotics remained a vital online platform for publishing top-tier research, with most of it readily available to the community via open access. Members of the Association for Robots in Architecture have unrestricted access to all publications.

In August 2022, our goal was to reinitialize the whole conference series, without knowing where the ROB|ARCH community has moved to in the last 4 years, where new talents and professorships have developed and new robotics workshops have been initialized. In order not to lose new talents and to include all future and well-established members, we decided to open a public call for conference hosts and received seven applications of exceptional quality from academic and commercial institutions that often collaborate. We were pleased to see applications from multiple geographic regions, including North America, Europe, Asia-Pacific, and the Middle East, demonstrating the global interest in hosting such an event, even after several years of dormancy.

The selection process involved several rounds of meetings and an on-site visit to two final institutions. Toronto was chosen for ROB|ARCH 2024 because of the great partnership between the Metropolitan University of Toronto and the University of Toronto and their many local partners.

As always, the decision was based on careful evaluation and mutual trust, especially for a conference such as ROB|ARCH, where our local hosts are committed to opening their doors for collaboration and providing access to their facilities for international institutions. Organizing such a large event would not be possible without the local help of our long-term sponsors to bring in additional equipment and materials. We are therefore very grateful for the long-standing trust of KUKA, ABB, Autodesk, and our other partners.

An event of this magnitude can only be realized through a collaborative effort, a shared vision, and an intrinsic motivation to create an exciting format that brings together the world's leading researchers and practitioners to discuss the future of robotic fabrication in architecture, art, and design.

We could not be more pleased with our collaboration with the multifaceted Toronto team, led by Jonathan Anderson for all the workshops, and Maria Yablonina, Paul Howard Harrison, and Brady Peters for this scientific publication.

Bringing back a conference series after such a long hiatus required considerable effort and dedication from all involved. We are deeply grateful to the editorial team, the entire large team of local organizers, the authors, and everyone who attended the conference.

We are thrilled to start organizing the next conference in Denmark in 2026!

Sigrid Brell-Cokcan
Johannes Braumann
Co-founder, Association for Robots in Architectures

The original version of the book has been revised. A correction to this book can be found at https://doi.org/10.1007/978-3-032-22607-5_25

Beyond Optimization

The growing presence of robotics in art, research, design, and construction has changed the way we practice and think through our tools. Since 2012, when the first ROB|ARCH conference was held, the role of robotics in creative practices and manufacturing industries has grown, shifted, and evolved. Robots are no longer a shiny novelty restricted to niche research proposals but are now a fundamental technical tool for interfacing between digital and physical worlds.

Even our understanding of what a robot can look like has evolved: the interdisciplinary discourse that has emerged around digital fabrication machines and industrial arms now includes myriad robot morphologies, scales, and applications—think aerial machines for construction, choreographed robotic swarms in performance, millimetre-scale robots for medical applications. In some ways, we have witnessed our machines grow up along with the discipline; no longer restricted to the structured environments of research labs, they have entered the real world and taken their place within it.

Acknowledging the momentum of these shifts within our broader social, political, and environmental contexts, ROB|ARCH2024 urged its community to reflect on this moment with a critical lens.

The ROB|ARCH2024 conference was presented by two host institutions: the **Daniels Faculty of Architecture, Landscape, and Design at the University of Toronto**, and **the Creative School at Toronto Metropolitan University**. The conference program included 8 workshops and 25 paper presentations. We were honoured to include keynote lectures by **Brian Ringley**, **Sougwen Chung**, and **Yusuke Obuchi** in our program schedule.

The conference theme, *Beyond Optimization*, was intended as a provocation: to reflect on the changes affecting the field of robotics in art, design, and architecture—and to respond, by shifting priorities and examining the criteria by which we evaluate our research. We aimed towards inclusion of a broad range of actors, voices, disciplines, and contexts. *Beyond Optimization*, then, implied a plurality of approaches and corresponding questions.

Any practice involving robotics engages multiple realms of knowledge and discourse, requiring an intuitive ability to manage interdependent parameters. As these knowledge-making practices are further encoded and augmented by other intelligences, ROB|ARCH2024 reflected on interdependence: both between the parameters that define our work and in the new relationships we must build.

One way to manage complex sets of interdependent parameters is top-down optimization: a range of computational methods for finding best-fit solutions within a mathematical design space we might call 'parametric'. As a metaphor, however, it implies that computational models can accurately represent the messy, interdependent systems that make up our physical world. Furthermore, it encourages a workflow wherein we model the physical to represent it digitally and then use this imperfect model to interact with the physical world through machine bodies. Staggered layers of abstraction—physical to digital and back—leave many of the intricacies of materiality, process, and politics lost in translation. Our optimizations can only be as accurate as the mathematical models that underpin them, so a fundamental problem of *resolution* both limits optimization's applicability and demands an alternative. As complex real-world environments require complex solutions, we must move beyond sterile parametric models and embrace the messy, stochastic reality of the incomputable.

Research Contributions

Papers in this volume are organised in six sections, each demonstrating a unique set of research contributions that respond to the conference topic through a shared lens of methodological, material, and technical approaches and tools.

Collaborations present four approaches to reconsidering robotic fabrication in architecture as a collaboration between multiple actors: physical machines, computational systems, human designers, and operators. Contributions in this section demonstrate a broad range of approaches, including novel developments in software, hardware and workflow methods, as well as conceptual reframing of machinic and human labour in design and fabrication processes.

Cyber-Physical Explorations of Collective Robotic Construction using Agent-Based Modelling details a novel bottom-up method for inter robotic communication, while Planning Non-repetitive Robotic Assembly Processes with Task and Motion Planning and A Loosely Coupled Observation Processing and Data Exchange System for Complex Teams of On-Site Construction Robots both outline comprehensive strategies for adapting robotic processes to real-world environments. Spontaneous Tensegrity: Exploring Improvisational Design and Robotic Fabrication in Tensegrity Structures extends the metaphor further, introducing a human-in-the-loop component to a complex structural balancing act.

Alternative Materials present four papers focusing on the development of novel robotic fabrication processes with materials, material systems, and fabrication methods that are currently underexplored in architecture and design. Contributions in this section include explorations of new forms that arise from natural materials in architecture, investigations of aesthetic expressions of unusual material combinations, and reframing of established materials for entirely new purposes and applications.

Additive Manufacturing with Graded Bio-Polymer Composites presents a novel method for robotic 3D printing with collagen-based materials while optimizing for part structure. Hollow Characters: Robotically Programmed Matter for Kinetic Behaviour and Guided Self-Assembly introduces a dynamic fabrication method wherein the robot repositions the mold to afford fabrication of complex behaving cast objects. Conformal Robotic 3D Printing and Post-fabrication Analysis of Mycelium-based Composites presents a method for utilizing mycelium as 3D printing material.

Other Robot Bodies present four research contributions exploring robotic processes beyond the use of existing industrial equipment. The papers propose new robot systems designed and developed for architectural fabrication and spatial adaptation. Each paper demonstrates novel robotic hardware, software, and design systems, promising new opportunities in architecture and construction where industrial machines are not sufficient.

Authors of Synergising Natural Material Complexity and Low Impact Automation propose a custom robotic system for winding of architectural components out of natural fibre materials. Differential Boundary Deposition: Mobile Robotic Construction with Amorphous Materials demonstrates an entirely new robot for 3D printing vertical structures by a mobile robot that climbs the structure as it prints, perpetually extending its own work envelope. A Suspended Cable-Driven Parallel Robot (CDPR) and Reorientation Mechanism for the Semi-Automated Positioning of Prefab Assemblies asks if lighter and more accessible cable robots are an overlooked actor in construction site automation. Authors of CHANDELIER'1244: Towards Cyber-Physical Experimentation With Structural Adaptation In Architecture demonstrate a workflow for adaptive building structures that applies to objects of variable scale: from a multistorey building to a light installation.

Out of Plane presents four contributions that develop novel additive fabrication processes that leverage complex robotic motion for spatial material deposition. The papers in this section explore 6-axis 3D printing, soft formwork for material extrusion, and novel slicing algorithms that afford material efficiency and new geometric possibilities for load bearing structures.

Non-planar 3D Printing of Double Shells introduces a novel slicing method for material-efficient and structurally superior 3D printing processes. Polymers, Fabric and Steel: A Hybrid Robotic Fabrication System for Bespoke Doubly-Curved Structural Concrete demonstrates a robotic method for producing steel and fabric formwork for cast concrete elements. 3D Printing Pattern Effects on SCF-PLA Extrusion on Pre-Stretched Textile Formwork explores robotic 3D printing on textile substrates, adding unique properties and variability to the elements produced. Design for Additive Construction: Uncovering Challenges and Opportunities through the Robotic 3D Printing of a Pedestrian Bridge exhibits the immense potential of new slicing algorithms for non-planar 3D printing with robot arms.

As-Built/As-Found presents five papers that focus on robotic processes for using existing materials and working in existing spaces. These contributions demonstrate novel methods that leverage robotics for fabrication with construction waste and reclaimed materials, informed disassembly for future material reuse, and new construction systems using as-found natural materials.

Rising From Rubble: Leveraging Existing Construction Tools for Upcycling Concrete Waste into Slender Walls illustrates a robotic fabrication process that utilizes discarded concrete rubble as construction material with minimal alteration and added material. *Mixed Reality Carpentry* leverages human sensing and agility to introduce new ways for leveraging non-standard wood elements in design and construction, while *Feasibility Check of a Workflow with Industrial Robots and Laser-Tracking for Timber Machining Across the Workspace* achieves a similar outcome through the integration of networked sensors. *Mobile Robotic Brickwork Disassembly for Material Reuse* proposes a new way to deconstruct existing building stock with intelligent machines to increase the amount of salvaged material for reuse. Finally, *Lost and Bound: Adaptive Detailing with Robotic*

Additive Joining for Reclaimed Steel uses complex sensing and machine learning to breathe life into discarded steel elements with new robotic assembly methods.

Models All the Way Down presents four papers that explore new possibilities afforded by Machine Learning methods for robotic fabrication. These research contributions offer novel and exciting approaches to ML that go beyond generating images with prompts, exploring fabrication processes and approaches that learn from material. Authors of *Reshaping Wood Waste: Creative Pathways to Enhance Resource Efficiency* demonstrate the use of machine learning strategies as tools for listening to the material's grain and its properties for fabrication. *Artifex Automata: Predicting Robotic Chiseling Strokes* offers a new method for combining human craft and machine precision through machine learning from material and craftsperson. *Toolpath Design Calibration for Robotic Spatial Printing with Supervised Learning* demonstrates the potential of machine learning approaches for unpredictable materials like clay, while *Towards Adaptive Additive Manufacturing with Semi-autonomous Robotic Binder Jet 3D Printing of Concrete* outlines a similar predictive strategy for binder-jet printing on arbitrary surfaces.

Overall, these research contributions suggest a compelling alternative (or antagonist) to top-down optimization methods. By discretizing, sensing, and distributing—by echoing the real world's messy complexity through a corresponding embrace of bottom-up understanding—we might overcome the un-computability of the real world.

Acknowledgements

As the Scientific Committee, we would like to express our gratitude to the Association for Robots in Architecture, Sigrid Brell-Çokcan and Johannes Braumann, for entrusting us with the development of ROB|ARCH2024. We would also like to thank our colleagues on the conference team, namely Zachary Mollica, Nicholas Hoban, and Jonathon Anderson, for their incredible dedication to the conference and for eight exceptional workshops.

We would like to extend our gratitude to the amazing support teams at both campuses, without whom none of the conference's events would have been possible. We want to thank the Daniels Applied Technology team for their support throughout the conference days: Alexander Gaskin, Amy Catherine George, Eugene Wang, Gabriel Li, Joseph Rogal, Lily Jeon, Paul Kozak, and Rahul Sehijpaul. We would also like to extend our gratitude to Sarah Whitehouse and Yuri Lomakin for their support throughout the conference. We want to thank the Creative School Workshop team for their support and assistance: Alda Escareno, Arnel Espanola, Adrian Kenny, Alex Verni, and Marcella Driver-Moliner. We want to thank the team of students who supported conference activities at Daniels: Ayaulym Yessengeldina, Cameron Manore, Caroline Guirguis, Huzaifa Chughtai, Kenny Vo, Liam Cassano, Mateusz Grabowski, Nicole Quesnelle, Olivia Carson, Rouqueya Romand, Sandy Nguyen, Sophia de Uria, Sean Lamb, Renee Powell Hines, and Brian Slocum.

Our sincere appreciation goes out to the Peer Review Committee, without whom this publication and the conference would not have been possible. We would also like to thank our paper session chairs and moderators: Kathrin Dörfler, Madeline Gannon, Kathy Velikov, Zachary Mollica, Shelby Doyle, and Daniela Mitterberger.

ROB|ARCH2024 would not have been possible without the generous support of our hosting institutions: Daniels Faculty of Architecture at the University of Toronto and the Creative School at Toronto Metropolitan University. It was further made possible by the generous support from our sponsors: Kuka, Mosaic, Trotec, Hoem, the University of Toronto Robotics Institute, RoboticPlus.AI, Springer, Giraffe, Haliburton Forest, Sparkbird, and WZMH. We would also like to extend our gratitude to the workshop sponsors: ABB, Autodesk, Shelley Automation, Schunk, Architectural Association School of Architecture, AtonAton, Carleton University School of Architecture and Urbanism, the Center for Augmented Computational Design in Architecture, Engineering and Construction at ETH Zurich, Chrysalis at the Creative School (TMU), the Institute of Technology in Architecture at ETH Zurich, the Institute for Computational Design and Construction at the University of Stuttgart, Iowa State University College of Design, Princeton University, Design+Technology Lab at the Creative School (TMU), the Department of Architecture at the TUM School of Engineering and Design (Technical University of Munich), Perkins&Will, and Zahner.

ROB|ARCH 2024: Beyond Optimization, Scientific Committee

Maria Yablonina
Paul H. Harrison
Brady Peters

Contents

Collaborations

Cyber-Physical Explorations of Collective Robotic Construction using Agent-Based Modelling ... 3
Samuel Leder, Lasath Siriwardena, Philipp Kragl, and Achim Menges

Planning Non-repetitive Robotic Assembly Processes with Task and Motion Planning (TAMP) ... 19
Pok Yin Victor Leung, Yijiang Huang, Caelan Garret, Fabio Gramazio, and Matthias Kohler

Loosely Coupled Observation Processing and Data Exchange System for Complex Teams of On-Site Construction Robots ... 34
Lukas Kirner, Marit Zöcklein, Jyrki Oraskari, Carsten Kamp, Fabian Claßen, and Sigrid Brell-Cokcan

Spontaneous Tensegrity: Exploring Improvisational Design and Robotic Fabrication in Tensegrity Structures ... 47
Isla Xi Han and Stefana Parascho

Alternative Materials

Additive Manufacturing with Graded Bio-Polymer Composites ... 65
Nicholas Paul, Eppinger Carl, Chiujdea Ruxandra, Sonne Konrad, and Ramsgaard Thomsen Mette

Conformal Robotic 3D Printing and Post-Fabrication Analysis of Mycelium-Based Composites ... 81
Özgüç Bertuğ Çapunaman, Alale Mohseni, Alireza Zamani, Natalie Walter, and Benay Gürsoy

Hollow-Characters: Robotically Programmed Matter for Kinetic Behavior and Guided Self-Assembly ... 98
Samim Mehdizadeh and Oliver Tessmann

Other Robot Bodies

Synergising Natural Material Complexity and Low-Impact Automation ... 115
Eszter Olah, Mehrdad Zareian, and Moritz Dörstelmann

Differential Boundary Deposition: Mobile Robotic Construction with Amorphous Materials ... 129
Matthew Johnson, Philipp Kragl, Michael Tucker, Nicolas Kubail Kalousdian, Samuel Leder, and Achim Menges

A Suspended Cable-Driven Parallel Robot (CDPR) and Reorientation Mechanism for the Semi-Automated Positioning of Prefab Assemblies ... 142
Steven Beites, Ethan McDonald, and Marc Arsenault

Out of Plane

CHANDELIER'1244: Towards Cyber-Physical Experimentation with Structural Adaptation in Architecture ... 161
Mathias Maierhofer and Achim Menges

Non-Planar 3D Printing of Double Shells ... 174
Ioanna Mitropoulou, Amir Vaxman, Olga Diamanti, and Benjamin Dillenburger

Polymers, Fabric and Steel: A Hybrid Robotic Fabrication System for Bespoke Doubly-Curved Structural Concrete ... 192
Jon Krähling Engholt and Dave Pigram

3D Printing Pattern Effects on SCF-PLA Extrusion on Pre-stretched Textile Formwork ... 210
Ehsan Baharlou

Design for Additive Construction: Uncovering Challenges and Opportunities through the Robotic 3D Printing of a Pedestrian Bridge ... 231
Peter Storey, Nathan King, James Donnelly, and Shadi Al-Ariss

As-Built/As-Found

Rising From Rubble Leveraging Existing Construction Tools for Upcycling Concrete Waste into Slender Walls ... 253
Maxence Grangeot, Qianqing Wang, Katrin Beyer, Corentin Fivet, and Stefana Parascho

Feasibility Check of a Workflow with Industrial Robots and Laser-Tracking for Timber Machining Across the Workspace ... 266
Marc Pantscharowitsch and Benjamin Kromoser

Mixed Reality Carpentry ... 285
Gwyllim Jahn, Cameron Newnham, and Nick van den Berg

Mobile Robotic Brickwork Disassembly for Material Reuse 299
Begüm Saral, Hanzhi Chen, Stefan Leutenegger, and Kathrin Dörfler

Lost and Bound: Adaptive Detailing with Robotic Additive Joining for Reclaimed Steel 313
Inés Ariza, Romana Rust, Vlad-Alexandru Silvestru, Andreas Taras, Fabio Gramazio, Matthias Kohler, and Catherine De Wolf

Models All the Way Down

Reshaping Wood Waste: Creative Pathways to Enhance Resource Efficiency ... 329
Maria-Eleni Papandreou, Vincent Huyghe, Joseph Naguib, Lucas Gottschild, Pit Siebenaler, Bora Çobanoğlu, Alexandre Dubor, and Aldo Sollazzo

Artifex Automata Predicting Robotic Chiseling Strokes 343
Sandro Siefert, Nadja Gaudillière-Jami, Tom Svilans, and Paul Nicholas

Towards Adaptive Additive Manufacturing with Semi-autonomous Robotic Binder Jet 3D Printing of Concrete 354
Linxiaoyi Wan, Jingyang Liu, Dana Cupkova, and Joshua Bard

Toolpath Design Calibration for Robotic Spatial Printing with Supervised Learning 371
Sulaiman Alothman, Hanspeter Pfister, and Martin Bechthold

Correction to: Robotic Fabrication in Architecture, Art and Design 2024 ...... C1
Maria Yablonina, Paul H. Harrison, Brady Peters, Jonathon Anderson, Nicholas Hoban, and Zachary Mollica

Author Index 389

Collaborations

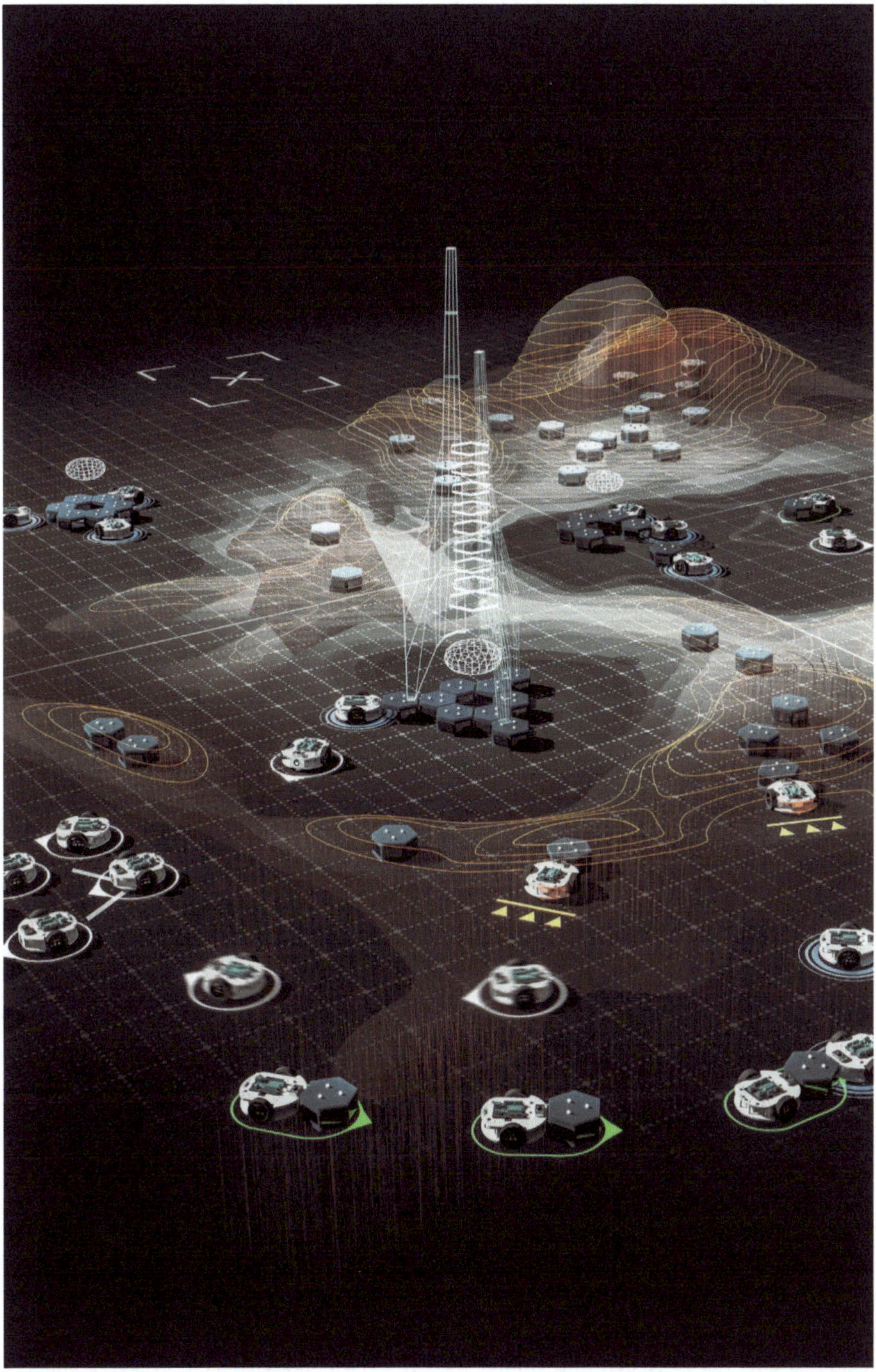

Cyber-Physical Explorations of Collective Robotic Construction using Agent-Based Modelling

Samuel Leder[1,2(✉)], Lasath Siriwardena[1,2], Philipp Kragl[1], and Achim Menges[1,2]

[1] Institute for Computational Design and Construction (ICD), University of Stuttgart, 70174 Stuttgart, Germany
samuel.leder@icd.uni-stuttgart.de
[2] Cluster of Excellence Integrative Computational Design and Construction for Architecture (IntCDC), University of Stuttgart, 70174 Stuttgart, Germany

Abstract. Collective robotic construction (CRC) is an emerging subset of on-site construction robotics, which, despite its interdisciplinary nature, remains predominantly centered within the field of engineering. The goal of this research is to create a highly accessible workflow for developing and working with CRC systems that allows for their cyber-physical control. To demonstrate this concept, a low complexity robotic system was created in conjunction with an agent-based model (ABM) and associated software framework, enabling direct design and control of a swarm of mobile robots. The proposed system consists of a team of homogenous mobile robots that can actively rearrange a set of identical custom digital materials. The developed robotic system was conceptually extended, and the provided workflow was experimentally tested in a workshop setting with architectural students. The results of the workshop, which included four different conceptualizations of the provided robotic system, investigated various research topics within CRC, demonstrating the ability of architects to aid further conceptualization of research in the field.

Keywords: Collective Robotic Construction(CRC) · Agent-based Modelling and Simulation (ABMS) · Behavioral Fabrication · Swarm robotics · Construction Automation

1 Introduction

Collective robotic construction (CRC) is a growing subdomain of on-site construction robotics. It builds upon principles of multi-robotic systems, focusing on the development of bespoke machines for construction of structures that are much larger in scale than the machines themselves (Petersen et al. 2019). Originating from space applications in the 1980s and 1990s (Bock and Linner 2016), CRC research has evolved to explore construction on Earth, employing diverse machines such as flying vehicles (Braithwaite

M. Yablonina et al. (Eds.): Robotic Fabrication in Architecture, Art and Design 2024, pp. 4–18, 2026.
https://doi.org/10.1007/978-3-032-22607-5_1

et al. 2017) and climbing robots (Leder et al. 2022). These machines can undertake various tasks, including superstructure assembly (Nigl et al. 2013), construction monitoring (Zhang et al. 2022), and building anchoring (Melenbrink et al. 2020), utilizing different material systems like discrete bricks (Petersen et al. 2012) and continuous fibers (Yablonina, Kalousdian and Menges 2020). Reviews on CRC systems can be found in works from Petersen et al. (2019), Ardiny et al. (2015), and Gerling and Von Mammen (2016).

Most CRC research is conducted from an engineering perspective, resulting in point solutions primarily within the engineering domain. This emphasis has led to the development of advanced decentralized control algorithms and modular hardware architectures for assembling pre-defined structures. As such, the question of architectural design, which is one of the major themes of CRC together with assembly algorithms, robotic systems and functional materials, is often left out (Leder and Menges 2023).

Despite this trend, a few projects have addressed the architectural design aspect within CRC. Heinrich et al. (2018) developed a digital setup within native architectural software for architects to design behaviors for self-organizing robotic construction processes. The implementation is limited in that it focuses mainly on the geometric placement of building elements and does not include robotic constraints such as the task and motion planning required to assemble the structure.

Two other research projects present entire workflows for the full deployment of CRC systems. Fiberbots employs a swarm robotic platform for independent yet interconnected fiber composite tube assembly, incorporating a flocking-based model for architectural design (Kayser et al. 2018). Ntzoufras et al. present a platform that unifies scanning, data analysis, G-code generation, and construction processes in a single workflow for CRC system deployment (Ntzoufras et al. 2020). In both cases, the architectural design is entirely separate from the assembly process, creating a distinct division between construction phases where design concludes once construction begins. The goal of presented research is to establish a workflow that will enable architects to explore bottom-up design and cyber-physically oversee structures' assembly by mobile machines, fostering more design interaction in CRC.

To do this, a CRC system was developed with a corresponding cyber-physical design workflow. The physical system is comprised of a two-wheeled robot that can actively rearrange custom digital material (Fig. 1). The cyber-physical workflow is centered around an agent-based model (ABM) that directly represents the components of the physical system. The workflow can be run (i) fully in simulation allowing for the development of new behaviors for the mobile robots and (ii) while the mobile robots are running which allows for direct control of the assembly process leading to varying design results.

The system was conceptually extended and experimentally tested in the context of a workshop with students from predominantly architectural training. During the workshops, four distinct concepts for the contextualization of the system were developed, along with corresponding digital behaviors for the robots. These concepts were then deployed on the physical robotic system. Each of the four conceptualizations discussed varying important aspects in the development of CRC systems, highlighting the importance of the incorporation of architects in the future development of CRC.

Fig. 1. Physical CRC system composed of 2-wheeled mobile robots and hexagonal digital material

2 Methods

The aim of this research is to introduce an accessible workflow for developing CRC systems that allows for their cyber-physical control. To showcase this, a simple, mobile robotic system was developed for easy comprehension and extension by non-specialists. Accompanying the physical system is an ABM and software architecture enabling direct design and control of mobile robots.

2.1 Robotic System

The proposed robotic system comprises a team of homogenous mobile robots and a set of identical custom digital material units (Fig. 2). The mobile robots can pick up and move the digital materials in space to assemble them into geometric configurations.

The design of the mobile robot is an adjustment of the one proposed in (Yablonina et al. 2019). It is a two wheeled mobile robot that is equipped with a LiPo battery and electromagnet. The wheels are powered by DYNAMIXEL motors, XL430-W250-T, which allow for the direct control of their velocity over time. An Arduino Uno Wi-Fi Rev2 is the microcontroller of the robot, as it is a low-cost and easy to use platform for developing hardware that is open for extension by others. A DYNAMIXEL shield was used to control the motors in conjunction with the Arduino.

Fig. 2. Two of the main components of the physical robotic system: two wheeled mobile robot and digital material.

The digital material is a hexagonal block that is held together by a ring of screws. The screws are evenly spaced to allow the electromagnet on the robot to pick up and release the block. The front of robot body was designed to aid in alignment between the material and robot.

On the top of both the mobile robot and the digital material are unique patterns of retroreflective spherical makers, which allow them to be tracked using an external vision system. This is important for the eventual control of the system using the developed cyber-physical workflow. The entire robotic system is, thus, composed of mobile robots, digital material and an external motion capture system. Camera setups from NOKOV[1] and OptiTrack[2] were both tested.

2.2 Agent-based Model (ABM)

For the modelling and control of the mobile robots, an ABM was created. Agent-based Modelling and Simulation (ABMS) is a computational method that aims at simulating complex systems from the perspective of autonomous agents. Each agent has behaviors that define how it interacts with other agents as well as their environment. Through the study of the interactions of the agents, the aim of ABMS is to gain insights on patterns exhibited by the system as a whole. As many aspects in architecture and construction can be understood as complex systems, ABMS has grown in popularity for research in these field (Stieler et al. 2022). The method has also been increasingly utilized in CRC due to the direct correlation that can be drawn between agents acting and reacting in simulation models within the digital world, and mobile robots assembling structures in the real world (Leder and Menges 2023).

[1] NOKOV Mocap: NOKOV Motion Capture System, https://en.nokov.com
[2] NaturalPoint Inc.: OptiTrack, https://optitrack.com/

The formalization of the ABM in this research was done through the creation of an extension to the agent-based modelling framework, ABxM (Nguyen 2022). ABxM is an open-source object-oriented framework that can be readily extended upon for application specific developments. ABxM allows for additional classes to be readily developed within Grasshopper, the visual programming environment for Rhinoceros 3D, or as add-ons to the existing plug in, allowing for generation and customization of novel agent systems. Parametric inputs to developed models enable architects to interactively modify them and in turn have a direct impact on the outcomes they generate.

The source code and a Unified Modeling Language (UML) diagram for the newly derived classes of the ABxM implementation is published open access (Leder, Siriwardena and Menges 2024). The source code includes (i) code for the customization of agent and behavior base classes of the ABxM core libraries, (ii) code for the creation of Grasshopper components for those classes as well as (iii) code for additional components, which were necessary for the implementation of the model in the cyber-physical workflow described in the next section. The following gives description to the main constructs of the model created.

Agent Systems. The development of agent systems generally requires the definition of three constructs: agents, their behaviors, and their environment. The model is designed to simulate the physical robotic system and to function as a mechanism for its cyber-physical control, necessitating two agent systems to interact. One agent system models the collection of mobile robots and another models the set of digital material. The former is modelled as a force-based system, specifically a Boid system in the ABxM framework, as the eventual control of the mobile robots is done by changing their velocity. The latter is a position-based system, or a Cartesian system in the ABxM framework, as the velocity of the digital material as it moves is not important to their modelling.

Environment. The environment in which the agents interact is a two-dimensional plane, which represents the area in which the robots can drive. In the ABM that was developed, the objective was defined to move all the digital materials to one of the colors stored in the mesh environment. Therefore, beyond the environment holding geometric information, it also holds additional attributes that influence how the agents act.

Agent Behaviors. The two agent types that correspond to the respective agent systems have different behaviors. Diagrammatic representations of the behaviors are in Fig. 3. The mobile robot agents have 7 behaviors: Separation, Containment, Wander, Pick Up/Push Material, Slow Down, Drop Material, and Update. The first three behaviors are derived from Craig Reynold's canonical boid model (Reynolds 1987; Reynolds 1999), while the others are specific to the robotic system. A brief description of the behaviors of the robots is as follows:

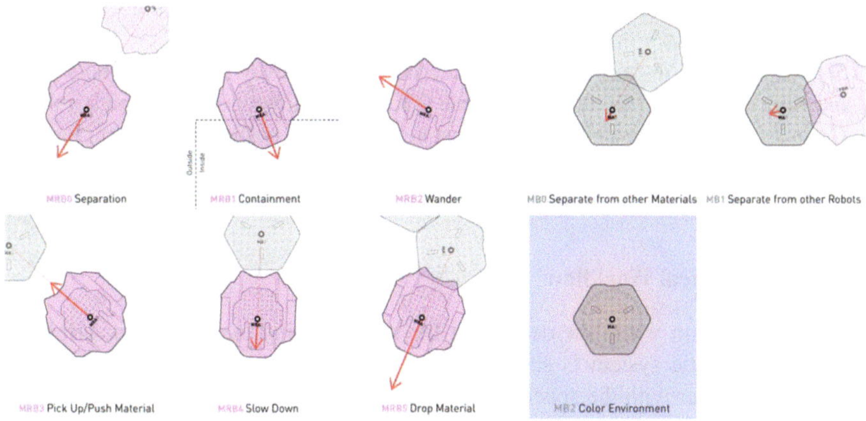

Fig. 3. Mobile robot behaviors (MRB0-MRB5) for mobile robot agents (MRA) and material behaviors (MB0-MB2) for material agents (MA). The update behavior is not displayed as it overwrites the position of the agent in the digital space and has no effect on the vector negotiation of the agent or effect on the environment.

- *Separation* steers the robots to avoid collision between each other.
- *Containment directs the robots to stay within a rectangular boundary.*
- *Wander* steers the robots randomly, adding some noise into their movement.
- *Pick Up/Push Material* tells the robot to pick up a digital material when it is close enough to it. When the model is run solely in simulation, the behavior also advances the digital material which is held by the robot in the same direction that it is traveling. This behavior triggers the electromagnet on the robot to turn on and stay on, when the physical system is being controlled by the model.
- *Slow Down* makes the robot decrease its velocity right before it picks up a digital material to aid in their alignment.
- *Drop Material* tells the robot to release the material it is holding and drive in the opposite direction. The condition for dropping the material is based on the color the robot reads from the environment based on its current position. This behavior triggers the electromagnet on the robot to turn off, when the physical system is being controlled by the model.
- *Update* is the behavior that updates the position of the digital agents to the real mobile robots using inputs from the external motion capture system. If the model is used in simulation only, the update behavior also adjusts the orientation of the robot relative to how it moves in the real world.

The digital material agents have a different set of 4 behaviors: Separation from other Materials, Separation from Robots, Color Environment and Update. The first two behaviors are used in simulation to model the physics of the material in the digital world. They are defined as follows:

- *Separation from other Materials* maintains that the materials do not overlap each other in the digital world.
- *Separation from Robots* forces the material to move away from the robots if they intersect at orientations other than those that would allow the robot to pick it up.
- *Color Environment* is a behavior that changes the color of the environment based on the position of the digital material.
- *Update* changes the position of the material based on its location in the real world.

2.3 Cyber-Physical Workflow

The communication workflow developed for implementing the described ABM with the physical robotic system is shown in Fig. 4. The entire workflow was developed within Grasshopper with the specific aim of embedding the development in a software platform utilized and understood within the architectural discipline. While the model possesses the capability to operate independently within Grasshopper and can be visualized in Rhinoceros to assess its functionality, communication between the ABM and components of the physical robotic system is essential during cyber-physical operation. Communication between the ABM and the mobile robots is facilitated through the MQTT protocol. Concurrently, interaction between the ABM and the external motion capture software is conducted utilizing the User Datagram Protocol (UDP).

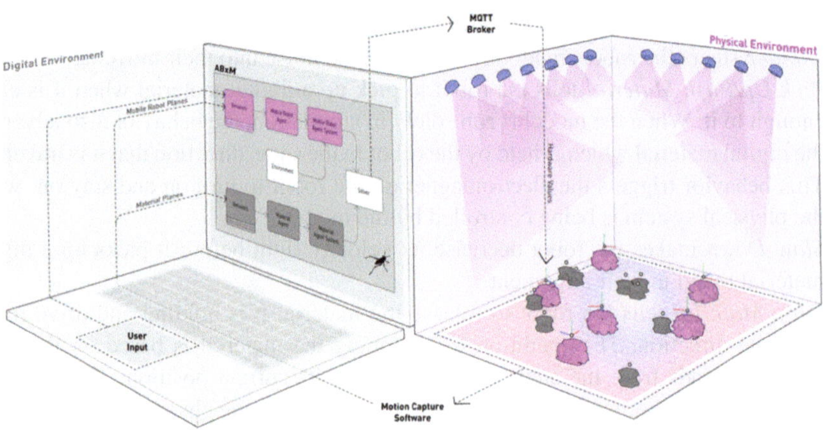

Fig. 4. Cyber-physical workflow diagram

The robots themselves have limited intelligence. The code on the robots, which can be found open source with the 3D models of the robots and digital material (Leder Kragl and Menges 2024), only allows for MQTT communication, and actuation of the motors and electromagnet onboard the robot. All the system decision making is happening in the ABM, which runs at the same time as the physical system. Position and orientation data collected by the motion capture system triggers the ABM to update. The ABM updates by synchronously running through the behaviors of the digital agents to calculate new actions for the robots based on that data collected on the robots and digital material.

These actions then are simultaneously translated to the robots through values for motor and electromagnet actuation. The system runs autonomously and iteratively based on the continuous updating of the ABM, which is tuned at a rate to match the velocity of the robots in the real world. Interaction with the ABM through the parametric inputs to the behaviors allows for adjustment to the physical actions of the robots in the real world or digital agents when the model is ran in simulation.

3 Workshop Results

The presented robotic system and cyber-physical workflow were tested in the context of the Architectural DigitalFUTURES Workshop series at the College of Architecture and Urban Planning (CAUP), Tongji University in summer 2023. Students were provided with the described methods and asked to conceptualize how the system could be further extended or applied to different use cases in architecture and construction. After an initial brainstorming, four concepts, entitled Collaborative Carrying, Resource Rush, LightHall and Skyline Scrubber, were selected for further development during the weeklong workshop. This entailed advancing and adjusting the provided methods to meet the needs of each system.

To test the four concepts, ten mobile robots were assembled with a large set of digital materials. The area in which the robots were able to move without interruption to the external motion capture system was roughly 2.5 m by 3.25 m. This area was augmented with a projection, which allowed for information from the ABM to be overlaid with the physical system. The projection aided in the debugging of the system and further allowed for architects interacting with the system or external viewers examining the system to have a better understanding of the processes that are otherwise not visible.

Description of the four concepts and their relationship to major concepts within research on CRC are described in the following subsections. The main objective of the workshop was to contribute to the development of new agent behaviors (Fig. 5). This was achieved by augmenting the base class for behaviors in ABxM using C# or Python. During the workshop, evaluation of the behaviors was conducted through visual testing in simulation on personal computers and then implementing them in the cyber-physical workflow.

3.1 Collaborative Carrying

Collaborative Carrying built upon principles of natural resource gathering behavior, a prominent topic within research in CRC. The agents or mobile robots take on the role of foragers, collecting digital material and relocating it to their home. The other concept inspired by nature was that of stigmergy. Stigmery is a form of coordination that involves leaving traces in the environment for the coordination of social insects. This principle, which has been used in other research on CRC (Petersen et al. 2012), was adapted in this formulation of the system. The robots left digital traces in the environment, which inform other robots where they have already traveled (Fig. 6).

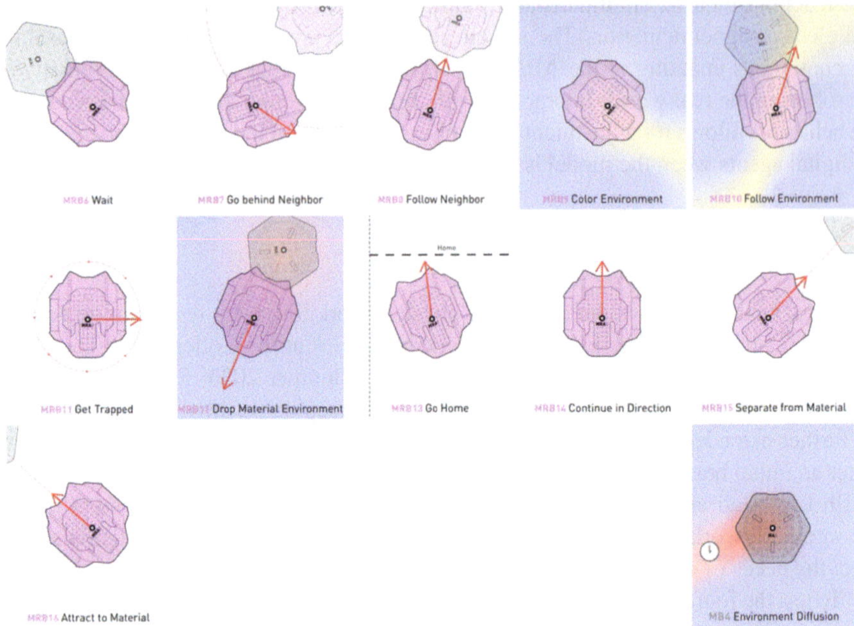

Fig. 5. Mobile robot behaviors (MRB6-MRB17) and material behaviors (MB4) developed in the workshop. Each behavior was tested in both the digital environment and using the cyber-physical workflow. The behaviors developed for each concept are as follows: Collaborative Carrying MRB6-MRB10, Resource Rush MRB11-MRB13, MRB14–16 and Skyline Scrubber MB4.

Fig. 6. Image sequence of two mobile robots forming a "chain" to combine strength to push the digital material to their home, marked in light purple in the upper right. Dark purple indicates the areas which all the mobile robots in the system have traveled in the environment.

Another research topic from CRC that was considered in the development of Collaborative Carrying is that of robot-robot collaboration. In order for the robots to be able to bring the digital material back to their home, they had to work together (Fig. 6). Within this implementation, a single robot is considered not strong enough to move the digital material on its own. Therefore, it must wait for another robot to help it push the material in a chain-like formation. The robots search for digital material on their own but collaborate when interacting with the material. Collaboration is an important aspect of CRC as the sum of the actions and reactions of the mobile robots is what results in the assembly of structures. Consideration of how and when agents need to collaborate, for example to combine strength to push material into place, inform what type of structures can be created.

The preliminary implementations of the model indicate that stigmergy can enhance collaboration speed, enabling robots to locate each other more quickly.

3.2 Resource Rush

Resource Rush involved mobile robots divided into two competing teams (labeled red and blue). Each team was assigned a goal to gather the most digital material on their side of the environment. (Fig. 7). Within each team agents were divided into types based on their task: a set of robots for collecting material and one robot for protecting the stored material, by staying within a smaller containment zone. An additional neutral robot roamed the environment and trapped agents by forcing them to rotate in a circle. This required the creation of new behaviors and adjustments to the mobile robot agent definitions to be able to assign roles at the model initialization phase.

Fig. 7. Snapshot from Resource Rush. The information on the side panels relates to one agent from each team

The important aspects from CRC that are considered in this system are heterogeneity in robot collectives, task allocation, and site organization. In this example, there are 3 different types of agents, some which collaborate and others which have opposing goals.

By playing with the balance of agent types, their goals and defining agent containment zones, different levels of success in obtaining prescribed goals can be achieved. The balance of these features defines which team was able to collect more of the digital material, potentially correlating to concepts of productivity on the construction site.

3.3 LightHall

LightHall merged the topics of lightweight construction and CRC, resulting in a dynamic and interactive environment that adapts to human presence. In LightHall, adjustments were made to the digital material (Fig. 8). A linear vault was constructed out of bending active elements. The anchor points of the vault were inserted into the digital materials. By moving positions of anchor points, mobile robots were able to adjust the height of the vault based on tracked human position within the space (Fig. 8). Robots were instructed to move closer to humans when they are standing to increase bending and thus height in the vault or moving away when a human is sitting to decrease height. Human positions were tracked by localizing one of the digital materials held by human users.

Fig. 8. Augmentation of digital materials to be an anchor point for a benching active vault in LightHall and representation of how the system can respond to humans

LightHall deals with questions relating to adaptivity in structures with CRC. The mobile robotic machines in CRC can assemble structures but can also live within the structure, giving them the ability to further adapt over time. As such, architectural spaces can be reconfigured to adjust to the needs of humans who inhabit them. Although in this setup the relationship between humans and robots only informs the height of the structure, a change to the building system and the integration of further behaviors could be used to explore more complex and conflicting relationships between humans, their desires and the spaces they occupy.

3.4 Skyline Scrubbers

In Skyline Scrubbers, the provided system was abstracted to represent flying vehicles that navigate a polluted city landscape (Fig. 9). The digital materials represented air purifiers that are moved throughout the city by flying vehicles to clean the air.

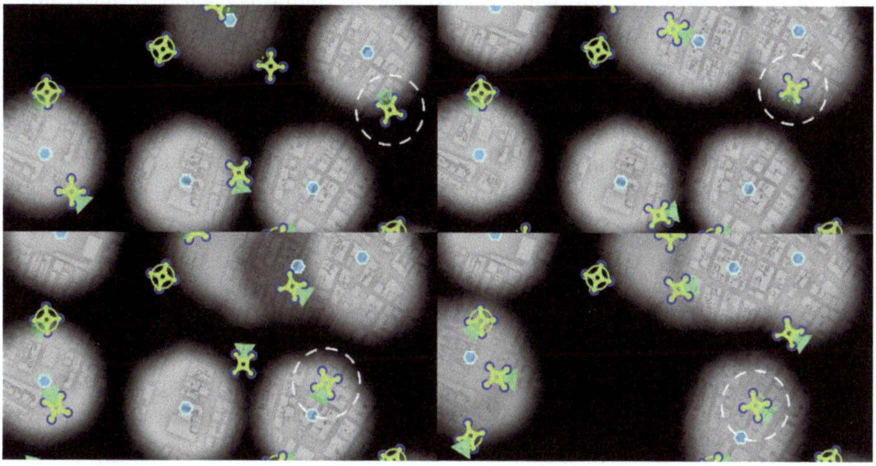

Fig. 9. Snapshots from running the ABM configured for Skyline Scrubbers

Like LightHall, Skyline scrubbers questioned how CRC systems can be utilized after the main construction of a building. In this example, mobile machines are utilized to offset emissions from construction or other human activities. CRC systems can be an active part of the architecture they assemble, potentially being involved in the maintenance, repair or other processes important to the lifecycle of a building.

4 Discussion and Conclusion

In summary, this paper introduced a CRC system with a cyber-physical design workflow that empowers architects to engage in bottom-up assembly processes with mobile machines. The research offers a compelling vision of the future of construction, where architects play a more active and integrated role in the development and deployment of novel construction automation systems. By combining advanced technology with architectural insight, CRC specifically has the potential to revolutionize the construction industry and create structures that are not only functional but also adaptable, efficient, and responsive to evolving needs and challenges in the built environment.

During the workshop conducted with the methods developed in this research, no analytical data was collected to quantitatively evaluate the four concepts and the corresponding adaptations of the ABM. Future work should consider collection of such data while the models run to be able to reach further conclusions about the behaviors developed for each concept. This can include the speed of assembly, the patterns or clustering of the digital material, and time spent assembly digital material versus wandering in the environment. This assessment would help in addressing the underlying questions that each concept poses, such as determining the circumstances where collaboration, competitive or individual action is essential and understanding how varying levels of heterogeneity in CRC contribute to a more efficient or exploratory construction processes.

Through the experimentation with the proposed workflow in a workshop environment, three key insights were gained. First, not all digital behaviors resulted in the successful translation into physical outcomes with mobile robots. This is largely due to updating of the ABM based on the motion capture date. Higher computation times of the ABM occur when many or complex behaviors are controlling the actions of the digital agents. Future work will focus on the fine tuning of the framerate of communication to solve this problem. Furthermore, the digital model neglects real world physical factors like wheel to floor friction and resistance when the robot pushes material, leading to varied results when the model runs solely in simulation.

The next insight is that by simulating CRC with ABM and additionally visually enhancing CRC systems in action, a more extensive exploration of the capabilities of such systems can be facilitated. Without architects having the tools to understand how many mobile robots can work together on construction process, it is nearly impossible for them to design structures that can be built using CRC.

Finally, the last major observation is that the behavioral approaches to developing CRC systems allow for exchange across varying robot system. Behaviors from different models were shared and built upon to fit different concept models developed in the workshop. In future work, the behaviors of the four models will be collected and made open source along with the rest of the source code. Although the behaviors in this research were developed specifically for implementation of wheeled mobile robots pushing digital material, the Skyline Scrubbers behavior model begins to suggest that the model can also be adapted for other types of robotic systems. In that example, a consideration of 3-dimensional space and a different translation of actions resulting from the ABM would be required to work with real flying machines. In conclusion, the wider community working on CRC can continue to grow by offering widely accessible and open-source ways for developing CRC systems as the one proposed in this paper.

Acknowledgement. The research has been supported by the Deutsche Forschungsgemeinschaft (DFG, German Research Foundation) under Germany's Excellence Strategy - EXC 2120/1-390831618.

The authors would like to especially thank Phillip Yuan and the FabUnion team of students, specifically Xingjie Xie, for their coordination of the 13th Architectural DigitalFUTURES Workshop at the College of Architecture and Urban Planning (CAUP), Tongji University. Furthermore, the authors would like to thank the participants of the workshop for their hard work and dedication to developing and deploying the system. The participants were Jules Wu, Jeremy Chen, Yiran Zhou, Zihao Zheng, Yuting Lei, Yuxuan Liu, Yujun Mao, Lianhan Huang, Fujian Zhao, Yilin Gao and Gloria Guo.

References

Ardiny, H., Witwicki, S., Mondada, F.: Construction automation with autonomous mobile robots: A review. In: 2015 3rd RSI International Conference on Robotics and Mechatronics (ICROM), pp. 418–424 (2015). https://doi.org/10.1109/ICRoM.2015.7367821

Bock, T., Linner, T.: Construction Robots: Elementary Technologies and Single-Task Construction Robots, vol. 3. Cambridge University Press (2016). https://doi.org/10.1017/CBO9781139872041

Braithwaite, A., Alhinai, T., Haas-Heger, M., McFarlane, E., Kovač, M.: Tensile web construction and perching with nano aerial vehicles. In: Springer Proceedings in Advanced Robotics, pp. 71–88. Springer International Publishing (July 2017). https://doi.org/10.1007/978-3-319-51532-8_5

Gerling, V., Von Mammen, S.: Robotics for self-organised construction. In: 2016 IEEE 1st International Workshops on Foundations and Applications of Self* Systems (FAS*W), pp. 162–167 (2016). https://doi.org/10.1109/FAS-W.2016.45.

Heinrich, M.K., et al.: Using interactive evolution to design behaviors for non-deterministic self-organized construction. In: 2018 Symposium on Simulation for Architecture and Urban Design (SIMAUD)., ISBN 9781510863156, pp. 213–220, Delft, Netherlands (2018). https://doi.org/10.22360/simaud

Kayser, M., et al.: FIBERBOTS: design and digital fabrication of tubular structures using robot swarms. In: Robotic Fabrication in Architecture, Art and Design 2018, pp. 285–296. Springer International Publishing (August 2018). https://doi.org/10.1007/978-3-319-92294-2_22

Leder, S., et al.: Leveraging building material as part of the in-plane robotic kinematic system for collective construction. Adv. Sci. **9**(24) (2022). https://doi.org/10.1002/advs.202201524

Leder, S., Menges, A.: Architectural design in collective robotic construction. Autom. Constr. **156** (December 2023). https://doi.org/10.1016/j.autcon.2023.105082

Leder, S., Siriwardena, L., Menges, A.: ABXM. distrubted robotics. RADr: Agent-based design and control of multiple roaming autonomous distributed robots (RADr). In: DARUS, vol. V1. https://doi.org/10.18419/darus-4058

Leder, S., Kragl, P., Menges, A.: Roaming Autonomous Distributed robot (RADr). In: DARUS, vol. V1, (2024). https://doi.org/10.18419/darus-4059

Melenbrink, N., Rinderspacher, K., Menges, A., Werfel, J.: Autonomous anchoring for robotic construction. Autom. Constr. **120** (2020). https://doi.org/10.1016/j.autcon.2020

Nguyen, Long, Schwinn, Tobias, Groenewolt, Abel, Maierhofer, Mathias, Zorn, Max Benjamin, Stieler, David, Siriwardena, Lasath, Kannenberg, Fabian and Achim Menges. ABxM. Core: The Core Libraries of the ABxM Framework, 2022., doi. 10.18419/DARUS-2994.

Nigl, F., Shuguang Li, J.E., Blum, and H. Lipson.: Structure-reconfiguring robots: Autonomous truss reconfiguration and manipulation. IEEE Robot. Autom. Mag. **20**(3), 60–71 (September 2013). https://doi.org/10.1109/mra.2012.2201579

Ntzoufras, S., Oungrinis, K.-A., Liapi, M., Papamanolis, A.: Robotic swarms in architectural design a communication platform bridging design analysis and robotic construction. In: Anthropologic: Architecture and Fabrication in the cognitive age - Proceedings of the 38th eCAADe Conference-Volume 2, vol. 2, pp. 453–462 (2020). https://doi.org/10.52842/conf.ecaade.2020.2.453

Petersen, K.H., Nagpal, R., Werfel, J.K.: Termes: An autonomous robotic system for three-dimensional collective construction. In: Robotics: Science and Systems VII, pp. 257–264. MIT Press (2012). https://doi.org/10.7551/mitpress/9481.003.0038

Petersen, K.H., Napp, N., Stuart-Smith, R., Rus, D., Kovac, M.: A review of collective robotic construction. Sci. Robot. **4**(28) (2019). https://doi.org/10.1126/scirobotics.aau8479

Reynolds, C.W.: Flocks, herds and schools: A distributed behavioral model. In: Proceedings of the 14th annual conference on Computer graphics and interactive techniques. ACM (August 1987). https://doi.org/10.1145/37401.37406

Reynolds, C.W.: Steering behaviors for autonomous characters. In: Game developers conference, vol. 1999, pp. 763–782 (1999, March)

Stieler, D., Schwinn, T., Leder, S., Maierhofer, M., Kannenberg, F., Menges, A.: Agent-based modeling and simulation in architecture. Autom. Constr. **141** (September 2022). https://doi.org/10.1016/j.autcon.2022.104426

Yablonina, M., Kalousdian, N.K., Menges, A.: Designing [with] machines. In: Distributed Proximities - Proceedings of the 40th Annual Conference of the Association of Computer Aided Design in Architecture (ACADIA). ISBN 978-0-578-95213-0,, pp. 330–339 (2020) https://papers.cumincad.org/cgi-bin/works/paper/acadia20_330

Yablonina, Maria, Leder, Samuel, Kubail Kalousdian, Nicolasand Achim Menges, "Choreographed space: Distributed mobile robotic system for temporal architectural change." In ICETAD [Proceedings of the ICETAD Conference 2019], Ryerson University, Toronto. (ISBN: 978-1-64713-769-4). 2019.

Zhang, K., et al.: Aerial additive manufacturing with multiple autonomous robots. Nature. **609**(7928), 709–717 (September 2022). https://doi.org/10.1038/s41586-022-04988-4

Planning Non-repetitive Robotic Assembly Processes with Task and Motion Planning (TAMP)

Pok Yin Victor Leung[1(✉)], Yijiang Huang[1,2,3], Caelan Garret[2,3], Fabio Gramazio[1], and Matthias Kohler[1]

[1] ETH Zurich, Zürich, Switzerland
yck011522@gmail.com
[2] MIT, Cambridge, USA
[3] NVIDIA Research, Santa Clara, USA

Abstract. Many recent examples of robotic construction are still planned using separate task and motion planning approaches. This separation makes it hard to create efficient action sequences, especially for non-repetitive assembly processes. This paper introduces Task and Motion Planning (TAMP) as a more advanced planning approach that integrates task planning and motion planning to optimize action sequences while ensuring collision-free robotic trajectories.

We demonstrate the benefit of TAMP by applying it on a robotically constructed timber pavilion that can benefit from an optimized tool change sequence. Robotic execution time was collected from the actual construction, and the results show an estimated 16% reduction in execution time using TAMP compared to traditional task planning approaches like flowcharts.

Our TAMP implementation relies on the Planning Domain Definition Language (PDDL) and the PDDLStream solver. The system allows a flexible approach to model actions and dependency and creates validation checks, such as collision. We explain our incremental programming approach for modeling assembly problems with PDDL. This approach can serve as a general guideline for applying TAMP to other robotic assembly problems. It will also promote the creation of modular and reusable code. Finally, we discuss the flexibility of TAMP with PDDLStream, its limitations, and future research opportunities.

Keywords: Robotic Assembly · Task and Motion Planning · PDDL

1 Introduction

Traditional paradigm for programming industrial robots is not suitable for architectural fabrication where non-repetitive tasks are often necessary for unique geometries. Instead, automated planning and validation are essential for non-repetitive robotic programs.

When discussing non-repetitive robotic programs, it is necessary to distinguish between the action template and the geometric targets. The action template (also known as a task template) is a list of sequentially executed actions, while geometric targets

M. Yablonina et al. (Eds.): Robotic Fabrication in Architecture, Art and Design 2024, pp. 20–33, 2026.
https://doi.org/10.1007/978-3-032-22607-5_2

are descriptions that complement some actions, such as robotic movements. Figure 1 shows an example of an action template with four actions and two geometric targets for a simple construction process. Note that the two movement actions are accompanied by geometric targets that specify the target pose (position and orientation) at which the end effector will reach.

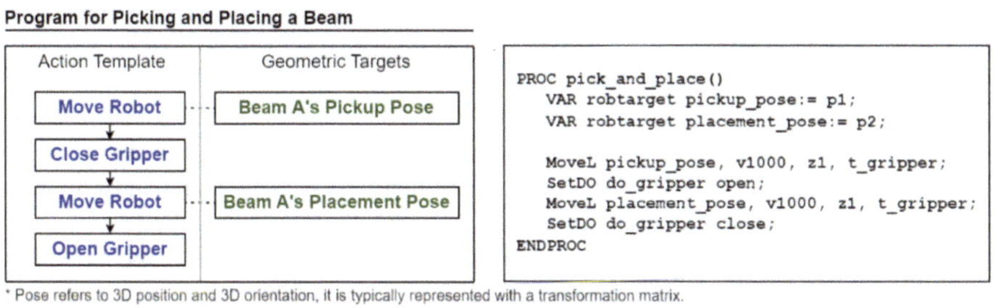

Fig. 1. Example of an action template and geometric targets represented symbolically (left) and in ABB RAPID language (right)

Using a single repetitive action template with a flexible geometric input allows the assembly of mass customized objects. However, structures created with this approach were often limited to a fixed topology. A more advanced customization approach is to use multiple snippets of action templates. For example, to construct a column-beam structure, two template snippets can be used. One snippet can be responsible for the "Column Assembly" while another for the "Beam Assembly". The selection and sequence of template snippets can be automated using computational logic based on the properties of a structure. In the computer science (CS) community, such process of selecting and scheduling actions is called task planning or symbolic planning.

Explicitly scripted conditional logic such as flowcharts (Huang et al. 2021) and behavior trees (Ögren and Colledanchise 2018) work well with construction tasks that have limited variations (Wagner et al. 2020). However, these approaches are not flexible enough when dealing with problems with many possible action sequences and interdependency (Garrett et al. 2020). Manually scripted action templates and the flowchart logic are often less than optimal. For these scenarios, new approaches that consider collisions and reachability are necessary (Hartmann et al. 2023).

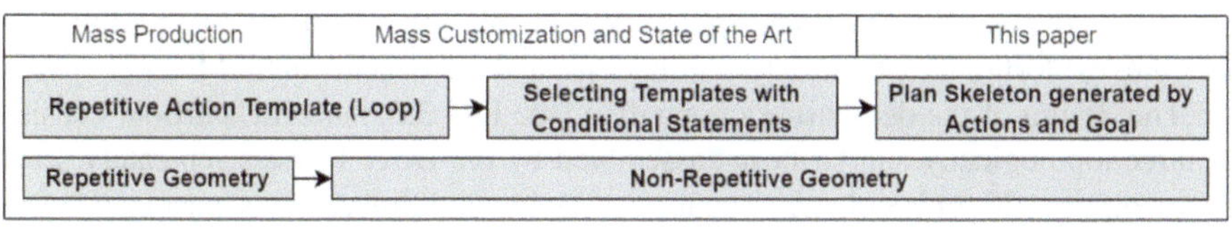

Fig. 2. Diagram showing the introduction of flexibility to the action templates and geometric targets.

The focus of this paper is to explore the application of Task and Motion Planning (TAMP) (Garrett et al. 2021), a state-of-art planning method developed in the CS community to a construction task. It integrates task planners (for deciding action sequences) and robotic motion planners (for finding and certifying collision-free trajectories). Proposed Integration allows the detection of robotic constraints by the motion planner to be used as feedback for planning a different and potentially more optimized action sequence (i.e., task plan). Fig. 2 highlights the difference of using TAMP vs. prior methods.

At this moment TAMP is not widely used in architectural robotics and digital fabrication due to its novelty and limited literature in the design field (Sherkat et al. 2023, Hartmann et al. 2020). Moreover, the Planning Domain Definition Language (PDDL) (Ghallab et al. 1998) is an action-based language for encoding state transitions, and is significantly different from the more common, sequentially executed languages. In other words, the users model their assembly process with PDDL, after which the TAMP solver will generate the conventional "robotic program" for execution. To make the modelling process easier and more transparent for the user, this paper will introduce an incremental programming approach for writing PDDL. **Specifically, this paper will demonstrate:**

1. How to encode an architectural assembly process for TAMP using Planning Domain Definition Language (PDDL) with an incremental programming approach.
2. How to apply PDDLStream, a state-of-the-art TAMP solver, which combines symbolic PDDL solvers with robotic motion planners.
3. The benefits of TAMP over prior planning techniques, with results showing a more efficient task sequence that can reduce execution times.

2 Case Study

To provide a tangible example, we will use a recently constructed timber pavilion to introduce the incremental programming approach we developed for writing PDDL and demonstrate TAMP's benefits. The robotic process used to assemble the pavilion is called Timber Assembly with Distributed Robotic Tools (DiRT) (Leung et al.). The assembly process is characterized by having a single industrial robotic arm (integrated with a large overhead gantry system) manipulating multiple tools capable of clamping and screwing tasks (see Table 1). One of the unique requirements in this process is that the distributed tools must be attached to and detached from the structure one by one before and after the assembly of each beam. The spatial and temporal complexity of the process is analogous to a chef managing many simultaneously cooking pots on a stove. A robotic fabrication process implies a multitude of feasible actions and ways of sequencing them at any moment in the process. Due to this plurality of choices, the robotic process is uniquely suited to exploring the possibilities, limitations, and the benefits of TAMP approaches.

The design of the demonstration pavilion (see Fig. 3a), named CantiBox, consists of three topologically similar boxes assembled by the DiRT Process separately. Each box comprises 20 timber elements connected by integral timber joints machined before assembly. More details of the pavilion's structural and architectural design can be found in (Tanadini et al. 2023). This paper will only focus on the planning of the robotic process.

Table 1. List of robotic tools and their purpose for assembling the case study structure

Tools	Type Name	Amount	Purpose
Robotic Clamp	`CL3`	2	Assemble *clamped joints*. The allowable joint intersection angle for CL3 is 25–90°, and CL3M is 90–155°.
	`CL3M`	2	
Robotic Screwdriver	`SL1`	3	Assemble *screwed joints*. SL1-G200 can also act as a timber gripper.
	`SL1_G200`	1	
Robotic Gripper	`PG500`	1	Grasp timber beams and bring them to the structure for assembly. Three different gripper lengths (500 mm to 1500 mm) are available to suit different timber lengths.
	`PG1000`	1	
	`PG1500`	1	
Scaffolding Bar	–	4	Stabilize structure during construction. (Ignored in this paper for simplicity)

Fig. 3. Photo and front elevation drawing of the CantiBox Pavilion

The planning of tasks and motions is performed separately for each of the three boxes, with a planning horizon[1] of twenty elements every time. Each of the twenty elements is labeled during design time to use one of three assembly methods: Ground Connection, Joint Clamping, and Joint Screwing. The following three paragraphs summarize the assembly methods (see Fig. 4 for related images). Note in advance that the Joint Clamping method is the focus of later discussions.

The Ground Connection Method is used for the first two timber elements to create a temporary connection to a ground platform. First, the robotic arm uses a gripper to position the timber element onto the construction platform. Then, a human operator fixes the timber to the platform using a fixture. Finally, the robotic gripper releases the timber, and the robot returns the gripper to storage.

The Joint Clamping Method is used for most of the timber elements (14 per box). First, the robotic arm attaches robotic clamps to the timber structure, one by one, at the

[1] Planning horizon refers to how far ahead the planning looks, it can refer to the number of actions performed or duration of time. In this work, it refers to the number of discrete timber elements being assembled.

location of the mating joints. Then, the robotic arm changes to a gripper tool to pick up a timber element and position it into the clamp jaws. Then, the robotic arm and the clamps move synchronously to close the joints. Finally, the robotic arm returns the gripper to storage and detaches the clamps, one by one, from the timber structure. The number of clamps used depends on the number of mating joints, and the clamp type (CL3 or CL3M) needs to match the joint's intersecting angle.

The Joint Screwing Method is used for the interlocking key elements (4 per box). First, the robotic arm picks up a timber element using a gripper tool. The operator manually attaches screwdrivers to the timber element held by the robot. Then, the robotic arm brings the timber element towards the structure, the screwdriver begins rotation, and the mating joints are pulled together by the screws. Finally, the robotic arm returns the gripper to storage and detaches the screwdrivers from the timber structure.

Fig. 4. Photos showing a snapshot of the three assembly methods: Ground Connection (left), Joint Clamping (middle), and Joint Screwing (right)

3 Challenge/Problem

Robotic planning without using TAMP can be generally broken down into two separate stages. In the first stage, a list of actions is composed manually or automatically using computational logic such as flowcharts. In the second stage, the robotic motions between each target are planned using motion planners.

Using the Joint Clamping Method of our case study as an example, if a flowchart is used to compose the action template, it would consist of three phases: Attach Clamps, Clamping, and Detach Clamps (see Fig. 5a).[2] While this logic may be simple to understand, the resulting list of actions could be more efficient. Consider the case when two or more beams are assembled. It would be inefficient to detach the used clamps one by one and bring them back to storage, only to pick them up from storage again and attach them to a new location (see Fig. 5b). In terms of saving the number of steps and assembly time, it would be more efficient to detach a clamp from the structure and bring it to the new location without putting it down (see Fig. 5c).

Unfortunately, different beams require different amounts of clamps and different clamp types. A direct transfer of clamps from the previous location to the next location is not always possible. Some of the clamps will have to stay on the structure for a longer period before it is used again. However, with the two-stage planning approach, there is

[2] The Attach Clamps and Detach Clamps actions would be considered sub-templates. They would be repeated as many times as the number of clamps needed for that specific beam.

no way to ensure that the clamps are still accessible by the robot at a later step.[3] If we attempt to create an action plan that will leave the clamps on the structure and retrieve them later, the motion planner in the second planning step has a high chance of getting stuck. This problem is analogous to a game of chess where a player makes a series of moves that result in a checkmate with no way out. The player must either concede defeat or re-evaluate their strategy and backtrack to avoid the hopeless situation. In TAMP, the ability to backtrack is the critical feature that sets it apart from prior work.

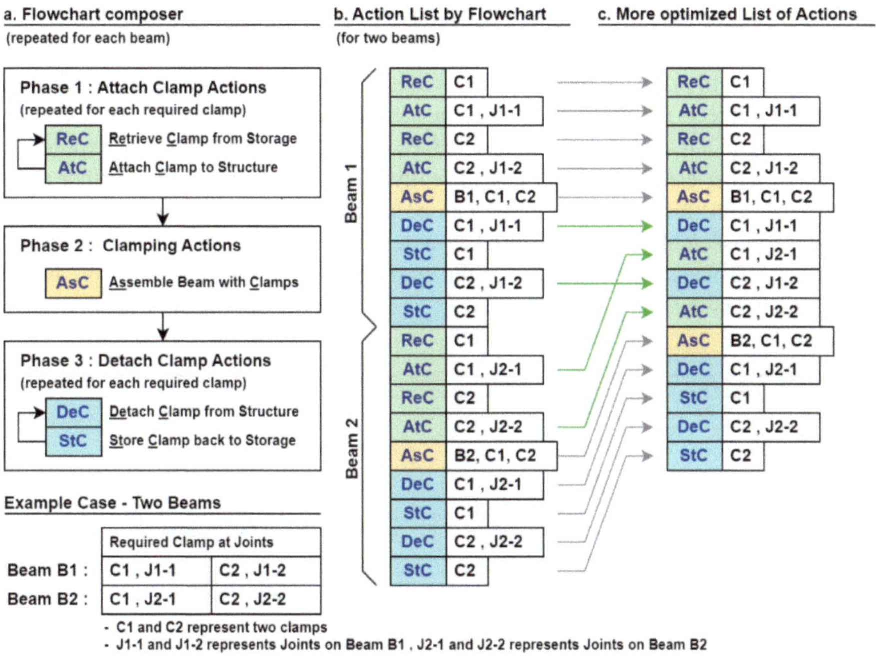

Fig. 5. Diagram showing the nonoptimal result of planning actions with a flowchart

4 Task and Motion Planning (TAMP)

Using a TAMP workflow allows the task planner to automatically choose the sequence of actions by verifying the feasibility of its robotic trajectory using a motion planner. If a trajectory cannot be found, the symbolic solver can backtrack and try a different sequence of action, for example, by transferring the clamp earlier or by moving the clamp back to storage after use. The key difference between TAMP versus flowcharts is that the actions do not have a predefined sequence (see Fig. 2).

[3] The addition of more timber elements may block the access of the clamps that are left on the structure. The clamps can be considered trapped or inaccessible when the motion planner cannot find any robotic trajectory to reach them.

4.1 Symbolic Planning

TAMP integrates task planning (symbolic) and motion planning. Symbolic planning involves using autonomous techniques to solve symbolic planning and scheduling problems. A *planning problem* is one in which we have some *initial starting state*, which we wish to transform into a desired *goal state* by applying a set of *actions* (Ghallab et al. 1998). In our case study example, the initial state is a list of statements asserting that (1) the timber elements are located at their storage location, (2) each tool is located at its storage location, and (3) the robotic arm is not holding any tool or timber elements. The goal state refers to the same set of statements, except that all the beams are in their assembled locations.

Predicate - In symbolic planning, states are expressed symbolically, using relationships among objects instead of geometry (such as a transformation matrix). For example, the following PDDL statement '`(BeamAtStorage b1)`' can be used to express the fact that a timber element called b1 is currently at the storage location. Another statement, '`(BeamAtAssembled b1)`' can be used to describe the same element but located at the assembled location. Each fact-asserting statement is called a *predicate* and expresses a Boolean statement. The initial starting state is simply a list of predicates describing the system at that time.[4]

a. (domain.pddl):
```
(define (domain beam_assembly)
   (:requirements :strips)
   (:predicates (BeamAtStorage ?beam) (BeamAtAssembled ?beam))
   (:action assemble_beam
      :parameters (?beam)
      :precondition (BeamAtStorage ?beam)
      :effect(and (not (BeamAtStorage ?beam)) (BeamAtAssembled
?beam))))
```
b. (problem.pddl):.
```
(define (problem minimum_working_example)
   (:domain beam_assembly)
   (:objects b0 b1 b2)
   (:init (BeamAtStorage b0)(BeamAtStorage b1)(BeamAtStorage b2))
   (:goal (and
      (BeamAtAssembled b0)(BeamAtAssembled b1)(BeamAtAssembled b2))))
```

Fig. 6. A minimal working example of a PDDL (a) domain and (b) problem describing beam assembly

Action In symbolic planning, users do not define a fixed order of actions. The user tells the system what it can do instead of what to do. Actions are composed of two main parts: Precondition and Effect. The *precondition* is a Boolean logical formula that allows the planner to decide if an action can be applied based on the current state. For example, a beam must be at the storage location for the "Assemble Beam" action to begin. The precondition can also prevent illogical actions, such as assembling an already-assembled

[4] In planning all unknown values are assumed to be false. That is to say that if we do not know the value of a predicate, then we assume it to be false. This is known as the "closed world" assumption. ("The AI Planning & PDDL Wiki" n.d.)

beam. If the planner decides to apply an action, the *effect* of that action describes the addition and removal of predicates from the current state.[5] For example, the effect of an "Assemble Beam" action should add the predicate `'BeamAtAssembled b1'` and remove `'BeamAtStorage b1'`. Therefore, the result of applying an action is a new state that is ready to apply another action. The symbolic planner can keep track of the changing states automatically and be smart about applying efficient action sequences to reach the goal state. However, it is the user's responsibility to ensure that actions do not create an incoherent state. For example, a beam cannot be in two places simultaneously.

Parameters PDDL actions are parametrically defined, using variables to express their preconditions and effects.[6] The example in Fig. 6a shows a simplified `'assemble_beam'` action from the case study with a parameter `'?beam'`. The same parameter is used in the precondition and effect.

4.2 Encoding Assembly Problem with PDDL

In PDDL, a problem is represented using a *domain file* and a *problem file*. The *domain file* contains the predicates and actions of the robotic process. Once established, this file remains reusable for different problems as long as the process logic is unchanged. In our case study problem, a single domain file is shared for the three problem files corresponding to each timber box. Table 2 shows the complete list of actions used in the case study, note that each action is very human-understandable.

Each of the three problem files in our case study contains objects, the initial state, and the desired goal specific to each timber box. This file contains a long list of predicates (see associated code repository) to declare the number of beams, the joints between the beams, the type of gripper required for each beam, and the type of clamp required for each joint. For integration with a parametric design workflow, this file was generated computationally via scripting.

4.3 Motion Planning with PDDLStream

The second step to achieving TAMP is integrating symbolic planning and motion planning. When used separately, motion planning ensures target reachability and avoids unforeseen collisions. However, as discussed in the problem statement, motion planners alone can get stuck if a trajectory cannot be found. TAMP handles this failure automatically by allowing the symbolic planner to try a different action sequence.

The TAMP implementation used in this paper is called PDDLStream. It introduces the concept of *stream*, an external function that a symbolic planner can call to certify whether a predicate is true. For robotic assembly processes, the checks include reachability (certified by motion planners) and collision (certified by collision checkers). PDDLStream simply acts as a framework that provides a link between the symbolic planner

[5] When a predicate is removed, it means that the statement is now false.

[6] The syntax shown in this paper is described in PDDL v1.2 (Ghallab et al. 1998). This paper uses a popular symbolic planner, FastDownward (Helmert 2006), which supports many common syntaxes.

Table 2. List of Actions used in the case study

#	Action Name	Parameters
1	`assemble_beam_by_clamping_method`	`?beam? gripper? grippertype`
2	`assemble_beam_by_screwing_method`	`?beam? gripper? grippertype`
3	`assemble_beam_by_ground_connection`	`?beam? gripper? grippertype`
4	`pick_gripper_from_storage`	`?gripper`
5	`place_gripper_to_storage`	`?gripper`
6	`pick_clamp_from_storage`	`?clamp`
7	`place_clamp_to_storage`	`?clamp`
8	`attach_clamp_to_joint`	`?clamp? beam1? beam2? clamptype`
9	`detach_clamp_from_joint`	`?clamp? beam1? beam2`

(such as FastDownward (Helmert 2006)) and the streams (the external functions). The interfaces to the stream samplers are defined in a stream file for PDDLStream (see associated code repository). Figure 7 shows an overview of the TAMP planning process.

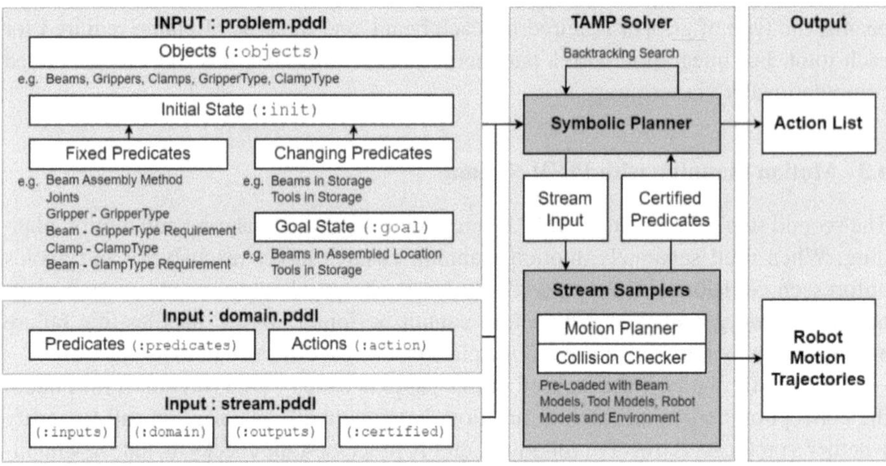

Fig. 7. Overview of the TAMP planning process showing input and output to the solver

Motion Planner (MP) Separate planners are needed for each action that can get stuck. They are specific to the actions and are preloaded with the necessary geometry for motion planning. For example, a stream for the `BeamAssembly` action should accept the?

beam parameter as input and check whether a trajectory exists to move that beam to the final assembled location. Only self-collision (the robotic arm, the gripper, and the timber beam) is considered during MP because collisions with other beams and tools are checked later. If the robot can reach the targets, the generated trajectory is stored as a predicate to assert that the corresponding action has a trajectory. For example, '(BeamAssembly-Traj? beam? traj)' can represent the fact that trajectory? traj is available for beam? beam.

Collision Checker (CC) Collisions between the movable beams and tools and the robot are checked separately using a CC stream.[7] Figure 8 shows the AssembleBeam action (as seen in Fig. 6) where MP and CC streams are added to the precondition. The 'exists(?otherbeam)' clause checks whether any other assembled beams are in collision.

```
:precondition(and
   (BeamAtStorage ?beam)
   (BeamAssemblyTraj ?beam ?traj)
   (not (exists (?otherbeam) (and
      (BeamAtAssembled ?otherbeam)
      (BeamAssemblyInCollision ?traj ?beam ?otherbeam)
   )))
```

Fig. 8. Precondition with motion planning and collision checking streams

5 Experimental Setup

Six experiments with increasing complexity were conducted to demonstrate how a PDDL problem can be formulated in an incremental manner. The first experiment is a minimal working example without considering tool and joint requirements while the final experiment considers all fabrication constraints and is equivalent to the case study process. Due to page limitations, the complete PDDL code can be found in the code repository associated with this paper.[8]

E01 Beam Assembly Minimal working example with only one action to assemble beams. No grippers or clamps. No order is enforced. Symbolic planning only.

E02 Joint Partial Order The joints are defined as an intersection between two beams. A precondition is added to ensure the joint's first beam is assembled before the second beam can be assembled (a.k.a. partial ordering). Symbolic planning only.

E03 Gripper Switch The gripper requirement is added to the assemble beam action. Additional actions for the robot to pick up and release grippers are added. Three gripper

[7] The movable objects are checked separately from the motion planning steam because the position of these objects is managed by the symbolic planner. More details related to this implementation can be found in (Garrett, Lozano-Pérez, and Kaelbling 2020).

[8] https://github.com/yck011522/robarch_pddl

types are available. Symbolic planning only. The planning result shows that tool change actions are inserted wherever necessary. The assembly order differs from the previous result as the planner tried to reduce the number of tool changes.

E04 Assembly Stream The symbolic logic is the same as E03. However, the `AssembleBeam` action is validated with MP and CC streams as described in the TAMP section above. This is the first TAMP example. The resulting action sequence is longer than E03. The interpretation is that E03 did not consider MP and CC checks, now that they are considered, the assembly sequence cannot be as optimized. The MP and CC checks can also be considered extra constraints for the TAMP solver, which means the solver has less freedom to choose a shorter action sequence.

E05 Clamp Transfer `AssembleBeam` actions are separated according to the three assembly methods of a beam. For the assembly action with *Joint Clamping*, clamps are required at the mating joints. The correct clamp type must be present on these joints before the beam can be assembled. Four clamp manipulation actions are added (the last four rows in Table 2). Symbolic planning only. Note that there are two clamp devices available for each of the two clamp types. The resulting action sequence shows the clamps transferred between storage and the joints. However, in many cases, the clamps are transferred directly between the joints.

E06 Clamp Stream The actions are similar to E05, but the designer defines the assembly order (manually, by intuition) to improve stability during construction. In addition, MP and CC streams validate the AssembleBeam, the AttachClampToJoint, and the DetachClampFromJoint actions. This experiment is the complete TAMP equivalent to the case study. The resulting action sequence is longer than that of E05. The rationale is similar to the one offered for E04.

6 Results and Discussions

Each experiment (E01 to E06) is performed with three problem files corresponding to the three boxes in the case study. The planner's goal is to minimize the number of actions to reach the assembled state. The results are broken down in Table 3. "Assemble Actions" include all three assembly methods (#1 - #3 in Table 2). "Gripper Actions" include picking and placing grippers (#4, #5 in Table 2). The main highlight is the "Clamp Actions", which include the four actions that pick and place the clamps (#6 - #9 in Table 2). Although the topology of the three boxes is the same, their geometry is different. Therefore, their geometrical constraints (such as robot reachability) that affect the choice of tasks are also different. This is reflected in the result where the number of planned tasks differs for each of the three boxes.

The optimization benefit of TAMP can be seen in Table 4, which compares the result of experiment E06 and what would have been required if a flowchart method was used. The execution time of each action was measured during the actual robotic construction. The estimated times listed in the table are calculated using the average value for each action type (calculated separately for each of the three box structures). The significantly reduced number of clamp transfer actions (see Fig. 9) resulted in a total time reduction of

Table 3. Number of planned actions in the six experiments with three problem sets. The results of the three boxes (left, mid, right) are separated by a + symbol.

Planning Experiment	Assemble Actions (Count)	Gripper Actions (Count)	Clamp Actions (Count)	Planning Time (s)
E01	20 + 20 + 20	–	–	<1 + <1 + <1
E02	20 + 20 + 20	–	–	<1 + <1 + <1
E03	20 + 20 + 20	4 + 8 + 4	–	<1 + <1 + <1
E04 (TAMP)	20 + 20 + 20	4 + 8 + 4	–	113 + 139 + 145
E05	20 + 20 + 20	28 + 28 + 26	56 + 56 + 56	2 + 3 + 3
E06 (TAMP)	20 + 20 + 20	34 + 38 + 32	56 + 56 + 56	54 + 32 + 52
Flowchart	20 + 20 + 20	40 + 40 + 40	96 + 96 + 96	<1 + <1 + <1

16%. Note that similarly optimized action sequences are almost impossible for a human to create manually because it would be highly prone to error.

On the other hand, we found that the TAMP solver planning time is highly sensitive to the number of beams to be assembled, especially for our process that has a high branching factor (many possible actions at each step). The partial ordering constraint introduced in E02 helps reduce the branching factor. However, in E06 we must specify the full assembly order to reduce the branching factor for the search to be completed in reasonable time. One explanation is to imagine the number of possible moments for a clamp to be attached to the structure before it is needed by a beam, and after the clamp is used, it can also be detached or relocated at any time after that. Future research can study how to reduce planning time by smarter problem formulation.

Table 4. Time comparison of experiment E06 (TAMP with a fixed order) and the flowchart planning method

	Execution Time for Actions (min)			
Planning Experiment	Assemble Actions	Gripper Actions	Clamp Actions	Total
E06 (TAMP)	179 + 194 + 185	93 + 114 + 107	194 + 189 + 191	1446 (−16.3%)
Flowchart	179 + 194 + 185	100 + 111 + 119	284 + 252 + 303	1729

Fig. 9. Image of a beam being assembled by screwdrivers. Note that some clamps remained on the structure from the previous assembly.

7 Conclusion

The integrated Task and Motion Planning (TAMP) workflow using PDDLStream overcomes the limitations of the previous two-stage planning methods such as flowcharts. It allows the planner to explore different task sequences while considering the motion planning constraints. The automatic generation of action plans results in a significant execution time reduction without manual programming efforts.

Using our case study, we have introduced an incremental programming approach to formulate complex assembly problems with PDDL. While the incremental addition of actions and constraints makes programming easier, it also creates modular PDDL code snippets that can be reused for different assembly problems. For example, future work can consider adding a stream for the planner to generate and validate grasp poses (i.e., where the workpiece is held), or checking the partially assembled structure to avoid instability.

Acknowledgments. The authors would like to acknowledge the contribution of Caitlin Mueller for her insightful discussions and scholarship support for Y. Huang; the design and engineering team behind the CantiBox Pavilion; and the Robotic Fabrication Laboratory team (ETH Zurich). This research was supported by the Swiss National Science Foundation (NCCR Digital Fabrication agreement number 51NF40-141853), and Y. Huang was supported by an ETH Postdoctoral Fellowship.

References

Garrett, C.R., et al.: Integrated task and motion planning. Annu. Rev. Control Robot. Auton. Syst. **4**(1), 265–293 (2021). https://doi.org/10.1146/annurev-control-091420-084139

Garrett, Caelan Reed, Tomás Lozano-Pérez, and Leslie P. Kaelbling. 2020. "PDDLStream: Integrating symbolic planners and blackbox samplers." In International Conference on Automated Planning and Scheduling (ICAPS). https://arxiv.org/abs/1802.08705.

Ghallab, Malik, et al. 1998. "PDDL the planning domain definition language." https://planning.wiki/_citedpapers/pddl1998.pdf.

Hartmann, V.N., Orthey, A., Driess, D., Oguz, O.S., Toussaint, M.: Long-horizon multi-robot rearrangement planning for construction assembly. IEEE Trans. Robot. **39**(1), 239–252 (2023). https://doi.org/10.1109/TRO.2022.3198020

Hartmann, V.N., Oguz, O.S., Driess, D., Toussaint, M., Menges, A.: Robust task and motion planning for long-horizon architectural construction planning. In: IEEE/RSJ 2020 International conference on intelligent robots and systems (IROS), pp. 6886–6893, Las Vegas, NV, USA (2020). https://doi.org/10.1109/IROS45743.2020.9341502

Helmert, M.: The fast downward planning system. J. Artif. Intell. Res. (JAIR). **26**, 191–246 (2006)

Huang, Y., Leung, V.P.Y., Garrett, C., Gramazio, F., Kohler, M., Mueller, C.: The new analog: a protocol for linking design and construction intent with algorithmic planning for robotic assembly of complex structures. In: Symposium on Computational Fabrication. SCF '21, pp. 1–17. Association for Computing Machinery (2021). https://doi.org/10.1145/3485114.3485122

Ögren, P., Colledanchise, M.: Behavior Trees in Robotics and AI: An Introduction. CRC Press, Boca Raton (2018). https://doi.org/10.1201/9780429489105

Sherkat, S., Skoury, L., Wortmann, A., Wortmann, T.: "Artificial Intelligence Automated Task Planning for Fabrication" Advances in Architectural Geometry 2023, pp. 249–260. De Gruyter (2023). https://doi.org/10.1515/9783111162683-019

Tanadini, D., Boller, G., Leung, P.Y.V., D'Acunto, P.: Plastic Design of Bespoke Interlocking Timber-to-Timber Connections for Robotic Assembly. In: World Conference on Timber Engineering (WCTE 2023), pp. 4399–4408. World Conference on Timber Engineering (WCTE 2023), Oslo, Norway (2023). https://doi.org/10.52202/069179-0573

Wagner, H.J., Alvarez, M., Groenewolt, A., Menges, A.: Towards digital automation flexibility in large-scale timber construction: integrative robotic prefabrication and co-design of the BUGA Wood Pavilion. Constr. Robot. **4**(3), 187–204 (2020). https://doi.org/10.1007/s41693-020-00038-5

"The AI Planning & PDDL Wiki." n.d. Planning. Wiki-The AI Planning & PDDL Wiki. Accessed 5 Oct. 2023. https://planning.wiki/.

ZEPPELIN
hier mieten!
ZEPPELIN
hier mieten!
GEDA
ndividualized
roduction
RWTH

Loosely Coupled Observation Processing and Data Exchange System for Complex Teams of On-Site Construction Robots

Lukas Kirner[1(✉)], Marit Zöcklein[2], Jyrki Oraskari[1], Carsten Kamp[1], Fabian Claßen[2], and Sigrid Brell-Cokcan[1]

[1] Individualized Production in Architecture, RWTH Aachen, Aachen, Germany
kirner@ip.rwth-aachen.de
[2] Center Construction Robotics, Aachen, Germany

Abstract. The construction sector has slow adaptation to emerging technologies and lagging productivity compared to other industries. Level of construction robot autonomy is impacted by complexity and variability of a conventional construction site environment. One approach to increasing robot autonomy is to integrate more environmental sensors. However, when faced with the reality of complex tooling and sensing systems used on a construction site, including rented equipment that frequently moves between sites, integration of multiple changing sensor data streams becomes a bottleneck for smooth operation. Machines that host sensory systems often become data silos that do not communicate across systems. To address this, we explore how to enable the sharing of sensor observations between actors on a construction site using loosely coupled communication and event processing between sensors and construction site machines. We present an exemplary system architecture and assess its performance.

Keywords: Construction Robotics · Linked Data · 5G Network · Loose Coupling

1 Introduction

Construction is one of the slower industries to adopt automation and digitalization [1]. One reason for this is the high variability of the construction environment [2]. Unlike in manufacturing, where the same process often repeats thousands of times to produce serialized output, in construction, each site is unique, occupied by workers from a broad range of trades, each performing unique construction tasks. Automating a system with this many parameters and unpredictable variables is thus challenging [3]. An approach to solving this problem is to equip construction machinery with more sensors for process monitoring and surveying. Although it might seem that the issue lies in sensing hardware and data collection, one of the fundamental challenges in integration is establishing shared data gathering protocols and exchanges between various construction site actors: human and machine. Typically, the network infrastructure can only handle

M. Yablonina et al. (Eds.): Robotic Fabrication in Architecture, Art and Design 2024, pp. 35–46, 2026.
https://doi.org/10.1007/978-3-032-22607-5_3

a limited number of participants transmitting at high data rates. Complicating matters further, the participants will change over time since machinery is usually rental equipment that moves between construction sites [4]. To address the highly variable, dynamic, and unpredictable construction site environment and the corresponding demand for high flexibility, we propose a fully modular solution for a containerized server infrastructure where sensors communicate over a 5G network. The sensors in this setup will be independent of the carrier to accommodate changing requirements. The overarching goal is to enable sharing sensor observations between actors on a construction site through a loosely coupled system, thereby increasing productivity and safety.

2 Background

While multi-robot setups and collaborative construction robots are still in their infancy, existing Internet of Things (IoT) construction site monitoring applications are already facing the challenges of digital construction site communication. These include the orchestration and management of data streams, the structure and formats of the data, and the underlying network technology for secure and efficient communication. However, the requirements for these systems vary widely, e.g. in terms of transmission rates and latency [5].

2.1 Loosely Coupled Systems

Coupling refers to the interdependence between systems. It encompasses various facets, such as Technology, which assesses the degree of shared tools and infrastructure; Location, which evaluates physical or network-related associations; Schema Coupling, addressing alignment in data structures; Temporal, focusing on timing dependencies; and Programming Model, which assesses reliance on specific programming paradigms. In [6], Altıntas and Ilal developed an approach that involved integrating data from two domains (BIM—Building Information Modeling and GIS—Geographic Information Systems) into a third, independent platform with its domain model. However, compared to systems that employ interface-based communication or message-oriented architecture to achieve enduring loose coupling, their approach can be characterized as process decoupling. Although a valuable approach, we focus more on efficient, cost-effective, and lightweight extension mechanisms in this work.

2.2 Linked Data for on-site Automation

Linked Data is a technique for sharing structured data on the Web. It enables interoperable access based on technologies such as Resource Description Format (RDF) and ontologies [7]. In computing, an ontology is defined as an explicit conceptual specification for a domain of interest [8]. In the last few years, many ontologies have been developed for the construction domain. Most of them are based on translating existing models [9] or developing new mapping techniques [10]. The majority of achievements in this area describe general geometric and topological information or describe product and component information [10, 11]. Linked Data workflows to describe and use general construction processes have been developed in various research projects like DiCtion or the Internet of Construction (IoC) [12, 13]. In addition, approaches for the semantic description of robotic systems [14] and their capabilities [15] have also been developed. However, these are not currently considered for use in on-site construction robotics, and are not well linked to relevant sensor or network ontologies such as the Semantic Sensor Network (SSN)/ Sensor, Observation, Sample, and Actuator (SOSA) [16]. As a result, the capabilities of a system that benefits from a combination of these advances in a loosely coupled fashion have not been explored to date.

2.3 5G for construction sites

In contrast to prior cellular network standards, 5G was developed for industrial use cases. It holds many potentials, such as increased bandwidth, lower latency and a larger number of participants in the network compared to the 4G standard for mobile communication or the current IEEE WLAN standard [17]. The latter were examined in [18] for their suitability for on-site machine-to-machine communication. They were insufficiently performant for the use case of a teleoperated mobile robot. In another analysis, Mendoza et al. conclude that construction use cases could explicitly benefit from using 5G [19]. Considering only necessary sensors and recorded data, reliable machine-to-machine communication requires high availability, reliability, low latency, or high bandwidth and stability. Usually, these parameters are mutually exclusive but could be united under a 5G service network.

3 Methods

To develop a system that merges loosely coupled observation processing, linked data, and 5G networking, we apply the Design Science Research Methodology (DSRM) [20]. DSRM-driven research aims to extend human and organizational capabilities by designing "artifacts". These can be represented by constructs, processes, models, or prototypes that are evaluated when implemented. The artifact consists of our experimental hardware and software setup and is evaluated using a case study in a construction site environment. Since the system seeks to loosely couple sensor and actuator, our evaluation is realized through queries that prove that, regardless of the sensor's host, observation events can trigger actions without a specific observation intent. We consider the system to behave as expected if actions can be performed by querying the semantic observation events that are dynamically created without the need for knowledge about the host or sensor system, which verifies the extensibility of the communication system.

4 Case Study: The Observation Processing Platform

To evaluate our proposed methodology, we have developed a prototype of the system and deployed it in a construction scale case study derived from the EConoM research project. Here, a construction intralogistics use case is to be researched and developed, where material is transported to a construction site, autonomously unloaded, stored, and later transported to the area in the building where it is to be installed. The use case involves a variety of machines: a mobile robot platform and a construction lift for material transport (see Fig. 1). These machines differ significantly in their degree of digitalization and automation. This might be a typical scenario for future construction in the foreseeable future, wherein we are more likely to expect a mix of sensor-heavy robots, analogue machines and human workers. In the depicted scenario, the robots – in our case, the mobile robot platform – are data silos. They use their sensory systems exclusively for their respective tasks, such as navigation, without the ability to share their observations, even though these observations could be of relevance to other actors on the construction site. The key element of the described intralogistics use case is a scenario where physical objects limit the actors' perception. To address this, we chose to decouple parts of the sensory equipment in a data stream regard. With this setup, we aim to explore if the loose coupling of the systems can extend capabilities where we face physical constraints.

Fig. 1. Mobile robot platform entering a lift for material transport on a construction site while a human worker exits a construction site container in the background.

4.1 Hardware

The mobile platform is based on Innok Robotics' Heros platform. It has been modified to carry additional sensors and is equipped with a Milesight UR-75 router and antennas for 5G networking. The mobile robot's primary sensing system is an Ouster OS0–128 3D LiDAR scanner.[1] It acquires 3D point clouds at 20 Hz, resulting in a constant data stream of about 30 MB/s. Therefore, a stable bandwidth of approximately 240Mbit/s is

[1] http://www.oxts.com/wp-content/uploads/2021/01/Ouster-datasheet-revc-v2p0-os0.pdf

required to transmit the Packet Capture (PCAP)/UDP stream generated by the scanner. The Geda 1200Z construction lift is an analogue construction machine.

4.2 Experimental Scenario

In our experimental scenario, one mobile robotic platform must reach a designated material drop-off location accessible via an elevator. The construction elevator is retrofitted with a control mechanism that can receive wireless commands from other actors (a mobile platform or human worker) to execute basic operations: raise, lower, open, close and emergency stop. The mobile platform can thus enter the elevator and dispatch a motion command that would lift it to the desired floor. However, during the rising and lowering operations, the elevator safety barrier limits the field of view of the mobile platform, resulting in insufficient data for assessment of the environment outside the platform. Such loss of valuable information can cause significant safety concerns if, for instance, a human worker was to enter the restricted area while the elevator was in motion (see Fig. 2).

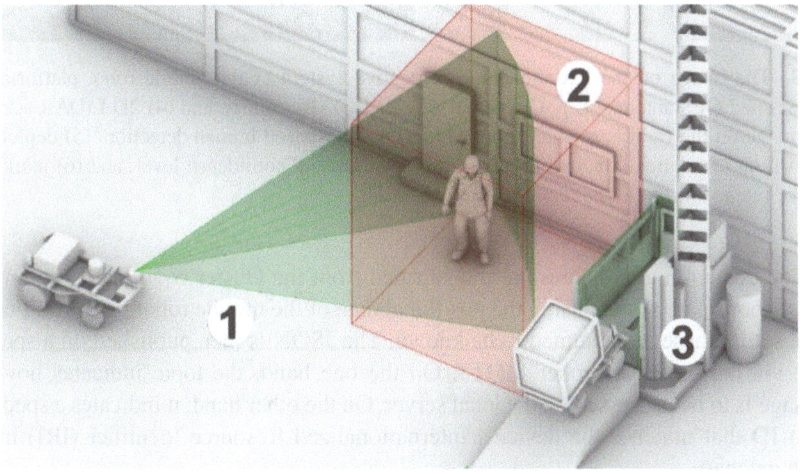

Fig. 2. Experimental scenario on the construction site: (1) Area with feature of interest in the field of view of the first mobile robot platform, (2) Restricted area around the lift when the lift is active, (3) Limited field of view of the second mobile robot platform when entering the lift.

4.3 Creating and Sharing Observations

The on-board LiDAR scanner requires stable wireless communication with high bandwidth, low latency, and high availability. To meet these criteria, we chose to roll out a private 5G network operating between 3.7 and 3.8 GHz. To grant the Ouster scanner access to the wireless 5G network, the mobile robot platform carries a 5G-capable industrial router to which the scanner is connected via ethernet. On the receiver side of the edge server, we are running a human detection software (based on YOLOv5[2]) that processes the data from the scanner. Each detected human is recorded as an observation JSON object, including detection time, detection location and confidence level (see Fig. 3).

Fig. 3. The image on the left shows the perceptive system of the mobile robot platform: (1) RTK refined GPS antenna, (2) 5G antenna, (3) 3D LiDAR scanner, and (4) 2D LiDAR scanner. The image on the right is a screenshot of the YOLO v5-based human detection. (5) depicts the observation of a human, including timestamp, location and confidence level, and (6) marks the zone of interest, the work zone of the lift.

The detection coordinates are transformed from the Ouster coordinate system to a global coordinate system using the positional data of the mobile robot platform. For the given setup, these are assumed to be known. The JSON is then published on a specific topic via the message broker MQTT. On the one hand, the topic indicates how the message is to be processed on the local server. On the other hand, it indicates a specified client ID that matches the device's Internationalized Resource Identifier (IRI) in the graph database.

In our setup, the client "`http://econom.one/CampusWeekDemo#OusterOS0-128_01`"publishes on the topic "`Refsite/HumanDetect`". The message is then handed over to Apache Kafka via kafka-connect. Apache Kafka[3] is an open-source, distributed, publish-subscribe messaging system designed for high throughput and low latency. Apache Kafka enables loose coupling by serving as an intermediary, enabling independent communication between data producers and consumers through asynchronous event publishing and reducing temporal and operational dependencies. In our configuration, Kafka adds the necessary flexibility to expand the system by further components such as OPC-UA [21] connectors or further stream processing applications in the future.

[2] *https://github.com/ultralytics/yolov5*

[3] *https://kafka.apache.org/*

Next, the central Linked Data Processing Core (IoC Core) instantly pulls the created JSON containing observation data and a connector to a time series database. For the latter, we use InfluxDB.[4] Independent of any following processing, all observations are stored persistently in a respective bucket belonging to the client IRI. Thus, all raw data is queryable by the edge system when using the timestamp as the key. The main component of the proposed edge system architecture is the IoC Core[13]. It is responsible for querying and updating the central triple store containing a linked data model of the construction site, including functional machine descriptions, spatial site and building topology, and dynamic process data. Furthermore, the IoC Core contains a configurable WebAPI and connectors to all relevant edge data repositories. In the IoC Core, the JSON containing the observation data is used to generate a SPARQL query. The query checks in various steps if the observation data leads to a relevant observation event that a) triggers an immediate machine actuation (in our case, an emergency stop) and b) creates an event in the triple store to be semantically described for other machines or agents to act or later analysis. While this is done in one query, the contents are hereinafter structured according to the underlying semantic constraints for modelling. The {x, y, z} coordinates of the observation are used to check if it occurred in a relevant zone. In the semantic construction site model in the triple store, machines can have workspaces classified as "`csto:HazardZone`". If the observation occurred in the workspace of the described construction lift, the HazardZone "`inst:ZeppelinLiftDZ_01`" will be returned. If no zone insertion is detected, no action is triggered. However, the observation data persists in the time series database (see Fig. 4).

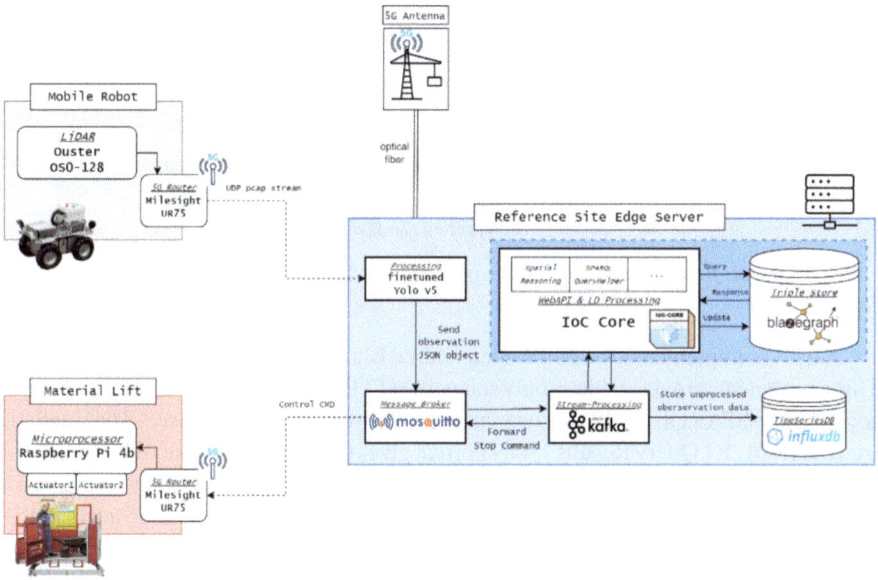

Fig. 4. Loosely coupled system architecture for sharing observations between actors on a construction site

---

[4] https://www.influxdata.com/

There are several approaches to storing geometries like zones in a linked data model. For use-case simplification, the zones queried in this artefact are described as boxes with boundary values {x, y, z}. These can be queried by checking if the given values are inside these bounds by a filter:

$$FILTER\left(?minX <\text{``}\{x\}\text{''}\hat{}\hat{}xsd : double\&\&\text{``}\{x\}\text{''}\hat{}\hat{}xsd : double <?maxX\right) \quad (1)$$

If the observation occurred in a HazardZone, the next step is to check whether this is relevant in the moment of observation. Therefore, it is checked if a condition of type "`eco:EmergencyStopCondition`" exists. In the modelled workspace of the construction lift, this condition is of type "`eco:CompositeANDCondition`". This means that it does not only consist of human presence as the stop condition but also of a running process. Both need to be met to trigger the emergency stop. This is typical behaviour for machine workspaces where danger is only imminent if the machine is running. In contrast, there are zones where no humans are allowed at any time, regardless of the processes performed. Lastly, the query takes the now-known IRI of the machine and lists its relevant actuators, their respective Endpoints (in this case, the MQTT topics) and the respective stop commands that need to be executed. For this critical functionality, data models can be checked via SHACL [22] to evaluate if all machines have this information available. An excerpt of the query can be seen in Fig. 5.

```
#FIND RELEVANCE HUMAN COND
?Zone eco:hasEmergencyStopCondition ?Condition.
?Condition eco:hasCondition ?Humancond.
?Humancond eco:isFulfilledBy inst:HumanPresence;
           eco:triggeredIfSimpleResult "true"^^xsd:boolean.
#CHECK IF PROCESS Condition is existent
OPTIONAL {?Condition eco:hasCondition ?Processcond.
          ?Processcond a eco:ProcessCondition.}
#CHECK OPTIONAL Process
OPTIONAL {?Condition eco:hasCondition ?Processcond.
          ?Processcond eco:startedCondition "true"^^xsd:boolean;
                       eco:finishedCondition "false"^^xsd:boolean;
                       eco:pausedCondition "false"^^xsd:boolean;
                       eco:cancelledCondition "false"^^xsd:boolean.
```

Fig. 5. SPARQL query checking for the presence of a human an possibly active workspace

A docker implementation of the open-source Blazegraphwas chosen as a triple store. This component is replaceable as only standard HTTP interfaces and Query Functionality according to SPARQL 1.1 [23], which most triple stores support, are used. The standardized SPARQL 1.1 Query Results JSON Format gives the server response. When receiving a positive response, as shown in Table 1, *STOPTOPIC* and *STOPCMD* are immediately used to execute the Emergency Stop Command as this is the most time-critical component to be activated. To this end, the IoC core sends out a Kafka message. This message is then published in MQTT by a sink connector that controls the construction lift's actuator. As a result, the lift stops.

Besides, an observation is created in the triple store, applying the concepts of the SOSA/SSN Ontology [16]. This describes the observation event semantically. Hence, not only do spatial relationships and conditions become machine-readable but also the

Table 1. Query Response as JSON

vars	values
Zone	<http://econom.one/CampusWeekDemo#ZeppelinLiftDZ_01>
Machine	<http://econom.one/CampusWeekDemo#ZeppelinLift_1200>
Processcond	<http://econom.one/CampusWeekDemo#ProcessCondition_01>
Processrunning	<http://econom.one/CampusWeekDemo#ProcessCondition_01>
STOPTOPIC	"ref_baustelle/maschinen/geda/1200z/"
STOPCMD	"EmergencyStop"

actions and events they trigger are accessible to all machines involved in processes on the construction site. When the sensor (origin) and timestamp are used, the time series database can additionally be triggered to retrieve the precise measurements or, for training purposes, to retrieve measurements close to the one that triggered the observation event Fig. 6 shows the instantiation of a "sosa:observation" in the graph model.

```
inst:HumanDetected_{GUID} a sosa:Observation, ioc:CurrentProcessDataState;
    sosa:observedProperty inst:HumanPresence;
    sosa:hasSimpleResult "true"^^xsd:boolean;
    sosa:madeBySensor inst:OusterOS0-128_01;
    sosa:hasFeatureOfInterest inst:ZeppelinLiftDZ_01;
    sosa:resultTime "2023-08-30T11:49:18+00:00"^^xsd:dateTime.
```

Fig. 6. SOSA observation

The resulting overall system response time to stop the machine is in the order of 70-100 ms running in a Docker environment on a PC with 16GB RAM and an AMD Ryzen 2700x processor. At this phase, the precise measurement of each network components' contribution to the overall response time is subject to ongoing research.

5 Discussion and Outlook

In the context of a construction site environment, we successfully demonstrated that practical loose coupling of sensors is achievable. While the artefact behaved as expected and the observation of a human presence in the defined hazard zone caused the lift to stop, the response time of the presented use case does not reach that of a primary safety system, as discussed in [19]. Therefore, the described implementation cannot replace a safety system. It can, however, aid in preventing safety problems before they occur. This can be done, for example, by scaling the danger zone to include the measured latency. It also enables preventive process logics, such as prohibiting the elevator from executing commands if a relevant observation has occurred. The artefact proves that observations created from sensor data can trigger commands to actuators. Further, it demonstrates that this is not subject to an initial intent. On the one hand, this means that the loosely coupled nature of the systems allows data from sensors to be reused for different purposes. On the other hand, multiple sensor inputs can be received from the same container and streamed to different processing applications. This means that other sensors that check for people in hazardous areas completely differently (e.g., infrared cameras or a light barrier) can be added to the system to augment it seamlessly. Future research should continue to optimize the system's latency, implement it for other use cases, and further modularize it. Processing logic, in the form of algorithms and scripts, should be able to be dynamically added depending on the use case or machine control to be implemented. In addition, the semantic description should be further developed so that the sensors and actuators added to the system are described in a machine-readable way. In this way, all participating robots and software agents can understand their capabilities, including what a sensor measures or a particular actuator manipulates, what protocols are required, and what constraints the participating machines inherit.

Acknowledgements. This work is part of the research project EConoM that is funded by the Federal Ministry for Digital and Transport of Germany within the initiative InnoNT (funding number 19OI22009F). It was supported within the framework TARGET-X, a project funded by the Smart Networks and Services Joint Undertaking (SNS JU) under Horizon Europe (funding number 101096614). The authors are responsible for the content.

References

1. *OECD*: Productivity and ULC by main economic activity (ISIC Rev.4), https://stats.oecd.org/Index.aspx?DataSetCode=PDBI_I4.
2. Abid Hasan*;* Bassam Baroudi*;* Abbas Elmualim *et al.:* Factors affecting construction productivity: A 30 year systematic review, Eng. Constr. Archit. Manag. Heft 25, 7 S. 916–937. doi. https://doi.org/10.1108/ECAM-02-2017-0035.
3. Snyman, T., Smallwood, J.: Improving productivity in the business of construction. Procedia Eng. **182**, 651–657 (2017). https://doi.org/10.1016/j.proeng.2017.03.175.
4. *Grand View Research:* Construction Equipment Rental Market Size, Share & Trends Analysis Report by Product (Earth Moving, Material Handling, Concrete & Road Construction), By Region, And Segment Forecasts, 2023–2030 (2023).
5. Kochovski, P., Stankovski, V.: Supporting smart construction with dependable edge computing infrastructures and applications. Autom. Constr. **85**, 182–192 (2018). https://doi.org/10.1016/j.autcon.2017.10.008.
6. Altıntaş, Y.D., Ilal, M.E.: Loose coupling of GIS and BIM data models for automated compliance checking against zoning codes. Autom. Constr. **128**, 103743 (2021)
7. Bizer, C., Heath, T., Berners-Lee, T.: Linked data—The story so far. Int. J. Semantic Web Inf. Syst. **5**, 1–22 (2009)
8. Gruber, T.R.*:* A translation approach to portable ontology specifications. *In:* Knowl. Acquis. 5 (1993), Heft 2, S. 199–220. doi:https://doi.org/10.1006/knac.1993.1008.
9. Bonduel, M., Oraskari, J., Pauwels, P., et al.: The IFC to linked building data converter—Current status. In: Proceedings of the 6th 2018 Linked Data in Architecture and Construction Workshop, p. 10

10. Rasmussen, M.H., Lefrançois, M., Schneider, G.F., Pauwels, P.: *BOT: the building topology ontology of the w3c linked building data group:* swj2224. Semantic Web. **12**, 143–161 (2021)
11. Wagner, A., Sprenger, W., Maurer, C., et al.: Building product ontology: Core ontology for linked building product data. Autom. Constr. **133**, 103927 (2022). https://doi.org/10.1016/j.autcon.2021.103927
12. Zheng, Y., Törmä, S., Seppänen, O.: A shared ontology suite for digital construction workflow. Autom. Constr. **132**, 103930 (2021). https://doi.org/10.1016/j.autcon.2021.103930
13. Kirner, L.*;* Lublasser, E.*;* Brell-Cokcan, S.*:* Internet of construction: Research methods for practical relevance in construction. *In:* Technol. Archit. Des. 2021 5, Heft 2, S. 146–152. doi:https://doi.org/10.1080/24751448.2021.1967053.
14. Fiorini, S.R., Carbonera, J.L., Gonçalves, P., et al.: Extensions to the core ontology for robotics and automation. Robot. Comput.-Integr. Manuf. **33**, 3–11 (2015). https://doi.org/10.1016/j.rcim.2014.08.004
15. Järvenpää, E.*;* Siltala, N.*;* Hylli, O. *et al.:* The development of an ontology for describing the capabilities of manufacturing resources. *In:* J. Intell. Manuf. 30 (2019), Heft 2, S. 959–978. doi:https://doi.org/10.1007/s10845-018-1427-6.
16. Haller, A.*;* Janowicz, K.*;* Cox, S.J. *et al.:* The modular SSN ontology: A joint W3C and OGC standard specifying the semantics of sensors, observations, sampling, and actuation. *In:* Semant. Web Interoperab. Usability Applicab.10 (2018), Heft 1, S. 9–32. doi:https://doi.org/10.3233/SW-180320.
17. Bellalta, B.: IEEE 802.11ax: High-efficiency WLANs. IEEE Wirel. Commun. **23**(1), 38–46 (2015). https://doi.org/10.1109/MWC.2016.7422404
18. Lee, H.J., Krishnan, A., Brell-Cokcan, S., et al.: Importance of a 5G network for construction sites: Limitation of WLAN in 3D sensing applications. In: Linner, T., García de Soto, B., Hu, R., et al. (eds.) Proceedings of the 39th International Symposium on Automation and Robotics in Construction. International Association for Automation and Robotics in Construction (IAARC), pp. 391–398 (2022)
19. Mendoza, J.; de-la-Bandera, I.; Álvarez-Merino, C.S. *et al.:* 5G for construction: Use cases and solutions. *In:* Electronics 10 1713; (2021), Heft 14. doi:https://doi.org/10.3390/electronics10141713.
20. Hevner, A.R., March, S.T., Park, J., et al.: Design science in information systems research. MIS Q. **28**(1), 75–105 (2004). https://doi.org/10.2307/25148625
21. Miriam Schleipen*;* Syed-Shiraz Gilani*;* Tino Bischoff *et al.:* OPC UA & industrie 4.0—Enabling technology with high diversity and variability. *In:* Procedia CIRP 57 (2016), S. 315–320. doi:10.1016/j.procir.2016.11.055.
22. *W3C* Shapes Constraint Language (SHACL), https://www.w3.org/TR/shacl/ (2023) Zugriff am: 01 Apr 2023.
23. *W3C* SPARQL 1.1 query language., https://www.w3.org/TR/sparql11-query/ (2023) Zugriff am: 01 Apr 2023

Spontaneous Tensegrity: Exploring Improvisational Design and Robotic Fabrication in Tensegrity Structures

Isla Xi Han[1,2](✉) and Stefana Parascho[1]

[1] Ecole Polytechnique Fédérale de Lausanne (EPFL), Laboratory for Creative Computation (CRCL), Lausanne, Switzerland
xihan@princeton.edu

[2] Princeton University, School of Architecture, Princeton, United States

Abstract. Over the last two decades, significant progress has been made in multi-robot systems (MRS) and human-robot interaction (HRI). In architectural applications, however, robots have mainly served as fabrication tools rather than design collaborators. Integrating robots into design and fabrication processes can yield numerous advantages, including enhanced adaptability at dynamic construction sites and more efficient and creative construction processes. This paper showcases how robots can cooperate with humans through visual computing to design and build X-module and T3-prism tensegrity structures. Instead of following predefined blueprints, the human-robot team relies on generalized procedures and spontaneous decisions, resulting in greater flexibility and an expanded design space. The experiments include: 1) Employing stigmergic mechanisms and visual servoing for robots to adapt to structural changes and human interventions during the fabrication of an X-module tensegrity structure. 2) Incorporating so-called robot design preferences to influence the final form of a T3-prism tensegrity structure. Audio feedback and direct human-robot interactions, such as material handling, make the robotic fabrication process more user-friendly and intuitive for human designers. The presented prototypes illustrate how robots can autonomously adjust during fabrication, promoting collective decision-making between robots and humans in the design process. The paper's contributions encompass four key aspects: 1) a novel robotic end effector design for manipulating X-module and T3-prism tensegrity modules; 2) a computationally inexpensive visual servoing method for robotic arms to track and interact with building elements; 3) human-in-the-loop improvisational workflows for the design and fabrication of tensegrity structures; 4) proof-of-concept prototypes.

Keywords: Collective human-robot construction · Human-robot interaction · Improvisation · Tensegrity · Stigmergy · Robotic fabrication · Multi-agent construction · Collective robotic construction · Construction robotics · End effectors · Visual computing

M. Yablonina et al. (Eds.): Robotic Fabrication in Architecture, Art and Design 2024, pp. 48–61, 2026.
https://doi.org/10.1007/978-3-032-22607-5_4

1 Introduction

Robotic technologies are adopted in the architectural field to enhance efficiency, enable new forms of expression, and automate the industry. Nonetheless, the currently dominant approach adheres to a linear design-fabrication workflow, where designers create digital models, programmers execute robotic path planning and sequencing, and fabrication takes place in a controlled factory or construction site. While this traditional approach has introduced notable advancements in robotic assembly processes (Parascho 2023), there remain drawbacks to the linear robotic fabrication workflow. First, traditional robotic fabrication processes often lack robustness against external factors such as material variations and structural deformations. Second, making design changes during the construction phase becomes prohibitively complex and costly due to the need to reprogram robots. Integrating these changes into a partially built structure also demands meticulous calibration, which is time-consuming and challenging. Third, the technical complexities surrounding robotic fabrication present a significant obstacle for architectural designers looking to fully explore the creative potential of this technology in their designs. Consequently, a substantial yet untapped potential exists for integrating robots into the design cycle in a user-friendly and intuitive manner (Han et al. 2021), enhancing efficiency and expanding the design space.

This research paper aims to elevate the agency of robots in architectural design and fabrication by positioning them as active collaborators rather than passive tools. Specifically, we introduced computationally inexpensive vision-based robot control to two ABB IRB 2600 industrial robotic arms (12 kg payload with 1.85 m reach). This technology enables these robots to make context-aware decisions rather than to adhere rigidly to predefined paths resulting from construction blueprints. Using a rule-based generalized procedure, robots and humans collaboratively design and fabricate tensegrity structures in a dynamic and flexible process.

Experiment I demonstrates how robots can adapt to external influences, such as structural deformations and human interventions, using local sensing and stigmergic mechanisms. Experiment II integrates various design influences, including robot preferences, human input, environmental variables, and structural constraints, throughout the construction process with an improvisational design-fabrication approach.

2 Related Work

2.1 Improvisation with Robotic Arms

Breaking away from the traditional linear workflow, we can draw inspiration from improvisation to envision a dynamic human-robot design-fabrication process. Improvisation is characterized by spontaneous decision-making and action devoid of pre-planning. This approach finds relevance in various creative domains, including theatre (Rond et al. 2019), dance (Jochum and Derks 2019), craft (Tokac et al. 2022), and music (G. Hoffman and Weinberg 2010). Notably, architecturally scaled projects, such as Improv-Structure (Han and Parascho 2022) and Tie a Knot (Mitterberger et al. 2022) experiment with human-in-the-loop co-design with robotic arms, marking a departure from conventional robotic construction practices.

Improvisational techniques can be categorized as open and closed skills (Pressing 2007), or external and emergent approaches (Troughton et al. 2022). The former involves responses to external stimuli, while the latter refers to self-produced movements or decisions. Various models for human-robot improvisation have been proposed. Troughton et al. combine both external and emergent techniques to create rule-based improvised robot movements that are playful and open-ended (Troughton et al. 2022). Hoffman builds upon related concepts such as inner monologue (Moore 1979) and responsiveness and maps such skills onto the application to robotics (Guy Hoffman, 2011).

In this paper, we integrate both internal (e.g., robot's design preferences) and external stimuli (e.g., human intervention, structural deformation, and environmental variables) to robotic arms to inform the improvisational design of tensegrity structures (see Section 4.2).

2.2 Collective Human-Robot Building Sequence

Research in multi-agent construction draws inspiration from nature to devise flexible assembly sequences. For instance, Werfel et al. replicated stigmergic construction principles inspired by termite behavior, where simple robots with minimal onboard sensing collectively achieve specific human-designed goals, such as 3D stacking, without human intervention (Werfel et al. 2014). Stigmergic construction means the agents do not directly communicate with each other but leave cues (physical or chemical) in the environment and make construction decisions according to such cues. Stigmergy is known for its robustness against external interventions. This concept serves as the foundation for our exploration of human-robot interaction in Experiment I (Section 4.1).

Conversely, another research trend focuses on a structured rhythm of human interventions, exemplified in projects like Tie a knot (Mitterberger et al. 2022), where human and robotic arms each contribute one rod to form a unit triangle to extend the existing structure. In improv-structure (Han and Parascho 2022), a new building cycle involves a robot placing rods in mid-air adjacent to the existing structure, with a human bridging the gap using bamboo rods to create organic weaving shapes. Experiment II (Section 4.2) delves into a more structured collaborative sequence.

2.3 The Assembly of Tensegrity Structures

Buckminster Fuller coined the term Tensegrity, derived from tensional integrity, to describe structures that combine compression elements (struts) and tension elements (tendons) to create stability in space (Pugh 1976). Unlike construction methods like stacking, casting, and spanning, where the relationship between each building block is tightly coupled, tensegrity structures exhibit strict topological rules while offering significant flexibility in geometric forms. This inherent flexibility fosters design versatility and expressive formal possibilities, making it an ideal candidate for improvisational structural design.

This paper focuses on two tensegrity unit types: the X-module, originating from Kenneth Snelson's *X-piece* sculpture of 1948 (Snelson 1996, 2012), and the T3-prism, the simplest prismatic tensegrity structure in 3D (Connelly and Terrell 1995). T3-prism also finds its root in Snelson's patented design, *the three-way tower* (Snelson 1965).

The assembly of a tensegrity structure is intricate. It typically involves a large team of human constructors holding elements until stability is achieved, as evidenced by projects like the MOOM Pavilion (KHUSHKARSTUDIO 2017) and Snelson's Soft Landing installation (Cohen 1999). There are a handful of robotic assembly processes developed for pre-defined tensegrity structures. Vasey et al. describe a process that utilizes the human's dexterity and the robot's precision to collaboratively fabricate pre-designed tensegrity units to form a large-scale structure (Vasey et al. 2016, Lafreniere et al. 2016). Nurimbetov et al. present a motion-planning strategy for robotic tensegrity assembly of a T3-prism unit with inverse kinematics (Nurimbetov et al. 2019). To the best of the authors' knowledge, our research paper marks the first endeavor in the field to improvise the design fabrication of a tensegrity structure through collaboration between robots and humans.

3 Methods

3.1 Material System and Gripper Design

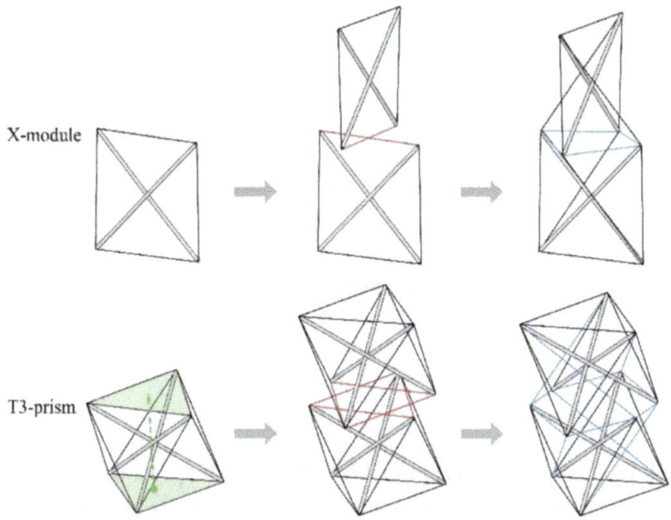

Fig. 1. Tensegrity modules and their aggregations: X-module and T3-prism

The tensegrity experiments covered in this paper consist of two unit types: X-module and T3-prism (Fig. 1). These units are individually stable but can also be combined. Once aggregated with additional tendons (marked in blue in Fig. 1), strategically removing selected tendons that support the original units (indicated in red in Fig. 1) reduces the redundancy of tension members in the combined structure. To facilitate this process, we employ two different materials for tendons. Nylon strings are used as temporary placeholders to indicate segments that can be removed after unit aggregation. Meanwhile, stainless steel ropes with diameters ranging from 1 to 1.5 mm serve as the more permanent tension members. The struts in the experiments are made of wooden rods with diameters ranging from 20-28 mm. One screw is drilled into each end of the strut with the stainless-steel rope wound around it to provide tension. Aluminum crip sleeves and tensioners are used to secure the tendons in place.

The customized end-effector we have designed (Fig. 2) is a pneumatic gripper-based tool engineered for handling X-module and T3-prism tensegrity units. The gripper features a unique perpendicularly positioned dual-jaw design with 3D-printed custom-shaped fingers that conform to rods of 24 ± 4 mm diameters. This perpendicular position is achieved by mounting two SCHUNK JGP 100–1 grippers onto an L-shaped aluminum profile, which is then affixed to the ABB IRB 2600 robot with a customized hardwood plate. A camera is mounted onto the L-shaped aluminum for 2D visual feedback. The control system utilizes computer vision algorithms to coordinate the robotic arm's movements (see details in Section 3.2).

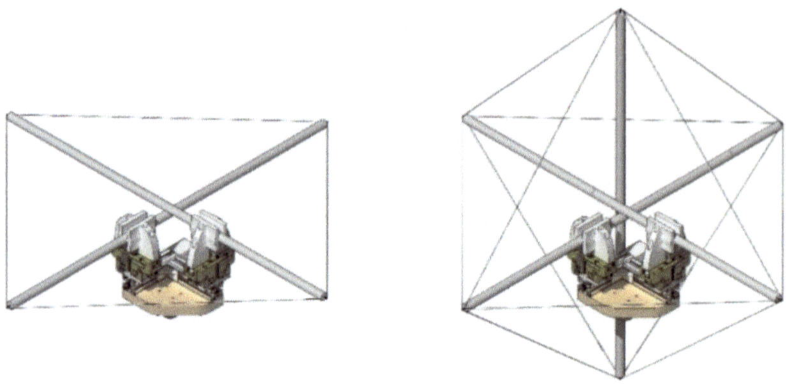

Fig. 2. Customized end effectors holding X-module and T3-prism tensegrity units

3.2 Visual Servoing

In the context of extending a tensegrity structure, knowledge of the location and orientation of a single rod at a specific topological position of an X-module or T3-prism unit can be sufficient to inform subsequent decisions. We mark these key struts in blue for ease of sensing and identification. The programs are written in Python using OpenCV (Bradski 2000) library for computer vision and COMPAS RRC (Van Mele et al. 2017) for robot movements. In order to balance context awareness and information overload, we process and analyze only the 2D visual data to track and inform the robot's relative positions to the key strut.

Calibration of the servoing system happens once at the beginning of the construction for each robot, given the lighting condition of the construction site doesn't go through drastic changes. To calibrate the system, the human operator needs to manually guide the robotic arm to a preferred relative position to the key strut for future reference. For example, in this paper, the calibrated relative location is for the robot to face the right-most end of a strut at roughly 35 cm away in **-z** direction of the tool-center point (TCP) coordinate system (Fig. 3, left), with the strut appearing entirely horizontal in the camera's view. We document the area of the key strut, denoted as $\mathbf{a_0}$, and the coordinates for the right-most strut end in the camera frame as point **T**. To create a target region **s** in the camera frame for tracking future key struts (Fig. 3, right), we assign a buffer distance of **p** for the horizontal direction and **q** for the vertical direction, resulting in a $\mathbf{2p} \times \mathbf{2q}$ rectangle **s** centered at **T**. Additionally, we define the upper and lower bounds of the HSV color range for the key strut's color (in this case, blue) as it appears in the camera for color filtering.

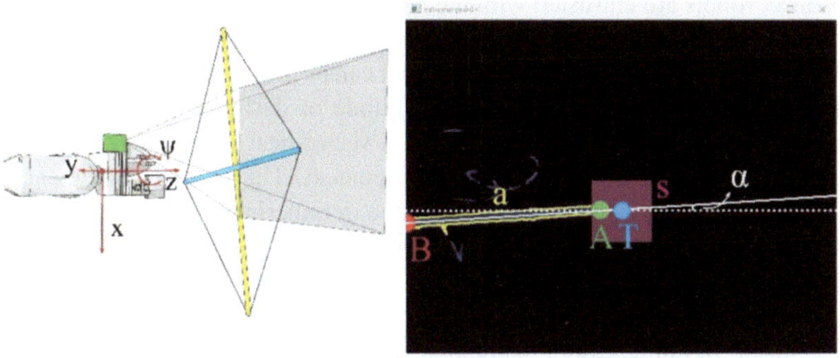

Fig. 3. Key parameters for visual servoing

With this information on hand, we can initiate the visual servoing process to track the key struts (in blue). This process begins with either manually positioning the robotic arm approximately 1 m away from the existing structure or directing the robot to move back 0.5 m in **-z** direction (TCP, Fig. 3, left) from its last building frame. These distances are relative to the scale of the building units. The main goal is to ensure the robotic arm has a clear view of the existing structure and sufficient space for location adjustments.

We isolate the struts of interest by employing a color filter with the preset HSV bounds. By selecting the region with the maximum area from the filtered regions, we focus exclusively on the nearest key strut. Our process begins by evaluating whether the right end of the key strut, denoted as point **A** (Fig. 3, right), falls within the designated region **s**. If it does not, the robotic arm makes a 1-step adjustment in the $\pm\mathbf{x}$ and/or $\pm\,\mathbf{y}$ direction (Fig. 3, left) to approach **T** in the next time frame, $\mathbf{t+1}$. If point **A** is within region **s**, the program proceeds to assess angle alignment, which is to minimize $\boldsymbol{\alpha}$ (the angular difference between the strut segment **AB** and the horizontal direction in the camera frame, Fig. 3, right). This is achieved by adjusting the value ψ (rotation around **z**-axis, Fig. 3, left). In the **z** direction, distance is estimated using the area of the closest strut, **a** (Fig. 3, right). $\mathbf{\Delta a = a - a_0}$ informs the displacement in **z** at the upcoming time frame. This method does not consider the rotation of the blue strut around the yellow strut for simplicity.

Through the iterative application of the steps above, $\boldsymbol{\alpha}$, $\mathbf{\Delta a}$, and distance **(T, A)** are gradually reduced to a preset tolerance (e.g., buffer **p, q**) over time. This process can be adapted to moving targets, allowing the robot to actively track a strut in a human's hand.

Eventually, the program exits the loop when the robot reaches the same relative location to the key strut as defined during calibration. This relative location forms the basis for any forthcoming robotic movements, whether it's proposing the next building frame or retrieving the key strut of interest from a human's hand (Sections 4.1 and 4.2).

3.3 Design Influencers: Human, Robot, and Environment

Even though tensegrity structures have specific topological rules (Fig. 1), there remains significant room for design freedom, with human, robot, and environmental factors collectively shaping the ultimate form of the structure. In Experiment II (Section 4.2), for instance, robots are entrusted with the autonomy to define two perpendicular struts within a T3-prism module (in blue and yellow), while humans dictate the placement of the third strut (in orange). Moreover, each robot can be programmed with distinct design characters (Fig. 7, right), resulting in a multitude of design proposals. Furthermore, the environment itself is preconfigured with a vector heat map (Fig. 7, left), exerting influence over each frame within the spatial context. The precise implementation of these influential factors will be expounded upon in Experiment III in Section 4.3.

3.4 Human-Robot Interaction

Multiple means of human-robot interaction are integrated into the experiments:

- **Vision:** Robots, with cameras and computer vision systems, can react to building elements within existing structures and to humans holding construction units.
- **Audio:** Human operators receive real-time updates about the robots' status through audio cues played via a laptop speaker, informing them of the robots' actions and providing instructions, such as Please feed rods into the gripper.
- **Direct Physical Contact:** Using computer vision, robots can physically interact with building elements, directly grasping them from a human's hand when instructed. This collaborative potential enhances design interaction; for example, in constructing tensegrity structures, humans can summon a robot for on-the-spot assistance.

4 Experiments

Two experiments in this paper explore building tensegrity structures with various complexities, including variations in geometry, agency distribution, and collaboration modes between humans and robots. To enhance clarity, a color-coding system was employed for the wooden struts: pink for the initial seed strut, blue for key struts guiding robotic actions, yellow for neutral struts aiding in forming X-modules with blue struts, and orange for manually placed struts by humans.

4.1 Experiment I: Stigmergic Construction of an X-module Tensegrity Structure

Experiment I drew inspiration from Kenneth Snelson's Self-Portrait in Pendleton, Oregon, in 1948. Experiment I explores the flexibility of construction sequences involving both humans and robots. This investigation relied on a stigmergic logic (see Section 2.2) applied to a thick vertical 2D structure. Within this framework, robots exhibit adaptive responses to existing structures and adjust to tolerate local inaccuracies.

The construction of this structure only requires the robot to execute a simple program repeatedly for the placement of each new unit. Initially, the camera undergoes calibration following the procedure in Section 3.2, with reference to one end of the blue strut in an X-module unit. Additionally, three predefined vectors, namely $\mathbf{v}_1$, $\mathbf{v}_2$, and $\mathbf{v}_3$, guide the robot in searching for struts within the general regions around $\mathbf{C}_1$, $\mathbf{C}_2$, and $\mathbf{C}_3$ (Fig. 4). At the beginning of the program, the robot initiates from one preset point (Fig. 4) and moves downwards following direction $\mathbf{v}_0$. While in motion, the robot scans for the presence of a blue key strut. Upon detecting a key strut, the robot fine-tunes the end-effector's location to align precisely with the strut's position (Section 3.2). The robot then checks regions $\mathbf{C}_2$ and $\mathbf{C}_3$. If a key strut is not detected, the robot places a new X-module unit at that location. If both $\mathbf{C}_2$ and $\mathbf{C}_3$ are occupied, the robot places a new strut unit at a displacement of $\mathbf{v}_3$ from $\mathbf{C}_3$. While the robot securely holds the X-module unit in place, the human builder inserts the connection rods (in orange) in the diagonal direction. The human also removes temporary nylon ropes and connects new steel cables according to the topological relationship illustrated in Fig. 1. Since the robot consistently fine-tunes its position in response to the presence of a key strut in the area of interest (e.g., $\mathbf{C}_1$, $\mathbf{C}_2$, $\mathbf{C}_3$), this process enables the robot to efficiently adapt to any deformations or minor changes in the existing structure, whether they result from material tolerance, human inaccuracies, or purposeful design intentions.

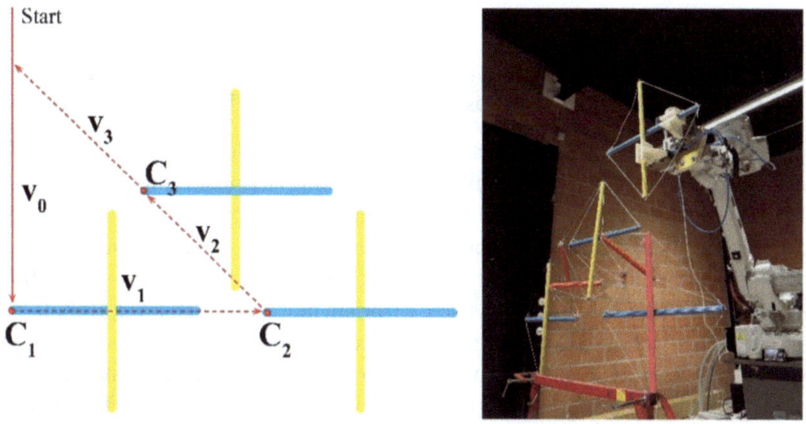

Fig. 4. Stigmergic construction for X-module tensegrity structure

Stigmergic construction exhibits robustness against external disruptions. For example, if a human inserts a new X-module, it will not adversely impact the robot's subsequent actions. The robot will proceed to add new pieces to the structure. This enhances design flexibility; for instance, humans can intentionally extend the structure in a particular direction, which the robot will seamlessly adapt to as structural tolerance.

Handover. In the same setting, we also test out an additional function that asks the robot to take a building unit from a human hand and hold it in place. This handover mode entails the robot tracking, approaching, and grasping the key strut of interest (Fig. 5), after which it maintains the unit in position for the humans to work on. By occasionally employing the handover function within the context of stigmergic construction, humans can introduce design interventions not only through the placement of orange connectors but also by directly positioning the blue-yellow struts with the robot's assistance.

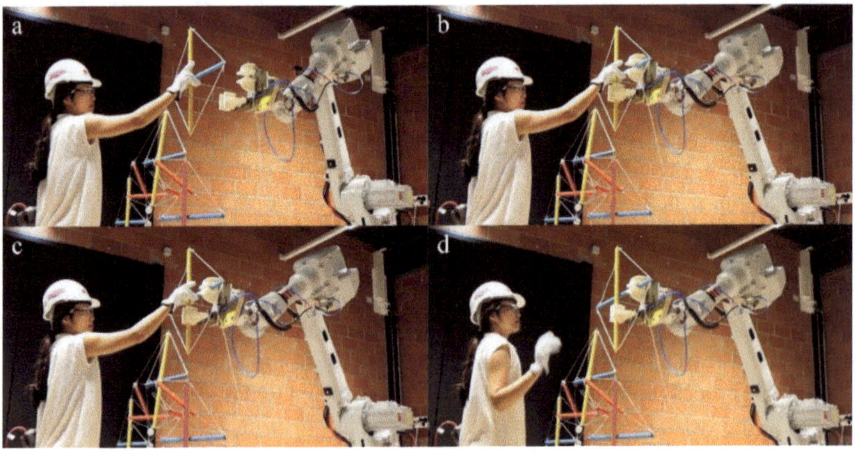

Fig. 5. Hand over a building unit to a robot. a) The robot detects the unit. b) The robot approaches the unit. c) Grippers close. d) The robot holds the unit in place

4.2 Experiment II: T3-Prism Tensegrity with Multiple Design Influences

Experiment II extends the design space from regular thick vertical 2D into irregular 3D tensegrity. The general goal is to build a cantilevering T3-prism tower (refer to Fig. 1 for T3-prism connection logic). Unlike a conventional T3-prism tower with uniformly sized components and consistent rotation angles between stacking units, Experiment II grants the robot autonomy to determine the orientation of two struts while allowing human placement of one strut within each T3-prism unit. On a high level, Robots #1 and #2 take turns extending the structure from the central seed unit towards opposite sides. They locate the nearest blue key strut, propose a relative displacement, and suspend two out of three struts (blue and yellow) in mid-air within a T3-prism unit. The human then inserts an orange strut at their preferred location and orientation to complete the unit and connects tendons to integrate the new unit into the existing structure, following the logic in Section 3.1 (Fig. 1).

The experiment setting involves two ABB IRB 2600 industrial robotic arms positioned 3 m apart, facing each other on elevated platforms 0.725 m high. The pink seed strut is 2.4 m tall with a 28 mm radius. It is anchored to a metal block on the ground, positioned equidistant between the two robots. Along with two additional struts (including one blue key strut), the pink strut forms the seed T-3 prism unit. The two robotic arms initiate the design process by utilizing this seed unit as a reference point for sensing and generating proposals for the placing of new units (Fig. 6).

Fig. 6. Human designer connecting steel cables

To start the building process, each robot first locates and finetunes its end effector's position in reference to the blue key strut of the last placed module using the method in Section 3.2. Once the key strut is located, the robotic arms advance toward a new frame. This frame is the result of three vectors, namely:

1. **A constant vector** addressing topological requirements, predetermined according to the height of a typical unit when uniformly stacked. This vector originates from the centroid of the bottom triangles and extends to that of the top triangles, highlighted in green in Fig. 1.
2. **An environmental vector** based on the end effector's location within a predefined 3D vector heatmap in the world coordinate system (Fig. 7, left). This environmental vector, continuously applied across the entire structure, ensures uniformity of design language. In Experiment II, the heatmap imposes fewer displacements in the central area of the site, thus aiding in aligning suggestion frames from two robots positioned on opposite sides. This heatmap can be easily adapted in future construction to reflect various design constraints.
3. **A robot preference vector** determined by the gripper's position relative to each individual robot's base frame coordinate system (Fig. 7, right). In this case, robot #1 adopts a uniform vector across its reachable zone, while robot #2 employs vectors pointing outward from the robot's base to create more space for movement and reduce the risk of collisions. These preferences, though relatively straightforward in this preliminary experiment, have the potential for enhancement in future iterations, for example, by incorporating visual inputs for more sophisticated design suggestions.

   The weight of the three influences can be tuned within the Python program to achieve the desired balance of design influences.

The design fabrication process emphasizes human participation by allowing human constructors to manually insert the third orange strut into the T3-prism unit at their desired position and orientation. Humans also have the flexibility to adjust the structure by tightening tendons.

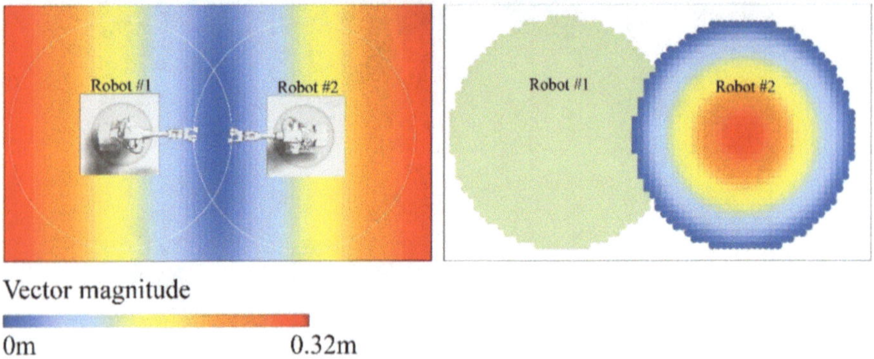

Fig. 7. Influencing vectors–a horizontal section view. Left: Heat map of the environment vectors. Right: Heat map of the robot preference vectors

In this experiment, a total of two T3-prism units are added by robot #1 and one unit by robot #2, with human involvement in cable connection, detachment, and placement of the orange struts. The resulting structure (Fig. 8), measuring 2.5 m x 1.6 m x 2.5 m, exhibits asymmetry due to angular variations and positional adjustments introduced by both humans and robots. Additionally, the system demonstrates a degree of flexibility in accommodating variances in rod radius and length.

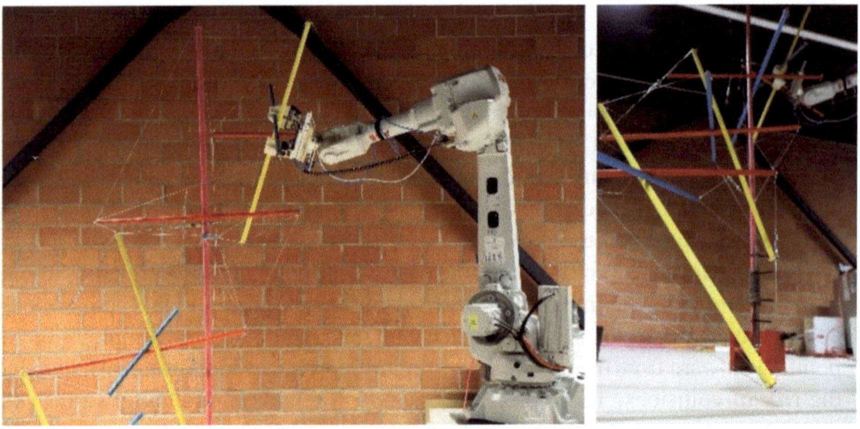

Fig. 8. Spontaneous Tensegrity - Prototype II

5 Conclusion and Outlook

Through two prototypes, this paper introduces novel ways for humans and robots to collaborate in designing and constructing complex tensegrity structures characterized by simplicity in logic rather than precise blueprints for element placement. Experiment I involves creating a thick vertical 2D tensegrity structure comprising X-modules, following a stigmergic logic, allowing room for human-driven variations while ensuring robotic adaptability to changes and sequencing disruptions. Experiment II elevates the agency of both humans and robots in design by constructing a T3-Prism structure, resulting in a blend of environmental, human, and robotic influences. Overall, this paper presents human-in-the-loop improvisational design and fabrication techniques specific to tensegrity structures alongside a novel end effector design and visual servoing methods for robotic arms, supported by preliminary proof-of-concept prototypes.

The current methods exhibit several limitations. As the robots only track selective local information (e.g., the nearest blue key struts) and lack an internal model of the existing structure, their proposed building frames are not considering the structural performance of the entire structure. Thus, human judgment is required to determine the viability of robot proposals. Despite Experiment II demonstrating humans and robots creating an irregular T3-prism tower, adherence to strict topological relationships persists. In experiment II, the structure quickly reaches its limits, as evidenced by the second unit proposed by Robot #2 nearly touching the ground and tilting at an angle that precludes adding another unit. Despite the prototypes laying the groundwork for integrating robot information into the design-fabrication process alongside human choices, there is a need for enhancements to enable robots to generate proposals that better consider surrounding obstacles and their own reachability constraints.

Looking forward, we envisage several potential advancements:

- The development of a computationally inexpensive structural simulator utilizing existing key strut data to enhance predictions of structural performance and limitations;
- The integration of a more context-aware environmental and robotic preference vector, for example, vector displacements in response to visual inputs instead of pre-defined heat maps;
- The potential development of a secondary mobile robot to adjust tension cables within the constructed structure. Given the directional nature of tensegrity aggregation, these minor adjustments can facilitate the extension of the structure in specific directions.

References

Bradski, G. 2000. The OpenCV Library. Dr. Dobb's Journal of Software Tools.

Cohen, M.D.: Kenneth snelson: The lyricism of structure. Sculpture. **1**, 1999 (1999) https://sculpturemagazine.art/kenneth-snelson-the-lyricism-of-structure/

Connelly, R., Terrell, M.: Globally rigid symmetric tensegrities, p. 21. Struct, Topol (1995)

Han, I.X., Meggers, F., Parascho, S.: Bridging the collectives: A review of collective human–robot construction. Int. J. Archit. Comput. **19**(4), 512–531 (2021)

Han, I.X., Parascho, S.: Improv-Structure: Exploring Improvisation in Collective Human-Robot Construction. In: Correia, A.G., Azenha, M., Cruz, P.J.S., Novais, P., Pereira, P. (eds.) Trends on Construction in the Digital Era Lecture Notes in Civil Engineering, pp. 233–243. Springer International Publishing, Cham (2022). https://doi.org/10.1007/978-3-031-20241-4_16

Hoffman, G., Weinberg, G.: Gesture-based human-robot jazz improvisation. In: In: 2010 IEEE International Conference on Robotics and Automation, pp. 582–587 (2010). https://doi.org/10.1109/ROBOT.2010.5509182

Hoffman, G.: On Stage: Robots as Performers. In: RSS 2011 Workshop on Human-Robot Interaction: Perspectives and Contributions to Robotics from the Human Sciences, vol. 1, p. 21, Los Angeles, CA (2011)

Jochum, E., Derks, J.: Tonight We Improvise!: Real-time tracking for human-robot improvisational dance. In: Proceedings of the 6th International Conference on Movement and Computing, pp. 1–11. ACM, Tempe AZ USA (2019). https://doi.org/10.1145/3347122.3347129

KHUSHKARSTUDIO: Case study: MOOM pavilion. Khushkar Design Studio (blog). (April 25, 2017) (2017) https://khushkarstudio.wordpress.com/2017/04/26/case-study/

Lafreniere, B., et al.: Crowdsourced fabrication. In: Proceedings of the 29th Annual Symposium on User Interface Software and Technology, pp. 15–28 (2016)

van Mele, Tom, et al. 2017. COMPAS: A framework for computational research in architecture and structures. http://compas.dev. doi:https://doi.org/10.5281/zenodo.2594510

Vasey, Lauren, et al. 2016. Collaborative Construction: Human and Robotic Collaboration Enabling the Fabrication and Assembly of a Filament-Wound Structure. In ACADIA//2016: POSTHUMAN FRONTIERS: Data, Designers, and Cognitive Machines [Proceedings of the 36th Annual Conference of the Association for Computer Aided Design in Architecture (ACADIA) ISBN 978–0–692-77095-5] Ann Arbor 27–29 October, 2016, Pp. 184–195. CUMINCAD. https://papers.cumincad.org/cgi-bin/works/paper/acadia16_184

Mitterberger, D., Atanasova, L., Dörfler, K., Gramazio, F., Kohler, M.: Tie a knot: Human–robot cooperative workflow for assembling wooden structures using rope joints. Constr. Robot. **6**(3), 277–292 (2022). https://doi.org/10.1007/s41693-022-00083-2

Moore, S.: Training an actor: The Stanislavski system in class. In: Harmondsworth. Penguin Books, New York (1979) http://archive.org/details/trainingactorsta0000moor

Nurimbetov, B., Issa, M., Varol, H.A.: Robotic assembly planning of tensegrity structures. In: 2019 IEEE/SICE International Symposium on System Integration (SII), pp. 73–78 (2019). https://doi.org/10.1109/SII.2019.8700342

Parascho, S.: Construction robotics: From automation to collaboration. Annual Review of Control, Robotics, and Autonomous Systems. **6**(2023), 183–204 (2023)

Pressing, J.: Improvisation: Methods and models. In: Physical Theatres: A Critical Reader. Routledge (2007)

Pugh, A.: An Introduction to Tensegrity. University of California Press. (1976). https://doi.org/10.1525/9780520338326

Puzatova, A., Shakor, P., Laghi, V., Dmitrieva, M.: Large-scale 3D printing for construction application by means of robotic arm and gantry 3D printer: A review. Buildings. **12**(11), 2023 (2022). https://doi.org/10.3390/buildings12112023

Snelson, Kenneth D. 1965. Continuous tension, discontinuous compression structures. U.S. Patent 3, 169,611, issued February 16, 1965

Snelson, K.: Snelson on the tensegrity invention. Int. J. Space Struct. **11**(1-2), 43–48 (1996)

Snelson, K.: The art of tensegrity. Int. J. Space Struct. **27**(2-3), 71–80 (2012)

Spencer, L., Peng, Y., Smallidge, P., Reiter, M.T., Zivkovic, S.: Slimlam method: Robotically bandsawn timber for glulam applications. Constr. Robot. **7**(2), 177–191 (2023). https://doi.org/10.1007/s41693-023-00102-w

Tokac, Iremnur, Benay Gursoy, Herman Bruyninckx, and Andrew Vande Moere. 2022. Craft-inspired digital fabrication: A study of interactive robotic clay carving. In: Proceedings of the 7th Annual ACM Symposium on Computational Fabrication, 1–14. SCF '22. New York, NY, USA: Association for Computing Machinery doi:https://doi.org/10.1145/3559400.3562003

Troughton, I.A., Baraka, K., Hindriks, K., Bleeker, M.: Robotic improvisers: Rule-based improvisation and emergent behaviour in HRI. In: 2022 17th ACM/IEEE International Conference on Human-Robot Interaction (HRI), pp. 561–569 (2022). https://doi.org/10.1109/HRI53351.2022.9889624

Werfel, J., Petersen, K., Nagpal, R.: Designing collective behavior in a termite-inspired robot construction team. Science. **343**(6172), 754–758 (2014). https://doi.org/10.1126/science.1245842

Alternative Materials

Additive Manufacturing with Graded Bio-Polymer Composites

Nicholas Paul(✉), Eppinger Carl, Chiujdea Ruxandra, Sonne Konrad, and Ramsgaard Thomsen Mette

Center for Information Technology and Architecture (CITA), Royal Danish Academy, Philip de Langes Allé 10, 1435 Copenhagen K, Denmark
{paul.nicholas,cepp,rchi,kson,mette.thomsen}@kglakademi.dk

Abstract. Bio-based materials are emerging a key alternative for carbon-intensive construction materials, which are becoming increasingly ill-suited to society's ambitions to decarbonize construction. Beyond already established low carbon biomaterials such as timber, locally sourced second and third generation waste and side-stream biomass is recognized as having significant potential to be used in architectural elements. However, the use of these materials requires processing routes based in contemporary design and fabrication technologies. Additive manufacturing is one such route. This paper outlines an approach for graded 3D printing with a biopolymer-composite that incorporates local waste and agricultural side stream residues. In this paper, we describe the underlying motivations, challenges and opportunities for multi-material grading, and detail the development stages of a custom-built robotic tool and feeding system that enables continuous grading of two or three material inputs. Further, we explain the complications involved with material viscosity changes and the correlation to the design of the printing toolpath. We demonstrate the use of the custom-built robotic tool and feeding system on a set of architectural-scale weather screen panels, using different biopolymer mixtures and grading strategies.

Keywords: Biopolymer composites · Robotic 3D printing · Material grading

1 Introduction

Architecture is engaging on a transformative shift towards biomaterials as a central part of its efforts to decarbonize the built environment. Biomaterials not only sequester carbon during their growth phase but also have a significantly reduced carbon footprint in their processing and production compared to more conventional construction materials such as concrete and steel. While timber, especially in the form of cross-laminated timber and other engineered wood products, is the leading biomaterial used in construction, it cannot be relied on alone, as increasing timber consumption would further challenge sustainable forestry practices (O'Brien and Bringezu 2017). Furthermore, different architectural applications require materials with different structural or visual properties and performances. It is therefore important to diversify the range of sustainable biomaterials

M. Yablonina et al. (Eds.): Robotic Fabrication in Architecture, Art and Design 2024, pp. 66–80, 2026.
https://doi.org/10.1007/978-3-032-22607-5_5

used in architecture and to develop processing routes for them that take advantage of contemporary design and fabrication technologies.

This paper describes the graded robotic 3D printing of a bio-based polymer composite. This material combines a collagen-based biopolymer with fibrous and filler materials that are also generated from biomass and the growth of microorganisms and plants. Grading different composites, allows to precisely distribute materials based on their mechanical performance. Rather than using virgin sources, the fibers and fillers are sourced from second and third generation feedstocks: biological residues and waste streams that would otherwise be burnt, landfilled, or left to decay. Valorizing them within 3D printed architectural elements stores and transfers their carbon into architectural material flows, helping to enable the built environment to become a store of atmospheric carbon.

However, these inherently local material resources introduce a new dynamic to the sourcing and flow of resources. Their quantity and quality are not constant – they are seasonally dependent, connected with local agricultural practice and terroir, and their properties vary in line with weather and climate. As new resource loops emerge within the circular bioeconomy they stop being wastes and raise in value. In summary, their continuous ready availability and perceived abundance cannot be assumed. To scale the use of this class of bio-based polymer composite within architectural 3D printing will therefore require modes of material specification and production that can address two criteria: material flexibility – the ability to interchange fillers and fibers in response to local material availability, and material optimization – the ability to specify the distribution of a graded material in response to design or performance criteria.

Our paper introduces the design of an extruder system that aims to address both of these criteria by enabling material grading. In the paper, we firstly describe state of the art in robotic 3D printing with bio-based materials for architectural applications, and then detail the specific challenges associated with our biopolymer composite material. We then outline the design and development of a system for the graded extrusion of a collagen-based bio polymer composite material. Through the description of four stages of design evolution, we address the key considerations that have informed the system design. These include material-based considerations, the mixing of multiple material feeds, the implementation of continuous feeding from a separated reservoir, and the grading control in the fabrication code. Lastly, the publication describes the application of our graded printing system to fabricate architectural scale wall panels that perform a weather screen function, and concludes with a discussion of these results and identification of the next steps that they imply.

2 State of the Art

The necessity to replace conventional building materials with more sustainable alternatives is a key driver for the construction industry (Pamu and Alugubelli 2023). Biopolymers and biopolymer composites are a particularly promising alternative. They can use less energy to produce and are in many cases by-products or waste products of other industries, activating cascading and recycling cycles (Plank 2004; Aaliya et al. 2021). Architecturally-oriented research has explored different sources and compositions of biopolymer (Bever and Duwez 1972) including chitin, cellulose or casein, as well as the use of different concentrations of these polymers with varying uses of fillers or fibers (Rech et al. 2021; Pradhan et al. 2021). The inherent properties of those materials have been linked to design explorations of transparency and translucency, color variation or structural value (Tusnim et al. 2020). Research that has created composites combining biopolymers with similar material states, viscosities and workability, but with different properties of the finished product (Lee et al. 2020), points to the vast further opportunities for research and application of biopolymers within architecture.

Realizing these possibilities within architectural elements is to a large extent dependent on fabrication processes. Much of the initial research has been achieved using small scale casting and molding approaches, with the aim of exploring recipes and producing simple geometries to enable assessment of material properties (Tarangini et al. 2023). 3D printing, an inherently scalable technology, has the potential to greatly enlarge the design space as well as the size of printed elements (Rossi et al. 2022, Sanandiya et al. 2018). However, this fabrication method also requires that a more complicated set of considerations need to be taken into account. These challenges include creating a printable material, controlling the transition of the material from viscous to firm across the printing process, and combining multiple material feeds of differing viscosities within a single print.

A small amount of literature reports multi-material robotic 3d printing with biopolymers. These investigations have focused on printing biopolymers without fillers (Mogas-Soldevila et al. 2015) or have developed approaches based on co-extrusion instead of material mixing within the printing process (Seibold et al. 2019). While co-extrusion enables two different materials, with different material properties, to be combined within one print, this method does not enable their continuous combination and mixing. A further approach to grading sequentially deposits different material recipes in a set order which then combine into a single graded result. In this approach, the exchange of tool or material during the printing process is highly time consuming and still limits the possibility of seamless material blending. (Lesna and Nicholas 2020; Mogas-Soldevilla et al. 2014).

3 Method

3.1 Collagen Glue Based Biopolymer

The biopolymer used in this investigation is composed of a mixture of collagen glue, water and glycerin. Compared to other common polymers used for 3d printing, this biopolymer enables significant adaptability to incorporate different fillers, as well as an ongoing thermoplastic response, however is also temperature sensitive and highly viscous. The biopolymer's adhesive quality enables strong interlayer fusion and bonding, as well as a highly consistent adhesion to the print bed. These properties favor the creation of more extensive prints, without the possibility of failure towards the end of the fabrication, however also create challenges both in the design of the extruder as well as the design of the printing path.

The high viscosity of the collagen-based biopolymer disrupts the mechanical components of valves, gates and monitoring devices that come in direct contact with the material. This impacts the design of the extruder and delivery system by limiting such flow-control devices to those that only have indirect contact with the material.

The temperature sensitivity of the collagen-based biopolymer requires a precisely controllable and closed heating system to control the viscosity of the material throughout the entire material feed. This requested the system to be able to react to the material's temperature, to guarantee a lower viscosity during the material feeding, but a higher viscosity at the point of extrusion to create a stable material bead.

At the level of the print path, the adherence of the polymer to both the extruder and itself also creates challenges. As the material is less deposited by the extruder, but more pulled out, by adhering to previously printed material or the print bed, free standing directional changes become challenging and must be approached through the design of the print path through methods such as short pauses at critical moments, as well as print geometry and direction control.

The introduction of different fillers introduces further considerations. To create the biopolymer composite, the biopolymer is mixed with varying combinations of three fillers – cotton, wood flour and bark. The grading of these fillers, such as a change from lower density materials like cotton to high density materials such as seagrass, impacts the viscosity of the material within the extruder, as well as the mechanical performance of the material once it is dried. As the composites are mixed by ratios of filler weight, cotton or wood flour drastically increase the volume of the material thus reducing the viscosity. The young's modulus varies between 2,51 GPa for a mixture of 6% cotton fiber by weight and 2,43 GPa for a 10% wood flour mixture.

3.2 Extruder Development

The custom extruder was developed in four stages, with each stage considering one major aspect of the intended goal of graded extrusion. The initial stage consisted of designing a single material extruder, without any grading capabilities, to understand the material and the implications of printing with different fiber ratios. The extruder was built out of a combination of aluminum and nylon. The majority of the extruder body was made of different sized aluminum tube extrusion, taking advantage of its high thermal conductivity and lightness, with custom parts manufactured using selective laser sintering of nylon. Key considerations in the choice of both materials was their ability to withstand the temperature of the biopolymer, which can reach up to 70 °C, robustness against abrasions from the fillers, reduction of the possibility of the collagen bonding with the extruder, and easy cleanability. The extruder was fed using a heated pneumatic pump. The second stage of the tool development (Fig. 1, left) tested a first incorporation of material grading. For this the initial extruder design was adapted to include two additional separate canisters. These could be individually controlled to add material accents to a continuously feeding base material.

Due to the different fiber ratios, the two additional materials needed to be kept at different temperatures, between 39 °C and 45 °C, to create an equal level of viscosity throughout the printing system. This included a separated heating system using resistance wires coated in silicone and separate heating controls, adjusted based on the optimal material temperatures.

In the third stage (Fig. 1, top right), the extruder was redesigned to implement insights gathered from the previous stages, most notably the difference in material flow between the base material feed and the accent material reservoirs. The additional reservoirs were removed and replaced by two more connections to external continuous feeding systems. To stop the residual flow and buildup of pressure, those connections were equipped with pneumatic cylinders acting as pinch valves. This constraint reduced the amount of buffer material to the material already in the extruder core and thus limited the time the secondary material needed to replace the initially printing primary material. Smaller adjustments were made to simplify the printing process, such as the integration of an overflow limit switch. This adjustment reduced the impact of the pressure used in the pump, as the limit switch closes the valve and cut the pressure to give the extruder time to process the material build up before restarting the material feed.

After testing with this tool, which greatly improved the consistency and effect of grading, the decision was made to develop a reduced version of this extruder (Fig. 1, bottom right). This design reduces the time taken to change between materials by reducing the material held within the extruder itself, in turn enabling more immediate grading strategies. In all stages the material mixing was controlled through the flow rate as the auger screw feeding the material also acted as a tool to homogenize the materials. The reduction in volume of this last stage omitted the necessity for the previously added pinch valves, as the volume where both material feeds would meet was severely reduced and used only to mix the material grade.

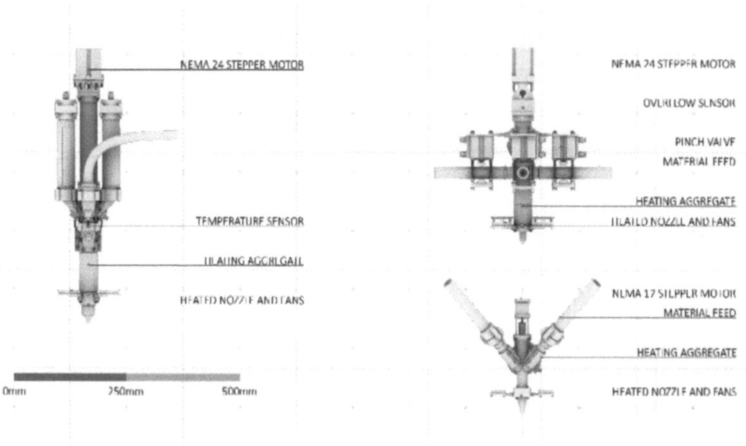

Fig. 1. Extruder development; from left to right: Stage 2 extruder with accent material reservoirs, Stage 3 extruder with pinch valves and additional material feeds, Stage 4 extruder with reduced extruder body.

3.3 Continuous Material Feeding Systems

During the process of extruder development it became clear that the use of one-liter reservoirs would greatly limit the extent of printing. Direct experience showed that replacing and refilling these reservoirs during the print resulted in inconsistent print qualities due to the fluctuation in material flow. To address this problem the small reservoirs were replaced with an external material feeding system that could hold three liters of material and be refilled without interruption to the printing process. This reservoir and pneumatic pump were developed by World Advanced Savings Project (WASP, 2024). During the initial grading tests with the stage 2 extruder it became clear that for controllable grading the materials need to be fed at a similar rate, which required a secondary pump. This additional pump and feeding system was built on the same basic principles as the WASP system, using pneumatics to feed the material (Fig. 2). Coordinated control for both pneumatic pumps was programmed into the robotic code, with Digital Outputs used to open or close the input of compressed air. The selected pressure was controlled by the operator to allow for a higher degree of intervention during the print.

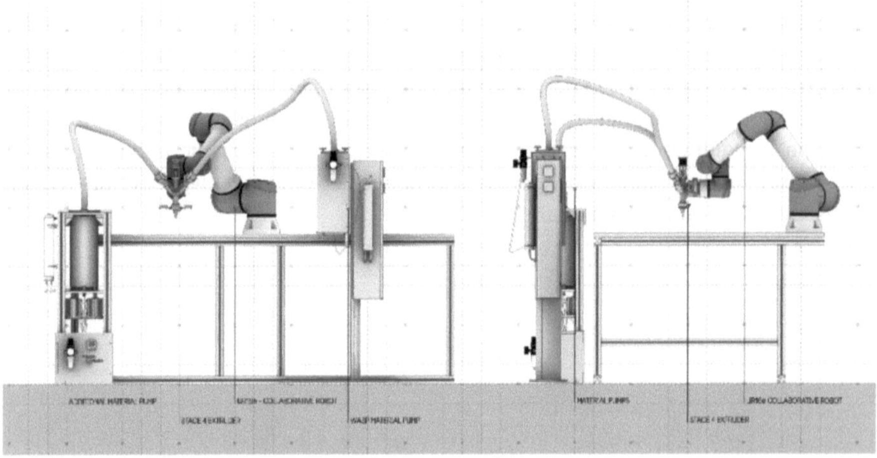

Fig. 2. Robotic fabrication setup; Universal Robots UR16e equipped with two material feeds and Stage 4 extruder.

3.4 Grading Test Strategy

Understanding the time necessary to switch between two materials is critical to be able to precisely plan and print graded geometries. To determine this parameter, a simple test geometry was developed where the frequency of switches between two separate material feeds was incrementally increased to understand the behavior of the material before, during and after switching. The test geometry consisted of a 30 cm by 30-cm panel, where lines in a five-millimeter offset were printed. Each test consisted of running the test twice, producing two layers on the same geometry. The first layer switched from an initial material, here called material "A" to a material with different filler content, here called material "B". The second layer switched back to the initial material. With the print path being regular, this allowed to measure the length of print, as well as the volume of material, necessary to switch from one material to the other. The result of these tests showed that, the previously mentioned, larger extruder, while able to switch from one material to another, took a lot of time and volume between material switches, which led to the development of the latest iteration. Furthermore, the tests revealed, that the choice of materials greatly impacted the form the gradient took while switching materials. These tests (Fig. 3) were done using the same two materials mixtures, but with different switching frequencies. The chosen material recipes were, for material "A", a mixture of collagen glue containing a filler of 10% wood flour by weight, and for material "B" a mixture of collagen glue containing a 6%bark fiber filler. The materials used in these experiments were sourced and processed as described in "Extending The Circular Design Framework For Bio-Based Materials: Reconsidering Cascading And Agency Through The Case of Biopolymer Composites" (2023). In test 1 (Fig. 3, far left), the extruder was purged of any traces of material "B" before the print started and then switched from material "A" to material "B" in the center of the print. On the second layer the same strategy was used but switching back to material "A". The resulting print was then easily analyzable, by identifying the point the material feeds were switched and measuring the distance to the first instance material "B" had purged out the entirety of material "A". In the second and third tests (Fig. 3, center and far right) the frequency of material changes was raised to one third of the print and one fifth of the print length between material shifts.

Fig. 3. Initial grading tests; Increasing the material switching frequency throughout iterations

3.5 Material Switching Delay

Throughout those tests the observation was made, that the volume of the extruder shared by both materials has a significant impact on the delay between the initiation of flow of the material to be printed and the time the material appears, homogenous at the nozzle. This volume translates into the length of the gradient within the printed geometry. This observation was the initial consequence for the design of the stage 4 extruder, which has a smaller volume in the extruder and can thus switch between materials faster. While the stage 3 extruder would need eighteen meters of print, using a four-millimeter nozzle, the reduced stage 4 extruder is able to switch within three meters, using the same nozzle diameter (Fig. 4). It was furthermore observed, that increasing the switching frequency to appoint where the secondary material does not have enough time to fully replace the initial material will create a continuous blend of the two materials, with little control over the exact composition. This can be especially observed in the third iteration of the grading tests (Fig. 3), where the material gradients become less defined, as well as more irregular throughout the print. This examination gave a clear design constraint for further explorations, and the design of the material feeding control throughout the print path.

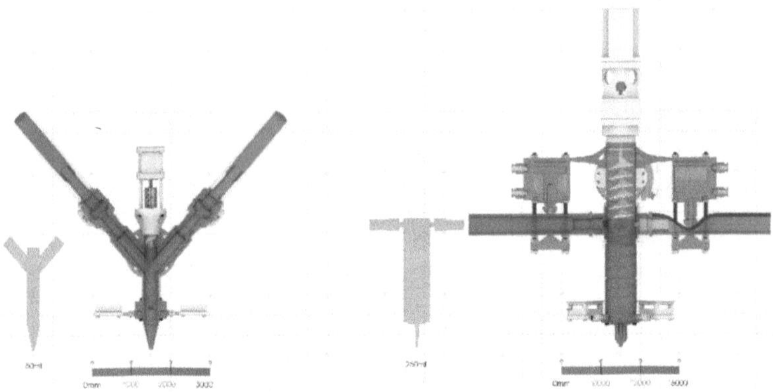

Fig. 4. Material Switching Delay comparison between Stage 4 and Stage 3 Extruders

4 Application of Graded Printing on Building Scale Panels

4.1 Three-Tiered Building Scale Panel Design Geometry

These observations were implemented in a set of three architectural scale panels, which were all designed with different grading strategies, applied to the same geometry. The panel is designed using a growth algorithm for variation of printing geometries as well as for aesthetic purposes, with the intended use as weather screens, modulating the impact of rain and wind. The panels combine three design features each with their own function – the base, the sacrificial layer and the stitching The first feature, the base, consists of four layers of collagen glue material, printed inside the boundary of a parallelogram of 350 mm by 700 mm. This base creates a perforated surface, adaptable for cladding a larger surface. It is separated into four larger fields of perforated print path and two solid end-pieces. The second design feature, the sacrificial layer, acts as a protective layer for the base. This geometry is designed as two horizontal extrusions, spanning the entire width of the panel. To minimize the footprint while still creating a large protective area, the sacrificial layer skews in a 30° overhang, which was considered the maximum inclination possible with the biopolymer composite using bark fiber. The last feature of the design, the stitching, creates a connection between the sacrificial layer and the base. This geometry uses the same growth algorithm, however with the branches growing in the opposite direction of the sacrificial layer and the base to connect multiple branches of the lower layers together.

This three-tiered panel geometry shows a wide variety of challenging print paths. From the growth algorithm-based branching, which includes small radii, fast directional changes, as well as long straight lines to the overhangs in the sacrificial layer and the bridging of the stitching layers over the base layer.

4.2 Grading Strategy and Expected Outcome

From what was learned through the grading control tests, three different grading strategies were applied to the panel geometry. The different grading strategies were applied with an understanding of the material change delay, switching material in different intervals to receive the desired outcome. In the first iteration (Fig. 5, far left), each field of the base was split in half and then shifted from field to field, creating a checkered pattern. The goal being to extend the time a material is printed before the switch of material occurs. The expected outcome was a global gradient from the bottom of the panel to the center and another from the center to the top of the panel. The sacrificial layer and stitching layer were divided, and each assigned a different material. The expected outcome was that the gradient would occur in the center of the individual sections, with the material distribution inverted between the sacrificial layer and the stitching layer. The second grading strategy (Fig. 5, center) splits the base in similar sized sections, however instead of shifting the material distribution between each field, the application of Material A is shifted less and less throughout the different fields. This approach was expected to allow for a gradient in a globally transverse fashion. The sacrificial layer and stitching layer were applied with grading strategies, that would allow for the initial material to be purged out and replaced by the secondary material towards the end of each layer, thus the expectation was, that the printed outcome would be similar to the grading strategy of the first panel. In the base of the last panel (Fig. 5, far right), the grading scheme steadily increased the delay of switching between the initial material and the secondary material. The expected outcome being a single, longitudinal gradient throughout the entire panel. On the sacrificial layer the grading strategy used the height of the section to increase the amount of the secondary material to create a gradient as the print rises. The stitching layer was sectioned in a similar fashion as in the second iteration, however the assigned materials were distributed in a way, that the expected

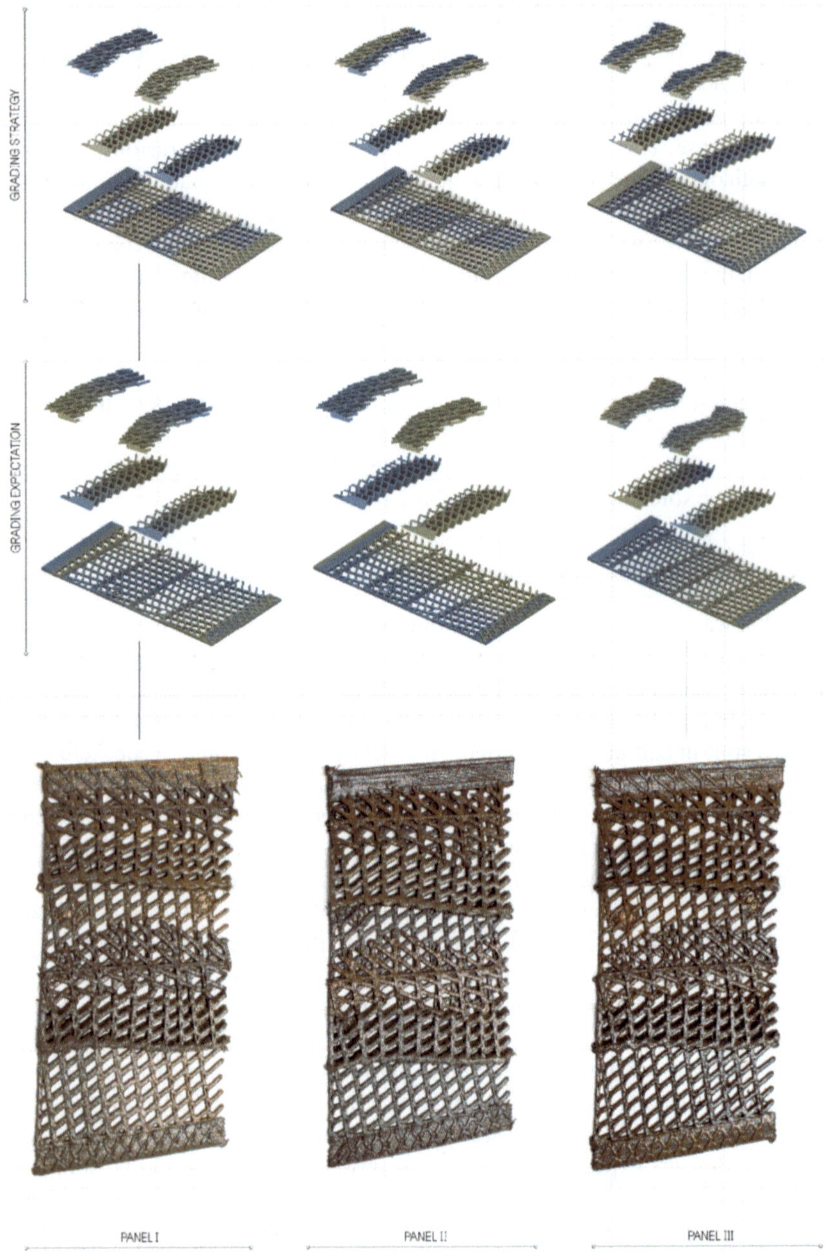

Fig. 5. Grading Strategy showing the distribution for 6% Cotton mixture (Blue) and 10%Wood-flour 1% Cotton mixture (Tan), Grading Expectations and Final Outcome of Building Scale panels

outcome would create a longitudinal gradient between the two parts of the sacrificial layer.

4.3 Chosen Materials

The three chosen materials were chosen for their similarity in viscosity, as well as their color variation, to create three easily comparable prints. The mixtures were all collagen based with different amounts and variations of fibers. The cotton mixture, recognizable by its distinct blue hue, consisted of a 94% base biopolymer mixture and a 6% cotton mixture. As an easily printable and recognizable material, it was used as a benchmark to be matched by the other materials in printability and viscosity. The darker brown material consists of a 91% base biopolymer with a 6% bark fiber, sourced from the forestry industry, and 3% cotton fiber addition, to match the viscosity of the cotton mixture. Lastly, a wood flour mixture was created, which consisted of 89% biopolymer, 10% wood flour and 1% cotton fiber. This mixture is very visibly detectable by its tan hue, as well as the higher grain size of the wood flour, compared to the other fillers, which was a visible marker within the gradient prints to detect residue of this mixture in the prints (Fig. 6).

Fig. 6. Printing Process of Graded Biopolymer Panels

4.4 Printed Outcome

The grading strategies were shifted along the print path to trigger the different feeding systems with a 3000 mm lead to the center of the gradient. This lead was determined through the aforementioned, initial grading tests (Fig. 3). This scheme allowed to print the panels, as expected with clearly defined gradients on a global scale. The three chosen materials behaved as planned, with their viscosity, tensile strength during printing and printing temperature being nearly identical, the changes between materials, and thus the gradients behaved similarly between all combinations of materiality. This gave the possibility to clearly identify the gradient lengths and placements (Figs. 7 and 8).

Fig. 7. Final Printed Outcome: Panel 1 & Panel 2

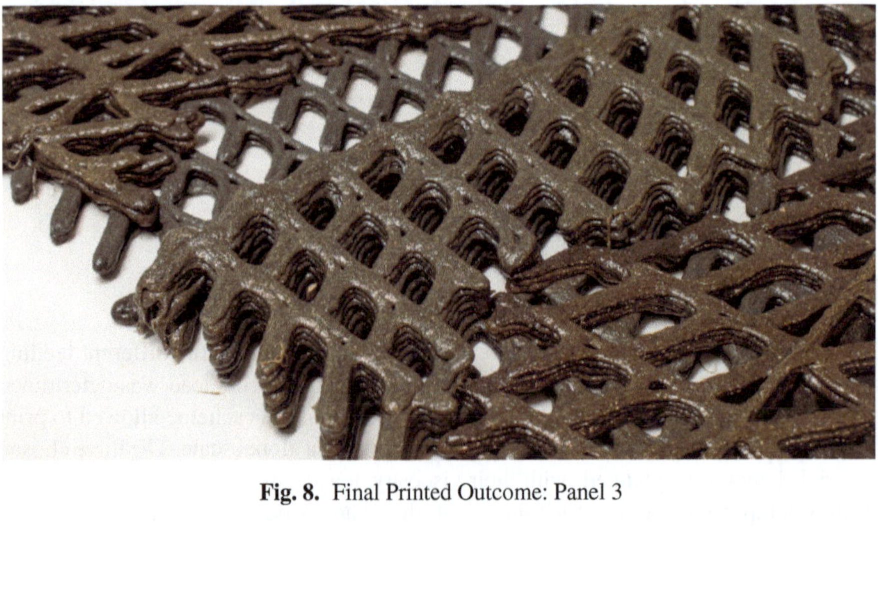

Fig. 8. Final Printed Outcome: Panel 3

5 Conclusion and Further Development

Our paper has introduced a graded 3D printing system for waste and side stream-based biopolymer composites. We are able to print with varying fillers and fibers, and therefore varying viscosities, making the process robust to variable quantities of local resource. By incorporating consideration of the influential parameters into the design of the overall print system and specific tooling, we have developed a printing approach capable of fabricating large-scale geometries with designed material distributions, material gradients and material properties to an architecturally acceptable degree of accuracy. Using multiple feeding systems with large material reservoirs, we are able to print graded architectural scale elements in one continuous process without the need to pause for material changes.

While printing different materials with varying degrees of fiber and filler creates a challenging condition of managing varying viscosities, our findings point towards the potential of fabrication approaches that are flexible with regard to these material parameters. The ability to maintain different temperatures for different material feeds, achieved through independent heating systems, underscores the flexibility in utilizing biopolymers and their mixtures at their ideal printing temperatures. The extruder we introduced in this paper can grade varied, heated materials within a short printing span and supports controlled grading at both a detail as well as a global geometry scale. This rapid grading facilitates continuous material variations and potential mixtures at varying rates during the print, achieving high-precision graded 3D printed geometries. It can be further adapted to ensure targeted material distributions based on the visual and structural traits of different biopolymer compositions.

Future research will focus on refining the heating system to adjust to in-process variations and handling material grading with a broader viscosity range for increased material versatility. Although our latest extruder design can transition between materials efficiently, further development is planned to reduce transition time, allow for shorter gradients and more controlled mixing, and grade three material variations of different compositions and viscosities to enable more complex grading strategies Further, we aim to extend our use of natural and recycled fillers and fibers to further broaden the material palette and incorporate local materials. Introducing fillers and fibers from different geographical regions, could further cement the adaptability of the material around the globe, using readily available materials.

Acknowledgements. This project has received funding from the European Research Council (ERC) under the European Union's Horizon 2020 research and innovation programme (grant agreement No 101019693).

The manuscript was written with the contribution of all authors. **All authors have approved the final version of the manuscript. Nicholas, P.** project conceptualization, methodology, design concept, writing – original draft, reviewing and editing, supervision, **Eppinger, C.,** design concept, 3D print hardware development, prototyping, fabrication, writing – original draft, reviewing and editing, **Chiujdea, R.,** design concept, prototyping **Sonne, K.,** prototyping, fabrication, **Ramsgaard Thomsen, M.** project conceptualization, methodology, design concept, writing – original draft, writing – review and editing, lab infrastructure, supervision, funding acquisition.

References

Aaliya, B., Sunooj, K., Lackner, M.: Biopolymer composites: A review. Int. J. Biobased Plast. **3**(1), 40–84 (2021)

Bever, M., Duwez, P.: Gradients in composite Materials. Mater. Sci. Eng. **10**, 1–8 (1972)

Lee, N., et al.: Sequential multimaterial additive manufacturing of functionally graded biopolymer composites. 3D Printing and Additive Manuf. **7**(5), 205–215 (2020)

Lesna, J., Nicholas, P.: De Gradus–Programming heterogeneous performance of functionally graded bio-polyerms for degradable agricultural shading structures. In: Holzer, D., Nakapan, W., Globa, A., Koh, I. (eds.) Anthropocene. Design in the Age of Humans, pp. 385–394. CAADRIA, Bangkok (2020)

Mogas-Soldevila, L., et al.: Designing the ocean pavilion: Biomaterial templating of structural, manufacturing and environmental performance. In: Proceedings of the International Association for Shell and Spatial Structures (IASS), pp. 1–13, Amsterdam (2015)

Mogas-Soldevilla, L., Duro-Royo, J., Oxman, N.: Water-based Robotic Fabrication: Large-Scale Additive Manufacturing of Functionally Graded Hydrogel Composites via Multichamber Extrusion. 3D Printing and Additive Manuf. **1**(3), 141–151 (2014)

O'Brien, M., Bringezu, S.: What is a sustainable level of timber consumption in the EU: Toward global and EU benchmarks for sustainable forest use. Sustainability. **9**, 812 (2017). https://doi.org/10.3390/su9050812

Pamu, Y., Alugubelli, S.: A comparative study of environmental impacts due to conventional and sustainable concrete. Mater. Today Proc. (2013)

Plank, J.: Applications of biopolymers and other biotechnological products in building materials. Appl. Microbiol. Biotechnol. **66**, 1–9 (2004)

Pradhan, R., Rahman, S., Qureshi, A., Ullah, A.: Biopolymer: Opportunities and challenges for 3D printing essay. In: Sabu, T., Gopi, S., Amalraj, A. (eds.) Biopolymers and Their Industrial Applications: From Plant, Animal, and Marine Sources, to Functional Products, pp. 281–303. Elsevier (2021)

Ramsgaard Thomsen, M., Nicholas, P., Rossi, G., Daugaard, A., Rech, A.: Extending the circular design framework for bio-based materials: reconsidering cascading and agency through the case of biopolymer composites. In: Design for Rethinking Resources, Proceedings of the UIA World Congress of Architects, Copenhagen (2023)

Rech, A., Nicholas, P., Ramsgaard Thomsen, M., Dougaard, A.: Waste-based biopolymer slurry for 3D printing with recycled materials. In: Nordic Polymer (2021)

Rosenthal, M., Henneberger, C., Gutkes, A., Bues, C.: Liquid Deposition Modeling: a promising approach for 3D printing of wood. European J. Wood and Wood Prod. **76**, 797–799 (2018)

Rossi, G. et al.: Integrated design strategies for multi-scalar biopolymer robotic 3D printing. In: Aviv, D., Jamelle, H., Stuart-Smith, R., Akbarzadeh, M. (eds.) Hybrids & Haecceities, pp. 346–355. ACADIA, Philadelphia (2022)

Sanandiya, N., Vijay, Y., Dimopoulou, M., Dritsas, S., Fernandez, G.: Large-scale additive manufacturing with bioinspired cellulosic materials. Sci. Rep. **8**, 8642 (2018)

Seibold, Z., Mhatre, S., López, J., Alhadidi, S., Bechthold, M.: Janus Printing: Co-Extrusion based Multi-material Additive Manufacturing for Ceramics. In: Bieg, K., Briscoe, D., Odom, C. (eds.) Ubiquity & Autonomy, pp. 576–585. ACADIA, Austin (2019)

Tarangini, K. et al.: Eco-friendly bioplastic material development via sustainable seaweed biocomposite. Ecological Chem. Eng. S. **30**, 333–341 (2023)

Tusnim, J., Hoque, M.E., Biswas, M.C.: Chapter 12—Biopolymers in Building Materials. In: Al-Oqla, F.M., Sapuan, S.M. (eds.) Advanced Processing, Properties, and Applications of Starch and Other Bio-Based Polymers, pp. 185–195. Elsevier (2020). https://doi.org/10.1016/B978-0-12-819661-8.00012-3

WASP. (2024, February 1). Stampanti 3D |WASP|Azienda leader nel settore della stampa 3D. Stampanti 3D|WASP. https: //www.3dwasp.com/

Conformal Robotic 3D Printing and Post-Fabrication Analysis of Mycelium-Based Composites

Özgüç Bertuğ Çapunaman(✉), Alale Mohseni, Alireza Zamani, Natalie Walter, and Benay Gürsoy

The Pennsylvania State University, University Park 16802, PA, USA
{okc5048,abm6706,aqz5247,nvw5160,bug61}@psu.edu

Abstract. Mycelium-based composites, heralded for their functional versatility in architecture, have emerged as a pivotal subject in sustainable biomaterials research in recent years. Within this context, conventional fabrication techniques for mycelium-based composites, predominantly relying on molding, are increasingly deemed inadequate, and a shift towards more sustainable and flexible fabrication processes is necessary. Within the larger body of work studying additive manufacturing of mycelium-based composites, our study presents a robotic additive manufacturing approach employing hybrid slicing and conformal 3D printing to fabricate geometrically complex architectural panels. The panels, however, geometrically deform as fungi colonize the 3D printed substrates post-fabrication. Addressing this, we have integrated a computer-vision-based post-fabrication analysis, adept at meticulously documenting and quantifying geometric deformations occurring during the material cultivation stage. By harnessing the precision of conformal 3D printing together with this comprehensive post-fabrication analysis, our research presents a novel framework for accurate materialization and geometric analysis of mycelium-based composites, signposting the direction for future architectural endeavors.

Keywords: Conformal 3D Printing · Robotic Additive Manufacturing · Computer Vision · Mycelium

1 Introduction

The shift towards sustainable practices in architecture has galvanized the pursuit of materials that align with ecological imperatives. Within this spectrum, mycelium-based composites, derived from the colonization of agricultural plant-based residues by the vegetative root of fungi, have garnered significant attention as a novel approach to repurposing waste streams into valuable resources (Ghazvinian and Gürsoy 2022). Instead of intensive material extraction and processing, which characterize many of today's construction materials, mycelium-based composites present a cost-effective and biodegradable alternative (Ghazvinian and Gürsoy 2022). Moreover, recent studies demonstrated

M. Yablonina et al. (Eds.): Robotic Fabrication in Architecture, Art and Design 2024, pp. 82–97, 2026.
https://doi.org/10.1007/978-3-032-22607-5_6

promising material properties of mycelium-based composites, presenting compelling alternatives to petrochemical, synthetic, or cement-based architectural materials (Pelletier et al. 2013; Appels et al. 2018; Elsacker et al. 2019; Walter and Gürsoy 2022). Despite such advantages presented in the literature, the need for sustainable and accurate materialization techniques poses inherent challenges to the large-scale adoption of mycelium-based composites in architecture due to this material's organic nature.

In this study, we leverage advancements in computational design and robotics to fabricate mycelium-based acoustic panels sustainably. A primary focus is placed on advanced robotic additive manufacturing strategies, emphasizing hybrid slicing and conformal 3D printing methods. These methodologies aim to improve the geometric accuracy of additively manufactured mycelium-based composite panels. In addition, mycelium-based composites can substantially deform post-fabrication due to heterogeneous loss of water content and subsequent shrinkage and deformation. We introduce a vision-based geometric reconstruction framework to monitor and analyze the geometric deformations observed in 3D-printed mycelium-based composite panels following fabrication. By detailing the alterations throughout the cultivation of mycelium-based composites, we seek to provide a granular understanding, facilitating a more strategic and adaptive approach to incorporating mycelium-based materials in architectural fabrication.

2 State of the Art

2.1 3D Printing of Mycelium-Based Bio-composites

In recent years, researchers developed materials and methods to 3D print mycelium-based composites (Bhardwaj et al. 2020, 2021; Soh et al. 2020; Elsacker et al. 2022; Mohseni et al. 2023), and designed and fabricated several experimental prototypes (Modanloo et al. 2021; Blast Studio 2023). While these studies paved the way for formwork-less, sustainable fabrication of mycelium-based components, few employed robots in 3D printing, and none explored advanced slicing techniques and conformal 3D printing strategies.

2.2 Advanced Additive Manufacturing Strategies

In additive manufacturing, the predominant method for discretizing digitally designed geometries into machine code has traditionally relied on planar slicing, wherein 3D objects are methodically constructed layer-by-layer by depositing material along parallel planar sections. Despite its widespread adoption, the *stair-stepping* phenomenon, a conspicuous consequence of this method, emerges prominently in objects with non-planar top surfaces, especially in large-scale fabrication scenarios.

In light of evolving fabrication demands, there is a revived shift towards advanced slicing methodologies, as Nayyeri et al. (2022) highlighted. These modern techniques, employing curved sectional planes, not only amplify geometric accuracy and reduce inter-layer fractures but also facilitate the creation of structures without external supports, even with significant overhangs (Krčma et al. 2022; Mitropoulou et al. 2020; Page

2018; Vele 2023). Their application has seen growth in adaptive fabrication, enabling fabricators to adapt to uneven build surfaces across a spectrum of materials, including thermoplastics (Kwon et al. 2019; Nicholas et al. 2021), concrete (Battaglia et al. 2019; Breseghello et al. 2023; Li et al. 2023; Naboni 2022; Schipper et al. 2017), and clay (Çapunaman et al. 2022; Çapunaman et al. 2023; Zamani et al. 2023). However, while revolutionary, non-planar slicing ushers in computational challenges and layer height inconsistencies especially in print volumes with highly variable cross-sectional height. Addressing these, advanced hybrid slicing techniques have surfaced as a compelling solution. As explained by Nayyeri et al. (2022) and exemplified by Ahlers et al. (2019), such hybrid slicing strategies merge the computational efficiency of planar slicing with the geometric precision of non-planar methods, marking a significant stride in the repertoire of additive manufacturing strategies.

2.3 Vision-Based Geometric Reconstruction and Analysis

With the increasing complexity in architectural fabrication, vision-based geometric reconstruction, combining optical sensing hardware and computational algorithms, has emerged as an essential tool to inform *the digital* of *the physical*. This transformative approach is the backbone for various applications, granting physical-to-digital feedback that ensures accurate representation and swift adaptation in advanced architectural fabrication scenarios.

Since the early 2010s, with devices like Microsoft Kinect and Intel RealSense becoming commercially available, architectural fabrication has widely adopted vision-based methodologies. This has led to notable research directions, such as the digitization of irregular stock materials and build environments, enabling conformal tooling operations across additive (Çapunaman et al. 2022; Çapunaman et al. 2023; Zamani et al. 2023; Ercan Jenny et al. 2020; Frangez et al. 2022; Naboni 2022; Nicholas et al. 2020; Breseghello et al. 2023), subtractive (Devadass et al. 2016; del Castillo and Lopez 2022; Larsen and Aagaard 2020; Zivkovic and Battaglia 2018; Çapunaman et al. 2022), and formative (Amtsberg 2015; Cui 2022; Nicholas et al. 2021; Raspall et al. 2014; Breseghello et al. 2023) fabrication techniques. Concurrently, real-time monitoring of the fabrication process is adopted to address anomalies and uncertainties, ensuring an alignment between design intent and realized constructs (Alvarez et al. 2019; Rossi and Nicholas 2018). Collectively, these strands highlight the increasingly essential role of vision-based geometric reconstruction in navigating the multifaceted challenges of contemporary architectural fabrication in the face of emerging material systems and complex tooling operations.

3 Methods

This research presents an integrated approach to robotic additive manufacturing through three targeted interventions at the design, pre-fabrication, and analysis stages, as highlighted in Fig. 1. The experimental design space assesses the relationship between surface intricacy and additive manufacturing constraints. At the core, our advanced additive manufacturing strategies harness geometric discretization, Differential Growth Curves

(DGC), and extruder pose optimization to enhance material deposition. This is coupled with our vision-based system, which meticulously captures and analyzes geometric changes. This section delineates the methodologies and insights underpinning each intervention.

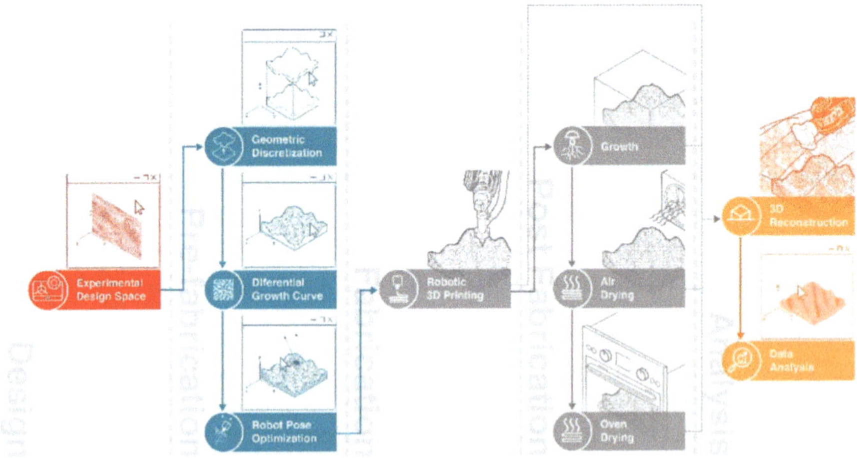

Fig. 1. Overview of the methodological workflow employed in this study, detailing the sequential stages of robotic fabrication and post-processing of mycelium-based bio-composites

3.1 Material Mixture and Handling

In this study, we made use of Ecovative's pre-inoculated Dehydrated Grow-It-Yourself Mushroom mixture (Grow It Yourself Material 2024), leveraging their patented technology for reliable cultivation of mycelium-based composites. Building upon the work of Bhardwaj et al. (2020, 2021), which outlined a six-step process – mycelium inoculation, primary colonization, mixing, 3D printing, secondary colonization, and drying – for 3D printing of mycelium composites, our approach sought to adapt this methodology for our fabrication setup. In order to improve the extrudability of the pre-inoculated Dehydrated Grow-It-Yourself Mushroom mixture, its particles were ground down to a finer size and sieved to ensure homogeneity. The ground mixture was mixed with flour, psyllium husk, and sterilized water in 5: 0.8: 1.2: 30 (ground mixture: flour: psyllium husk: water, by weight) ratio. Different than the methodology presented by Bhardwaj et al., the initial inoculation and primary colonization phases were strategically omitted, and the resulting material mixture was directly extruded to reduce the overall material preparation time.

3.2 Experimental Design Space

Within our experimental framework, we have delineated an exploratory geometric domain tailored to methodically assess the intricate interplay between geometric complexity and the inherent limitations of additive manufacturing. The foundational underpinning of this domain is an interference pattern synthesized from a 2D simplex noise

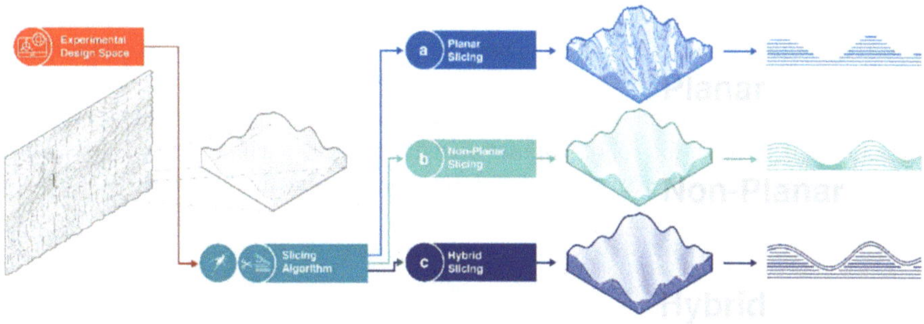

Fig. 2. The adopted experimental design space highlights the stair stepping and excessive layer height variance issues inherent in (a) planar and (b) non-planar slicing approaches, respectively, alongside the adopted (c) hybrid slicing approach

and a 1D sine wave, aiming to validate the proposed fabrication framework through a challenging case of materialization as a benchmark.

The experimental space presented here showcases an array of geometric variations, highlighting the stair-stepping challenges of traditional planar slicing approaches and the layer height variability issues found in newer non-planar slicing methods, as shown in Fig. 2.b and 2.c, respectively. Such artifacts compromise the geometric fidelity and detract from the desired aesthetic of the printed constructs. From this design space afforded by the interference pattern, a region characterized by its elevated geometric feature frequency was chosen for our fabrication tests. This selection represents a rigorous test case, where we would expect heightened geometric inaccuracies and operational limitations using planar or non-planar slicing approaches, especially when 3D printing with large nozzles diameters, and serving as a metric for assessing the efficacy of our proposed hybrid slicing strategy.

3.3 Robotic Work Cell

Within the context of our study, we have established a sophisticated robotic work cell, leveraging the capabilities of the ABB IRB 2400 industrial robotic arm over conventional gantry or delta-style 3D printers due to its 6° of freedom, essential for intricate non-planar material deposition. The robotic task planning and simulation processes are streamlined using the Robots plugin in Grasshopper for Rhinoceros 3D, ensuring precision and optimization.

To ensure the fidelity of the material deposition process, especially when working with mycelium mixtures, we developed a customized extruder End-of-Arm Tool (EoAT) (Fig. 3), which is made available to the broader research community through a GitHub repository (Çapunaman and Mohseni (2023)). Since sterilization is crucial in working with mycelium-based composites to mitigate contamination, we purposefully designed the interference plates for the LDM WASP Extruder XL 3.0 and the WASP 5 L pressurized tank for ease of disassembly, which facilitates thorough sterilization. Within this setup, the WASP 5 L pressurized tank also provides a secure and airtight container for transporting living mycelium mixtures from sterile lab environments directly to our robotic work cell. Furthermore, our extrusion system is augmented by a custom motor speed regulation circuit. It dynamically interacts with the 24 V robot group output signals, modulating the speed of the stepper motor associated with the extrusion screw and facilitating on-demand start and stop functionality and reversal of print direction for material retraction, all through the robot's digital signal interface. Such active synchronization assures a consistent and precise deposition rate for the mycelium mixtures.

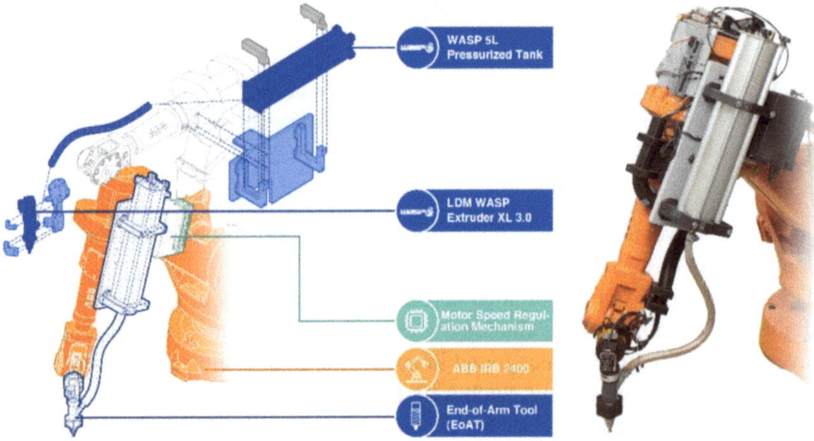

Fig. 3. ABB IRB 2400 robotic arm integrated with a customized LDM WASP Extruder XL 3.0 and WASP 5 L pressurized tank for precise non-planar deposition of mycelium mixtures.

3.4 Advanced Additive Manufacturing Strategies

To amplify the precision and capabilities of additive manufacturing for mycelium-based composites, our computational strategy fuses three core computational approaches: geometric discretization of the print volume, toolpath generation employing DGC, and extruder pose optimization, as illustrated in Fig. 4. Notably, the methodologies for toolpath generation and extruder pose optimization are grounded on a preceding work in clay 3D printing (Çapunaman, Iseman, and Gürsoy 2023), which offers a comprehensive detailing of the implementation in more depth.

Geometric Discretization.
At the core of our computational methodology is geometric discretization, tailored to incorporate both planar and non-planar slicing methods. By cohesively merging these methods, we seek to exploit the computational efficiency of planar slicing while capitalizing on the geometric precision of non-planar approaches.

Utilizing the target print volume, represented as a NURBS solid, we identify the top surface based on surface normals that align with the intended material deposition direction. The surface then undergoes an offset adhering to the surface normals, calculated as $h_{offset} = h_{layer} \times n_{conformal}$, where h_{layer} and $n_{conformal}$ are the desired layer height and

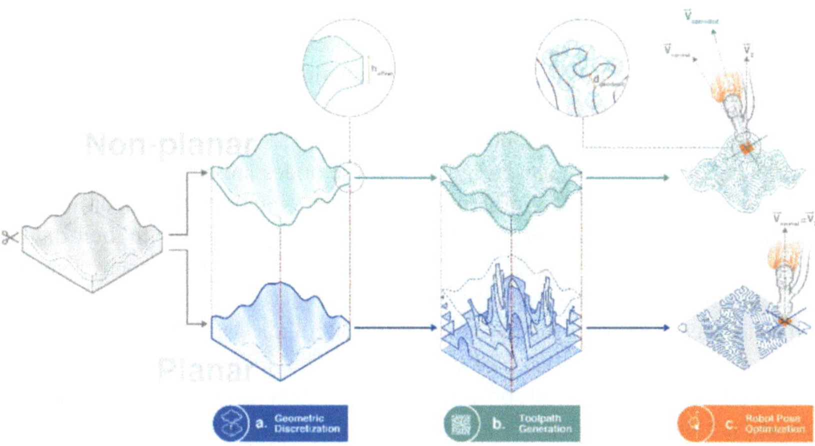

Fig. 4. Computational Workflow for Additive Manufacturing of Mycelium-based Composites: (a) Geometric Discretization, (b) Toolpath Generation employing DGC, and (c) Extruder Pose Optimization

the number of conformal layers, respectively. During this phase, a simple boolean logic is incorporated to ascertain that the minimum height of the panel along the Z direction, h_{min}, surpasses the total offset by at least one layer height, $h_{min} - h_{offset} \geq h_{layer}$. This measure ensures the offset surface consistently remains above the build surface, averting potential collisions.

Using the offset surface, the target print volume is divided into two separate volumes, each specifically tailored for either planar or non-planar slicing methodologies as illustrated in Fig. 4.a. For the base volume, equidistant mathematical planes, separated by h_{layer}, are utilized to generate planar NURBS surfaces through systematic slicing. Concurrently, the non-planar volume undergoes a distinct slicing method. By intersecting the corresponding volume with the extended and offset top surface at each h_{layer} increments, a series of non-planar NURBS surfaces are obtained, which serve as the basis for the DGC toolpath generation in subsequent stages.

Toolpath Generation.
In the second stage of our computational approach, DGC's adoption functions as a critical element in facilitating continuous material deposition, markedly attenuating perturbations stemming from travel movements between distinct print islands. Additionally, DGC's intrinsic nature as a surface-filling curve allows the successful generation of planar layers with holes and non-planar layers alike, while maintaining a consistent geodesic distance, $d_{geodesic}$, between proximate material beads.

For every print surface we obtain from the previous geometric discretization stage, we identify the largest boundary curve to serve as the initial seed. This seed curve undergoes systematic discretization, equally dividing up the input curve into a continuous spring system with a segment length of $L_{segment} = L_{curve} / n_{nodes}$ where L_{curve} represents the length of the seed curve and the number of nodes in this system, denoted by n_{node}. In our implementation, n_{node} was derived using Thue's theorem on circle packing (Chang and

Wang 2010) based on the surface area. This strategy enabled us to enhance the efficiency of our computational resources by minimizing unnecessary calculations.

Building upon this, the Kangaroo Physics plugin in Grasshopper is employed to derive a physics-based approximation of DGC. Adopting a simple spring model to ensure path continuity, the simulated spherical nodes are anchored to the designated print surfaces to ensure geometric accuracy. These nodes are designed to exert forces on their neighboring nodes, culminating in a winding, continuous toolpath (Fig. 4.b). As the simulation progresses, each node is systematically grown, with their diameters gradually incremented over $L_{segment} \leq R_{node} \leq d_{geodesic}$ in a set number of iterations. The increments follow an exponential growth function, and each growth iteration persists until the summed lengths of the motion vectors within our particle system drop below a set threshold, indicating a state of equilibrium. This toolpath generation approach is consistently applied across all print islands, regardless of their planar or non-planar nature, and the subsequent deposition sequences are sorted out along the print direction.

Robot Pose Optimization.

The third stage in our methodological framework is centered around optimizing the extruder orientation during the deposition process. Given the geometric intricacies of the panel, a strategic approach was required to determine the optimal extruder orientation for material deposition. Each generated toolpath from the previous stage, coupled with the respective evaluation of print surface normal at discrete toolpath nodes, serves as a foundation for determining the robotic extruder's initial orientation.

However, the undulating nature of the surfaces and the subsequent toolpaths introduces complexities in directly using surface normals, especially to avoid collisions and maintain the fluidity of the deposition process. To address this, a simple optimization strategy is integrated, formulating a new vector, $\vec{v}_{optimized}$, between the surface normal vector, $\vec{v}_{normal}$, and the world Z-axis vector, $\vec{v}_{Z}$ as illustrated in Fig. 4.c. This optimization ensures a collision-free deposition and maximizes the alignment of the extruder with the underlying print surface, thereby enhancing the fabricated component's geometric fidelity and overall quality.

3.5 Geometric Reconstruction and Analysis

Vision-based Geometric Reconstruction Framework.

As living fungi colonize the 3D-printed components and bind the layers, the overall geometry of the components goes through a change. This change is more prevalent as the 3D-printed components are air-dried and oven-dried to cease the mycelial growth. In this study, we employed a vision-based sensing framework to systematically gather data and assess the impacts of fungi colonization, air-drying, and oven-drying stages on the panel geometry. The vision-based reconstruction framework employed in this study was initially developed and exhaustively validated for accuracy and fabrication feasibility in an earlier study (Çapunaman et al. 2022) and has since been applied in complex manufacturing scenarios, successfully enabling additive manufacturing of clay on digitally undefined surfaces like sand (Çapunaman et al. 2023) and adaptable tensile formworks (Zamani et al. 2023).

Adopting the vision-based framework, we utilized 8x8 robot poses derived from a hemispherical trajectory with 90- and 60° angular ranges on the YZ and XZ planes of the robot coordinate system, respectively. The set distance between the camera and the panel's center was maintained at 1000 mm, while the voxel resolution for the TSDF integration was set at ~1.9 mm (1/512 m). These specific parameters were chosen based on our thorough testing documented in a prior publication, where readers can find a more comprehensive breakdown (Çapunaman et al. 2022).

Geometric Deformation Analysis.
To analyze the data collected at various post-fabrication stages of the mycelium-based composite panel, we employed the signed Hausdorff distance method (Aspert et al. 2002), a computational approach often used to quantify geometric similarity. The signed Hausdorff distance provides a robust measure of the maximum discrepancy between two point clouds—in this case, between reconstructed geometries of consecutive stages. Importantly, this metric's *signed* aspect captures the direction of the deviation, helping differentiate whether the reconstructed geometry expands beyond or retracts within the idealized one. Within this definition, we have set the signed attribute based on the dot product of $\overrightarrow{v}_{normal}$ and the vector from each sampled point to its closest match on the other geometry. This approach effectively captured the geometric transformations resulting from mycelium growth and the subsequent drying phases. For precise insights, the panel was scanned post-printing, post-colonization, post-air-drying, and post-oven-drying, with each stage's geometric deviations thoroughly assessed in comparison to the previous stages.

4 Results

4.1 Robotic 3D Printing of Mycelium-Based Bio-Composite Panels

Figure 5 provides a visual representation of our proof-of-concept mycelium-based panel fabricated using an 8 mm nozzle before and after the cultivation stages. The panel, with dimensions 250 mm by 250 mm, was 3D printed in approximately 45 min, utilizing around 3.5 l of the mycelium mixture. A notable observation from this fabrication process was the consistent material deposition across all layers. This uniformity underscores the efficiency and reliability of the advanced additive manufacturing strategies described in this paper, particularly when deployed for panels with high geometric complexity.

4.2 Robotic 3D Scanning of Growth and Analyzing Changes

By comparing the 3D scan data gathered from different stages, namely post-printing, during mycelium growth, after air-drying, and after oven-drying; we were able to track the changes occurring on the 3D printed panel over time. This approach yields a comprehensive view of the geometric deformations of the panel at each step of the process. Figure 6 visualizes changes between each subsequent stage using the signed Hausdorff distance method introduced earlier.

The comparison of the post-printing scan with the original CAD model revealed an average over-extrusion error of 2.59 mm and a considerable volume difference of

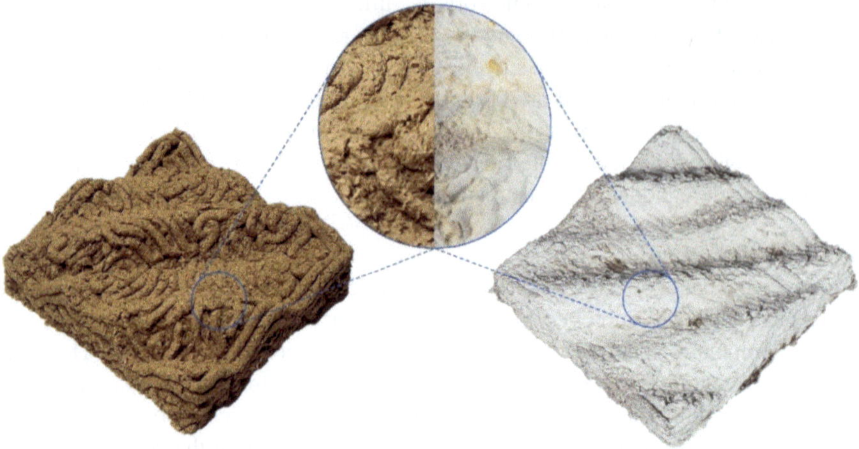

Fig. 5. Robotically 3D printed mycelium-based bio-composite panel; (Left) Panel immediately following the printing process, (Right) Panel after successful cultivation of the substrate.

14.96%. This finding underscores the challenges encountered during the deposition of the material, indicating potential areas of improvement in the precision of the 3D printing process. The subsequent Growth stage involves the mycelial colonization of the panel and presents a mean shrinkage of −0.83 mm with a smaller volume reduction of 3.97%. This modest change suggests that the growth stage exerts a minor influence on the overall geometry of the panel. The air-drying stage resulted in an average deformation of −1.09 mm and a total volume difference of −5.45%, indicating a significant volumetric change. Moreover, since we can observe a localized reversal of the geometric deformation trend from shrinkage to expansion around the corners, we can also conclude that this process introduces a considerable amount of localized warpages concentrated around the corners. The concluding oven-drying stage led to further shrinkage, with the total volume reducing by an additional −9.38% and an average shrinkage of −1.88 mm. Similar to the previous air-drying stage, we observed localized reversal of the geometric deformation trends, further underlining the material indeterminacy during dehydration stages. This significant reduction during the drying stages, particularly oven-drying, confirms that these processes contribute substantially to post-fabrication dimensional changes. Together, the combined impact of all the post-fabrication stages resulted in a significant −17.72% reduction in total volume.

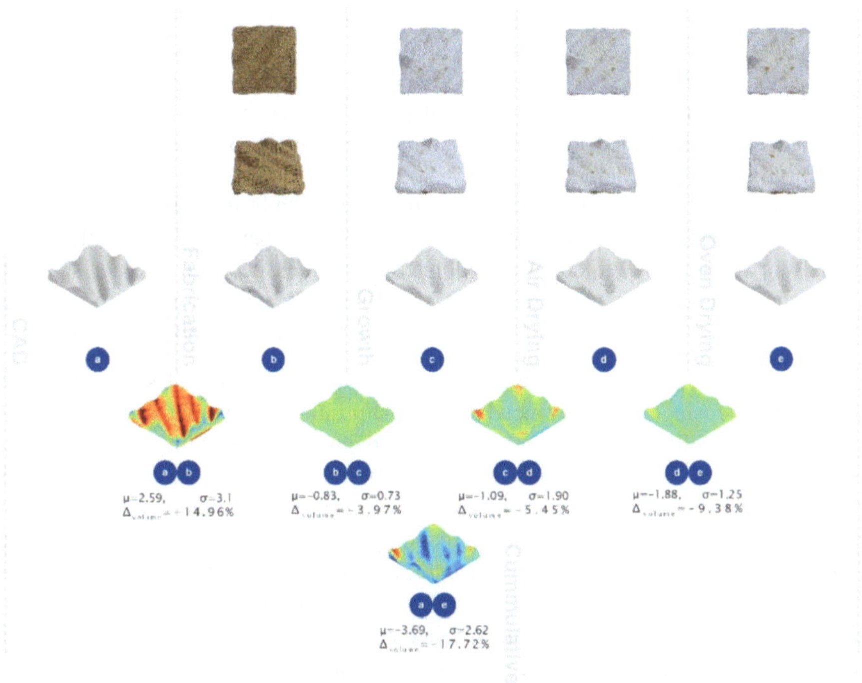

Fig. 6. Post-fabrication deformation analysis of the fabricated panel, illustrating the comparative differences between each successive stages: (a) digitally designed, (b) 3D printed, (c) colonized, (d) air-dried, (e) oven-dried panels

5 Discussion and Conclusions

In this paper, we presented an experimental study on robotic 3D printing of mycelium-based composites, exploring advanced additive manufacturing strategies for non-planar geometries and employing a vision-based sensing framework tailored explicitly for the geometric quantification of post-fabrication deformations. This research is timely as mycelium-based materials are gaining an increasing interest in architecture and design due to their sustainable features, evoking a need to explore sustainable and advanced manufacturing methods for mycelium-based building parts and structures.

5.1 Contributions

The contributions of this research are multifold:

- Developing a custom extruder end-of-arm tool greatly improved robotic 3D printing with mycelium mixtures, allowing for precise control over material extrusion rates. With components that can be easily disassembled and sterilized, the proposed extrusion implementation significantly mitigates contamination risks.

- By integrating non-planar slicing methods, leveraging DGC, and optimizing robotic extruder orientations, our methodology facilitates improved surface finishes and superior geometric fidelity, addressing the challenges inherent to mycelium-based composite 3D printing. We believe the strategies presented here can also be adopted beyond mycelium-based composite mixtures, contributing to extrusion based robotic 3D printing efforts at large.
- Incorporating a vision-based sensing and reconstruction framework enabled meticulous documentation of post-fabrication changes in the panels. Through the signed Hausdorff distance method, we pinpointed deviations between distinct stages of mycelium-based bio-composite cultivation, offering deeper insights into the material's behavior and transformation during fabrication and post-fabrication phases.

In conclusion, our research aimed at pushing the boundaries of sustainable construction by focusing on the advanced robotic 3D printing of mycelium-based composites. We successfully developed and implemented cutting-edge additive manufacturing strategies tailored for complex non-planar geometries. Furthermore, we innovatively applied a vision-based sensing framework to quantify the geometric transformations of mycelium-based composites post-fabrication precisely. Collectively, these strategies facilitate superior material deposition accuracy, which is imperative for the successful application of mycelium-based composite 3D printing. Our contributions lay the groundwork for future explorations, offering advanced manufacturing techniques that not only cater to mycelium's unique properties but can potentially extend to other extrusion-based robotic additive manufacturing workflows aiming to achieve higher levels of geometric fidelity.

5.2 Future Work

In the pursuit of advancing the fabrication of mycelium-based composites, our study opens several avenues for future research that promise to further refine the precision and efficiency of this sustainable manufacturing process. A primary focus of forthcoming investigations will revolve around enhancing the predictability and control over the material's behavior during and after the fabrication process. We believe that our geometric deviation analysis framework can be employed to proactively predict and counteract warpage and shrinkage, enhancing the accuracy of the finished product. By incorporating machine learning models trained on discrete 3D printed sample datasets, we can more accurately model the material's behavior, including the effect of material composition in observed deformations. Such models can then be used to compensate for deformations by optimizing the toolpaths and material extrusion rate, ensuring the desired final volume is achieved after the colonization and drying stages. Additionally, integrating a real-time vision-based feedback system during the fabrication process can monitor and correct discrepancies, such as under or over-extrusion or material springback, leading to consistently high-quality outcomes.

Within the scope of application space enabled by the proposed advanced additive manufacturing strategies, intentionally varied cross-sectional areas can serve as a means to optimize structural integrity. By modulating the material distribution, we can strategically balance load-bearing capacities, potentially enhancing the panel's overall durability and longevity while minimizing the material used. This methodical variation ensures that materials are used where they are most needed, effectively distributing stresses and optimizing weight.

Another promising avenue for future studies is investigating the potential interplay between mycelium growth and infill patterns. Different infill densities and arrangements might influence mycelium colonization speeds and patterns, which can have implications for the final composite's structural and material performance. By controlling infill densities, there is potential to manipulate the inner micro-environments of the printed volume, thereby optimizing conditions for faster, more uniform mycelial growth. This understanding could be critical in scaling up the fabrication of mycelium-based structures, ensuring both material efficiency and performance consistency.

Acknowledgements. This research is funded by the American Institute of Architects (AIA) Upjohn Research Initiative Grant (2022) and the Stuckeman Center for Design Computing Interdisciplinary Faculty Research Grant (2023).

References

Ahlers, D., Wasserfall, F., Hendrich, N., Zhang, J.: 3D Printing of Nonplanar Layers for Smooth Surface Generation. In: 2019 IEEE 15th International Conference on Automation Science and Engineering (CASE), pp. 1737–1743 (2019). https://doi.org/10.1109/COASE.2019.8843116

Alvarez, M.E., et al.: Tailored structures, robotic sewing of wooden shells. In: Willmann, J., Block, P., Hutter, M., Byrne, K., Schork, T. (eds.) Robotic Fabrication in Architecture, Art and Design 2018, pp. 405–420. Springer International Publishing, Cham (2019). https://doi.org/10.1007/978-3-319-92294-2_31

Amtsberg, F., Raspall, F., Trummer, A.: Digital-Material Feedback in Architectural Design. In: Emerging Experience in Past, Present and Future of Digital Architecture, Proceedings of the 20th International Conference of the Association for Computer-Aided Architectural Design Research in Asia (CAADRIA 2015), pp. 631–640. CUMINCAD (2015) https://papers.cumincad.org/cgi-bin/works/paper/caadria2015_202

Appels, F.V.W., Dijksterhuis, J., Lukasiewicz, C.E., Jansen, K.M.B., Wösten, H.A.B., Krijgsheld, P.: Hydrophobin Gene Deletion and Environmental Growth Conditions Impact Mechanical Properties of Mycelium by Affecting the Density of the Material. Sci. Rep. **8**(1), 4703 (2018). https://doi.org/10.1038/s41598-018-23171-2

Aspert, N., Santa-Cruz, D., Ebrahimi, T.: MESH: Measuring errors between surfaces using the hausdorff distance. In: Proceedings. IEEE International Conference on Multimedia and Expo, vol. 1, pp. 705–708 (2002). https://doi.org/10.1109/ICME.2002.1035879

Battaglia, C.A., Miller, M.F., Zivkovic, S.: Sub-additive 3D printing of optimized double curved concrete lattice structures. In: Willmann, J., Block, P., Hutter, M., Byrne, K., Schork, T. (eds.) Robotic Fabrication in Architecture, Art and Design 2018, pp. 242–255. Springer International Publishing, Cham (2019). https://doi.org/10.1007/978-3-319-92294-2_19

Bhardwaj, A., et al.: 3D printing of biomass–fungi composite material: Effects of mixture composition on print quality. J. Manuf. Mater. Proc. **5**(4), 112 (2021). https://doi.org/10.3390/jmmp5040112

Bhardwaj, A., et al.: 3D printing of biomass-fungi composite material: A preliminary study. Manuf. Lett. **24**(April), 96–99 (2020). https://doi.org/10.1016/j.mfglet.2020.04.005

Blast Studio. 2023. Our Process. https://www.blast-studio.com/our-innovation Accessed 16 Oct

Breseghello, L., Talaei, A., Florenzano, D., Naboni, R.: Shape-Env-Camera-Enhanced Robotic terrain-shaping for complex 3D concrete printing. In: Dokonal, W., Hirschberg, U., Wurzer, G. (eds.) Digital Design Reconsidered - Proceedings of the 41st Conference on Education and Research in Computer Aided Architectural Design in Europe (eCAADe 2023)-Volume 1, Graz, 20-22 September 2023, pp. 539–548. CUMINCAD (2023) https://papers.cumincad.org/cgi-bin/works/paper/ecaade2023_403

Çapunaman, Ö.B., Dong, W., Gürsoy, B.: A vision-based sensing framework for adaptive robotic tooling of indefinite surfaces. Construct. Robot. (2022). https://doi.org/10.1007/s41693-022-00081-4

Çapunaman, Ö.B., Iseman, E.K., Gürsoy, B.: Material-in-the-loop fabrication: A vision-based adaptive clay 3D printing workflow on indeterminate sand surfaces. In: Turrin, M., Andriotis, C., Rafiee, A. (eds.) Computer-Aided Architectural Design. INTERCONNECTIONS: Co-Computing Beyond Boundaries Communications in Computer and Information Science, pp. 410–424. Springer Nature Switzerland, Cham (2023). https://doi.org/10.1007/978-3-031-37189-9_27

Çapunaman, Özgüç Bertuğ, and Alale Mohseni. (2023). ForMat-Lab/WASP-XL-Extruder. C++. ForMat Lab. https://github.com/ForMat-Lab/WASP-XL-Extruder

Chang, Hai-Chau, and Lih-Chung Wang. 2010. A Simple Proof of Thue's Theorem on Circle Packing. arXiv. doi:10.48550/arXiv.1009.4322

Cui, Q.: Forming strategies for robotic incremental sheet forming. In: van Ameijde, J., Gardner, N., Hyun, K.H., Luo, D., Sheth, U. (eds.) POST-CARBON-Proceedings of the 27th CAADRIA Conference, Sydney, 9–15 April 2022, pp. 171–180. CUMINCAD (2022) http://papers.cumincad.org/cgi-bin/works/paper/caadria2022_294

Devadass, P., Dailami, F., Mollica, Z., Self, M.: Robotic fabrication of non-standard material. In: ACADIA//2016: POSTHUMAN FRONTIERS: Data, Designers, and Cognitive Machines [Proceedings of the 36th Annual Conference of the Association for Computer Aided Design in Architecture (ACADIA) ISBN 978–0–692-77095-5] Ann Arbor 27–29 October, 2016. CUMINCAD (2016) http://papers.cumincad.org/cgi-bin/works/paper/acadia16_206

Elsacker, E., Peeters, E., De Laet, L.: Large-scale robotic extrusion-based additive manufacturing with living mycelium materials. Sustainable Futures. **4**(January), 100085 (2022). https://doi.org/10.1016/j.sftr.2022.100085

Elsacker, E., Vandelook, S., Brancart, J., Peeters, E., De Laet, L.: Mechanical, physical and chemical characterisation of mycelium-based composites with different types of lignocellulosic substrates. PLoS One. **14**(7), e0213954 (2019). https://doi.org/10.1371/journal.pone.0213954

Jenny, E., Selen, E.L.-F., Gramazio, F., Kohler, M.: Crafting plaster through continuous mobile robotic fabrication on-site. Constr. Robot. **4**(3), 261–271 (2020). https://doi.org/10.1007/s41693-020-00043-8

Frangez, V., et al.: Geometric Feedback System for Robotic Spraying. PFG–J. Photogramm., Remote Sens. Geoinf. Sci. **90**(6), 531–542 (2022). https://doi.org/10.1007/s41064-022-00219-7

Garcia del Castillo y Lopez, Jose L. 2022. The digital touch-towards novel modeling frameworks for robotically-enhanced marble sculpting. In Pak, B, Wurzer, G and Stouffs, R (Eds.), Co-Creating the Future: Inclusion in and through Design-Proceedings of the 40th Conference on Education and Research in Computer Aided Architectural Design in Europe (eCAADe 2022)-Volume 1, Ghent, 13–16 September 2022, PP. 37–46. CUMINCAD. http://papers.cumincad.org/cgi-bin/works/paper/ecaade2022_250

Ghazvinian, A., Gürsoy, B.: Mycelium-based composite graded materials: Assessing the effects of time and substrate mixture on mechanical properties. Biomimetics. **7**(2), 48 (2022). https://doi.org/10.3390/biomimetics7020048

Grow It Yourself Material. (2024). Grow.Bio. https://grow.bio/products/grow-it-yourself-material. Accessed 2 March 2024

Krčma, Martin, David Paloušek, David Škaroupka, Johannes Braumann, and Daniel Koutný. 2022. Method of Multiaxis Three-Dimensional Printing with Intralayer Height Variation for Stairstep Effect Compensation. 3D Printing and Additive Manufacturing, September. Mary Ann Liebert, Inc., publishers. doi:https://doi.org/10.1089/3dp.2022.0097

Kwon, H., Eichenhofer, M., Kyttas, T., Dillenburger, B.: Digital composites: Robotic 3D printing of continuous carbon fiber-reinforced plastics for functionally-graded building components. In: Willmann, J., Block, P., Hutter, M., Byrne, K., Schork, T. (eds.) Robotic Fabrication in Architecture, Art and Design 2018, pp. 363–376. Springer International Publishing, Cham (2019). https://doi.org/10.1007/978-3-319-92294-2_28

Larsen, N.M., Aagaard, A.K.: Robotic processing of crooked sawlogs for use in architectural construction. Constr. Robot. **4**(1), 75–83 (2020). https://doi.org/10.1007/s41693-020-00028-7

Li, S., Nguyen-Xuan, H., Tran, P.: Digital design and parametric study of 3D concrete printing on non-planar surfaces. Autom. Constr. **145**(January), 104624 (2023). https://doi.org/10.1016/j.autcon.2022.104624

Mitropoulou, Ioanna, Mathias Bernhard, and Benjamin Dillenburger. 2020. Print paths key-framing: Design for non-planar layered robotic FDM printing. In Symposium on Computational Fabrication, 1–10. SCF '20. New York, NY, USA: Association for Computing Machinery. doi:https://doi.org/10.1145/3424630.3425408

Modanloo, B., Ghazvinian, A., Matini, M., Andaroodi, E.: Tilted arch; Implementation of additive manufacturing and bio-welding of mycelium-based composites. Biomimetics. **6**(4), 68 (2021). https://doi.org/10.3390/biomimetics6040068

Mohseni, A., Vieira, F.R., Pecchia, J.A., Gürsoy, B.: Three-dimensional printing of living mycelium-based composites: Material compositions, workflows, and ways to mitigate contamination. Biomimetics. **8**(2), 257 (2023). https://doi.org/10.3390/biomimetics8020257

Naboni, R.: Environment-aware 3D concrete printing through robot-vision. In: Pak, B., Wurzer, G., Stouffs, R. (eds.) Co-Creating the Future: Inclusion in and through Design-Proceedings of the 40th Conference on Education and Research in Computer Aided Architectural Design in Europe (eCAADe 2022) - Volume 2, Ghent, 13–16 September 2022, pp. 409–418. CUMINCAD (2022) http://papers.cumincad.org/cgi-bin/works/paper/ecaade2022_384

Nayyeri, P., Zareinia, K., Bougherara, H.: Planar and nonplanar slicing algorithms for fused deposition modeling technology: A critical review. Int. J. Adv. Manuf. Technol. **119**(5), 2785–2810 (2022). https://doi.org/10.1007/s00170-021-08347-x

Nicholas, P., Chiujdea, R.S., Sonne, K., Scaffidi: Design and fabrication methodologies for repurposing end of life metal via robotic incremental sheet metal forming. In: Stojakovic, V., Tepavcevic, B. (eds.) Towards a New, Configurable Architecture-Proceedings of the 39th eCAADe Conference-Volume 2, University of Novi Sad, Novi Sad, Serbia, 8–10 September 2021, pp. 171–180. CUMINCAD (2021) http://papers.cumincad.org/cgi-bin/works/paper/ecaade2021_178

Nicholas, P., Rossi, G., Williams, E., Bennett, M., Schork, T.: Integrating real-time multi-resolution scanning and machine learning for conformal robotic 3D printing in architecture. Int. J. Archit. Comput. **18**(4), 371–384 (2020). https://doi.org/10.1177/1478077120948203

Page, James Sherwood. 2018. Systems and methods for improved 3D printing. United States US10005126B2, filed March 19, 2015, and issued June 26, 2018. https://patents.google.com/patent/US10005126B2/en

Pelletier, M.G., Holt, G.A., Wanjura, J.D., Bayer, E., McIntyre, G.: An evaluation study of mycelium based acoustic absorbers grown on agricultural by-product substrates. Ind. Crop. Prod. **51**(November), 480–485 (2013). https://doi.org/10.1016/j.indcrop.2013.09.008

Raspall, F., Amtsberg, F., Peters, S.: Material feedback in robotic production. In: McGee, W., de Leon, M.P. (eds.) Robotic Fabrication in Architecture, Art and Design 2014, pp. 333–345. Springer International Publishing, Cham (2014). https://doi.org/10.1007/978-3-319-04663-1_23

Rossi, Gabriella, and Paul Nicholas. 2018. Re/Learning the Wheel: Methods to Utilize Neural Networks as Design Tools for Doubly Curved Metal Surfaces.

Schipper, R., Costanzi, C.B., Bos, F., Ahmed, Z., Wolfs, R.: Double curved concrete printing: printing on non-planar surfaces. SPOOL. **4**(2), 17–21 (2017). https://doi.org/10.7480/spool.2017.2.1920

Soh, E., Chew, Z.Y., Saeidi, N., Javadian, A., Hebel, D., Le Ferrand, H.: Development of an extrudable paste to build mycelium-bound composites. Mater. Des. **195**(October), 109058 (2020). https://doi.org/10.1016/j.matdes.2020.109058

Vele, J.: Improving buildability of overhangs of 3D printed objects through non-planar slicing informed by force flow analysis. In: Dokonal, W., Hirschberg, U., Wurzer, G. (eds.) Digital Design Reconsidered-Proceedings of the 41st Conference on Education and Research in Computer Aided Architectural Design in Europe (eCAADe 2023) - Volume 1, Graz, 20–22 September 2023, pp. 519–526. CUMINCAD (2023) https://papers.cumincad.org/cgi-bin/works/paper/ecaade2023_282

Walter, N., Gürsoy, B.: A Study on the sound absorption properties of mycelium-based composites cultivated on waste paper-based substrates. Biomimetics. **7**(3), 100 (2022). https://doi.org/10.3390/biomimetics7030100

Zamani, A., Mohseni, A., Bertug, Ö., Çapunaman, G.: Reconfigurable formwork system for vision-informed conformal robotic 3D printing. In: Dokonal, W., Hirschberg, U., Wurzer, G. (eds.) Digital Design Reconsidered-Proceedings of the 41st Conference on Education and Research in Computer Aided Architectural Design in Europe (eCAADe 2023) - Volume 1, Graz, 20–22 September 2023, pp. 387–396. CUMINCAD (2023) https://papers.cumincad.org/cgi-bin/works/paper/ecaade2023_317

Zivkovic, S., Battaglia, C.: Rough pass extrusion tooling. CNC post-processing of 3D-printed sub-additive concrete lattice structures. In: Recalibration. On Imprecisionand Infidelity. CUMINCAD, Mexico City, Mexico (2018) http://papers.cumincad.org/cgi-bin/works/paper/acadia18_302

Hollow-Characters: Robotically Programmed Matter for Kinetic Behavior and Guided Self-Assembly

Samim Mehdizadeh(✉) and Oliver Tessmann

Digital Design Unit (DDU), Technical University of Darmstadt, Darmstadt, Germany
mehddizaeh@dg.tu-darmstadt.de

Abstract. This paper describes a design to fabrication framework for robotic sequential casting of architectural objects: "RotoForm". RotoForm utilizes simulation, thermal imaging, and robotic fabrication for a responsive casing process that allows to control material distribution through robotic adaptation to incoming sensor data. The RotoForm robotic fabrication framework materializes large hollow objects, prioritizing activated performance over traditional optimization criteria. The targeted results are robotically fabricated hollow-concrete elements with specific motion behavior. The scope of this research extends beyond optimizing the weight of construction to creating objects/building elements capable of specific kinetic behaviors by manipulating their physical attributes through robotic fabrication.

We ask: "What if the architecture moves, becomes dynamic, and animated? What if architectural space transforms from being static into reconfigurable compositions that adapt themselves to the use and the users? What, if seemingly massive pieces rock and move at the touch of your hand?"

This research paper focuses on integrating advanced robotic fabrication methods with the design and assembly of hollow-concrete elements, notably focusing on the "RotoForm" robotic fabrication setup. Central to this investigation is the methodological approach to activating desired motion behaviors of building elements for guided self-assembly by tailoring their center of mass. A critical challenge addressed is translating digital data into tangible objects, ensuring precision in fabrication and adherence to desired motion patterns. Our method in this paper employs an array of innovative tracking techniques, such as thermal imaging, to increase the awareness during the robotic fabrication process and also assess the bespoke motion of the elements in interaction. We utilize this approach, materializing programmed building elements for kinetic performances, showcasing the potential of robotics to enable more circularity in architectural design. This research highlights a method for linking digital simulation and robotic fabrication and emphasizes how digital simulations relate to real-world matter.

Keywords: Guided Self-Assembly · Robotic Fabrication · RotoForm · Sequential Dynamic Casting · Architectural Choreography · Motion Tracking

M. Yablonina et al. (Eds.): Robotic Fabrication in Architecture, Art and Design 2024, pp. 99–112, 2026.
https://doi.org/10.1007/978-3-032-22607-5_7

1 Introduction

1.1 Robotic Fabrication as a Mediate

In the last two decades, two prominent yet interrelated directions have been distinguishable in robotics in art and architecture (Bechthold and King 2013). The first direction focuses on leveraging robotics for industrial optimization, addressing construction and manufacturing challenges while targeting to optimize the engineering processes. The second direction involves robots in creative and artistic contexts, such as generating artwork, interactive performances, choreographing a swarm of robots, or dancing with robots. There is a third direction, which links the design innovation to the industrial manufacturing realm by novel materialization methods. Hence, the advanced and easily accessible digital communication tools merging these two directions are acutely needed. This research focuses on the third direction between fabrication and performance, and a medial platform to use the object specific kinetic behavior through a robotic fabrication method called RotoForm.

The presented approach is close to the third direction, by implementing intelligence into the matter and the physical world using robotics and digital fabrication. We aim at a method that applies intelligence to the performance of the physical robotically fabricated objects, which can act and interact.

1.2 Activating the Kinetic Behavior Through Robotic Fabrication

The massive-looking yet lightweight, hollow, robotically materialized concrete elements are designed to join together, but not permanently. A single person can move and assemble them with their dry-joint, interlocking connection, and charged Center of Mass (CoM). The elements are programmed for guided self-assembly, disassembly, and reassembly through embedded geometry and material attributes (Mehizadeh and Tessmann, 2023) (see Fig. 1). The reconfiguration allows for a more circular approach and reuse of concrete elements in the architectural context.

This paper gives a detail insight to the robotic fabrication technique applied to designing and materializing kinetic architectural hollow-concrete elements, enabling self-assembly and reassembly. The research focuses on robotically fabricated building elements with programmed motion behaviors and potential for future reuse. It provides insight into the properties of lightweight, dynamic concrete elements. Despite their heavy appearance, these hollow objects can move, roll, and be assembled or disassembled through robotic fabrication and dynamic casting. Within this framework, the fabrication method programs the kinetic behavior of building elements for both autonomous and guided self-assembly.

2 Background

2.1 Material Programming and Robotic Fabrication

Using robotic arms to prefabricate concrete elements is a growing research area in the academic and practice sectors. The conjunction of automation, robotics, and digital design frameworks allows a seamless pipeline from the initial design phase to the

Fig. 1. The installation (blinded for review) to choreograph the architectural assembly with robotically fabricated hollow-concrete architectural characters.

on-site/off-site construction to the maintenance and eventually to the assembly and disassembly of elements and parts or demolishing (Baines, 2004). In the past two decades, computational design and robotic fabrication advancements revolutionized traditional construction methods. Concrete 3D printing-mostly adding layers of mortar as within (FDM) 3d-printing method-uses robotic arms to materialize prefab concrete elements (Wangler et al., 2019). While 3D printing concrete has transitioned from research to practical applications, there are different researches questioning how robots can be more intricately involved in material handling by using the robot trajectory and gravity to position the concrete.

M. Bechtold and N. King from Harvard GSD presented a robotic fabrication method for making customized concrete objects at the Robots in Architecture Conference 2012 workshops (King, 2012). The Smart Dynamic Casting (SDC) project at ETH Zürich explores the material-efficient production procedure for freeform columns by combining slip forming and robotic fabrication (Shahab et al., 2013). A small-scale slip form is attached to the 6-ax robotic arm. The Robotic arm gradually shifts the form and dynamically changes the geometry during the extrusion.

This Robotic fabrication methods like RotoForm, focuses on creating of hollow concrete structures with varying internal cavities (Tessmann and Mehdizadeh 2020). This technique leverages the fluidity of concrete and robotic motion to craft these structures. RotoForm facilitates multi-layered casting, leading to distinct weight distributions within the hollow elements, thereby altering their center of mass (CoM). Dunaway at University of Pennsylvania presents a robotic clay slip casting method for bespoke material distribution over the mold surface by incrementally injecting different amounts of colored slip into the mold while it is rotated (Dunaway et al., 2022)

2.2 Autonomy and Assembly of Parts

Automation in architecture largely relies on discrete solutions to design, fabrication, and assembly of part into complete structures. Process of translation between digital design and physical assembly is complex and has to consider unpredictable material behaviors and accumulating material tolerances at every stage. Claypool and Retsin describe this issue as the question of discreteness in design and offer a perspective on how the components could be gathered and assembled (Retsin 2016). The physical aggregation of discrete assemblies comes with a series of challenges concerning the imperfections of materials, gravitational forces and assembly processes. Tibbits at MIT demonstrates the potential of programming material for a specific assembly. In his work Tibbits activates the assembly of physical parts by manipulating the physical and geometrical attributes of parts. (Tibbits, 2012). The concept of "designing for self-assembly" underscores the potential to incorporate assembly instructions directly into building components by tailoring their intrinsic properties during the design and fabrication stages (Lendlein and Trask 2018). In our case, robotic fabrication is defined as a framework for designing and programming parts with a certain degree of autonomy for assembly. In the late 1950's, Lionel Penrose presented his proposal for self-replication machines and articulation of geometries for self-assembly (Penrose, 1959). The work of Penrose sought to program the geometry of discrete pieces so that the parts self-assemble (Griffith, 2004).

2.3 Monitoring and Motion Tracking

A motion Tracking system allows us to compare the motion results of simulation and materialized objects (Kiss and Nagymáté 2018)

3 Method

This study addresses four technical challenges, emphasizing the connection between mathematical and physical properties, elements for self-assembly and the robotic fabrication method. Central to these challenges is translating digital data to its physical objects and the system's awareness within the robotic fabrication framework (see Fig. 2).

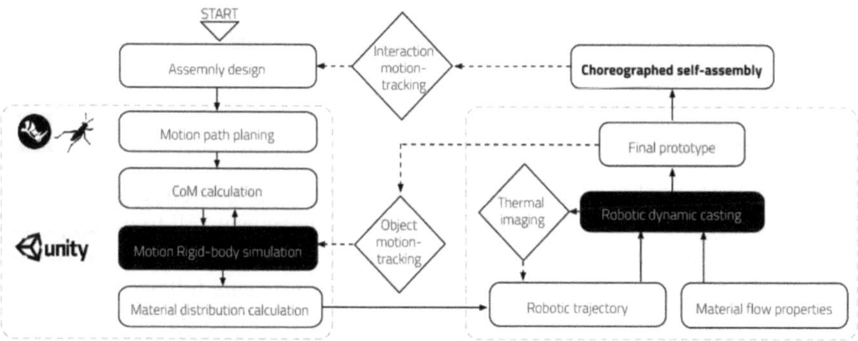

Fig. 2. The workflow from Design Simulation to Fabrication with the Feedback loops through Thermal Imaging and Motion Tracking.

We systematically detail the design and fabrication challenges across four main areas and describe the challenges in part c and d for robotic fabrication in detail. The following sections follow the order of design tasks presented in our developed framework:

(a) Motion through shifting the CoM within hollow-concrete elements.
(b) Strategizing assembly and interlocking of hollow-concrete elements based on the motion planning.
(c) Implementing robotic sequential casting to produce hollow-concrete elements with a manipulated CoM.
(d) Introducing a monitoring and evaluation system that encompasses material distribution tracking and the motion behavior of the element.

3.1 Motion Through Manipulating the CoM

To activate the motion of objects in any direction, it is essential first to determine the material distribution within the objects, ensuring they remain balanced in any desired position. Then, the object can be charged for motion in desired direction by pouring material in the motion's direction. We employ an algorithmic method for weight balance that represents any geometry subdivided into a voxel model (Bächer et al., 2014). The object's balance over its ground anchor point is achieved by iteratively removing voxels. This method is related to topological optimization but with different objectives. The same algorithm can also determine an element's motion from one spot to another (Mehdizadeh and Tessmann 2024) (see Fig. 3). Each movement is depicted as a sequence of stable positions and the transitions in between. The weight optimization algorithm facilitates the computation of mass distribution within an object, allowing it to have two or more stable positions. The added mass to trigger the motion determines the motion's speed but compromises the element's stability. Thus, an initial force is required to initiate the motion; once the force is applied, the material rolls to its next stable position.

All calculations are done within the Rhino and Grasshopper CAD environments. We employ rigid body simulations via Nvidia Physics within the Unity game engine to validate our computational approach. The seamless integration between these software tools ensures real-time assessment of the simulated object behavior. The robotic trajectory algorithm iteratively defines the motion of the robotic arm and the amount of cast material for each material layer. Similar to 3D printing, RotoForm allocates material in a series of consecutive layers. However, these layers follow the three-dimensional form of the formwork (see Fig. 2).

3.2 Assembly and Interlocking Through Motion

The planning of assembly in this research is connected with the motion path planning and the interlocking of the elements at the same time. As the objects roll from one position to the other, the motions in this research are rotational and not transitional (Gilibert et al., 2022). Subsequently the interlocking of rolling elements follows the rotational directions instead of transitional directions (Mehdizadeh and Tessmann 2024) use the rotational-locks for assembly of several element, that allow for sequential assembly of several elements, with dry joints. The dry joint connection of elements with rotational parts are designed as reversible assemblies that allow for dis-assembly and re-assembly in different configurations (Mehdizadeh and Tessmann 2024). The geometrical principle cycloid describes a calculation method for defining the rolling motion of the elements, their stable positions and the direction of rotational interlocking, all within the same framework (Barra 1975) (see Fig. 4).

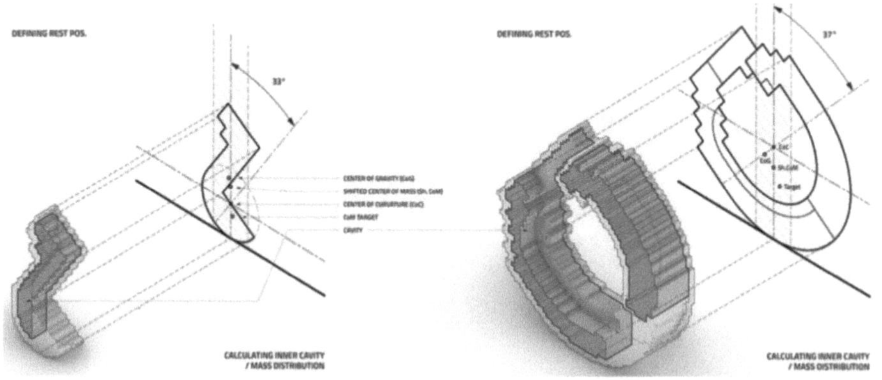

Fig. 3. Left: The generative model based on the mathematical principle of Cycloid for rolling and interlocking. Right: Optimization algorithm results for the weight distribution inside the Concrete Element

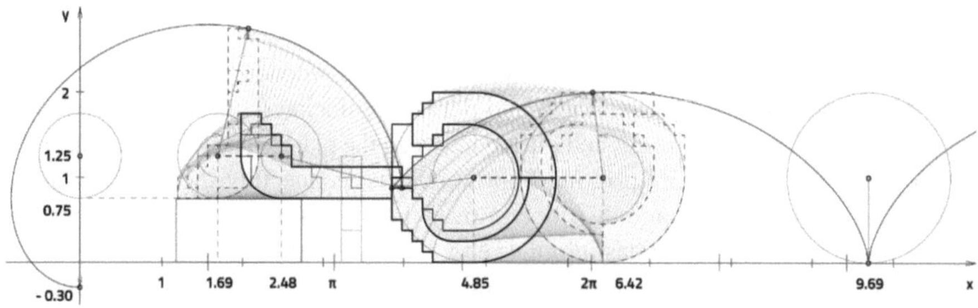

Fig. 4. The assembly path and rotational interlocking planning with the geometrical principle of Cycloids

3.3 Robotic Sequential Casting Setup

This research seeks to activate the specific kinetic behavior of building elements through the robotic fabrication method **"RotoForm."** In the sequential robotic casting process, small amounts of liquid material are cast in a formwork and then slowly rotated by a robot. The material spreads along the formwork surface, creating a hollow body (Mehdizadeh and Tessmann 2024). The robot arm rotates the formwork in a defined path with a variable velocity, distributing the material to create a variable wall thickness and mass distribution (see Figs. 5 and 6). RotoForm synthesizes elements by merging the robot's motion trajectory with the material's flow properties (Tessmann and Mehdizadeh 2019). After designing the motion-assembly scenario of concrete elements, the CoM and the material distribution are calculated and optimized in Grasshopper (Mehdizadeh and Tessmann 2024).

The robotic setup consists of three stages: formwork assembly, infill-rebars assembly, rotationally casting, and monitoring feedback. All three stages are planned and tested in a continuous step framework for materializing hollow-concrete building elements with shifted CoM on a large-scale (see Fig. 5).

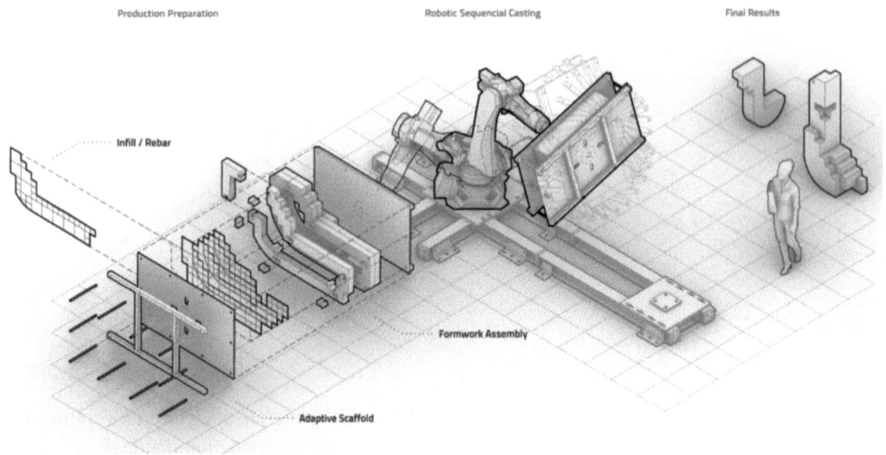

Fig. 5. The Fabrication pipeline of hollow concrete building elements, robotic sequential casting (RotoForm)

The adaptive formwork consists of two plates and a wooden frame, which allows us to adjust the distance between them and define the thickness of building elements. The sections of formwork are cut and placed between the plates. The rebars in this setup function like the infill of 3D printing hollow objects, connecting the surfaces of the element and allowing for force distribution over the outer surface of the element (see Fig. 5). During the rotational casting, the rebars are covered by the cast material, and their thickness increases with more layers of concrete, which allows bridging the two surfaces of the concrete with the same material.

Robotic path for Dynamic Casting. In order to materialize the large-scale hollow-concrete building elements with the shifted CoM, the calculation of the robotic motion path and the amount of material for each casting layer is required. After calculating the material distribution inside the formwork for the desired motion, an algorithm slices the layers inside the formwork with a specific thickness, much like the slicer for 3D printing with the FDM procedure that calculates the machine flow for each layer. Rotoforming a series of sequential layers gives precise control over material quantity at any given area of the formwork. Using too little material inside the formwork results in certain parts being uncovered over the surface of the formwork. Using too much material would result in very high dynamic forces inside the formwork as the material flows inside the formwork, resulting in uneven distribution of material inside the formwork and the restriction of material flow properties. After initial testing, we have established that layer thickness of 5 ± 2 mm gives the best results by flow property. After 20 min the material is cured enough to not move in the formwork anymore. Subsequently, the next layer can be added and rotationally cast (Mehdizadeh et al., 2022).

Two variables adjust the robotic motion path for the rotational casting: the motion end position and the duration of material flow. Despite the many fabrication techniques, such as 3D printing, that the motion path of the robot coincides with the material deposition, in this procedure, the robot defines the end target of material deployment from point A to B, given the time and gravity forcing the material to flow over the specific position and after the next position. The velocity and acceleration are constant parameters in this procedure. However, controlling the material distribution inside the closed formwork with variable wall thickness requires a monitoring feedback system.

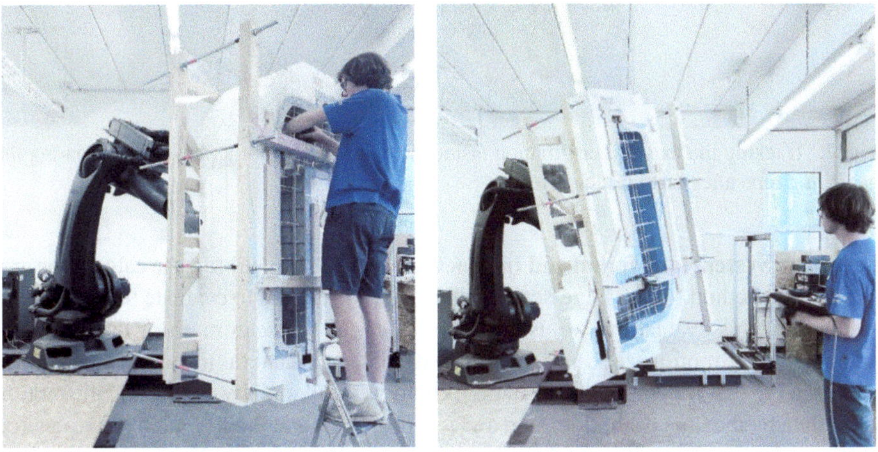

Fig. 6. The Robotic Fabrication Setup for Rotoforming Large-Scale elements up to 250 cm long

3.4 Feedback System for Material Distribution

The monitoring and feedback system for this procedure is divided into two scopes. Firstly, using the thermal cameras to track the material distribution inside the formwork as the robot is rotationally casting the parts and secondly, using the motion tracking system with Opti-track to evaluate the bespoke motion of elements with shifted CoM.

Monitoring through thermal Imaging. The Rotoforming in the closed formwork is challenging since it is impossible to visually track the material distribution inside the formwork. The exact concrete mixture's rheology properties and the formwork's shape do not help estimate the exact material flow path inside the formwork in reality. This challenge addresses the difference between the data simulation and the reality of material properties since many environmental parameters impact the material flow. In our previous work we have investigated using an endoscope camera to track the material and fluid behavior inside the formwork (Mehdizadeh and Tessmann, 2024). However, this method is restricted to a specific camera view. The (Fries-Briggs, 2014) introduces a technique that uses a thermal camera to track the material distribution inside the volumetric cavities of the Styrofoam formwork. The Thermal camera can detect the material since the concrete produces a considerable amount of heat during the curing process. In this phase, the material reaches temperatures up to 33 °C (see Fig. 7), which makes the position and density of the material inside the formwork visible. Using this technique and robotic sequential casting allows us to recalibrate the following sequence of the rotational cast to cover the uncovered areas of formwork duo to the thermal imaging process.

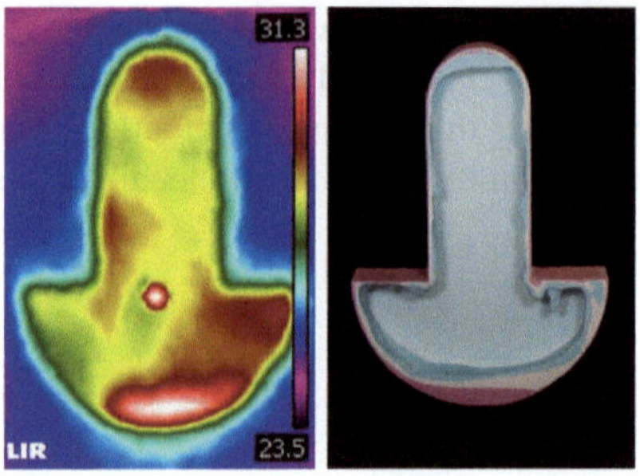

Fig. 7. Tracking the material distribution inside the formwork with thermal imaging during the hardening time after casting each layer

Feedback System for Motion and Interaction. In order to calibrate the calculation and simulation method for the large-scale application, it is crucial to evaluate the simulation results. For this evaluation, we use the overlay of the 3D-scan mesh of the element's geometry to the motion tracking data using the Opti-track cameras to quantify the motion data (see Fig. 8). We compare the results of motion simulation by comparing the motion track data by the time of motion, start and rest positions, and the deviation of the desired motions. The results on the small scale are very similar to the rigid body simulations in the game engine Unity, using Nvidia Physics. However, the collision of parts and their interaction need more accurate adjustment of the physical parameters, such as the surface friction. By scaling the system, this parameter becomes more important to adjust.

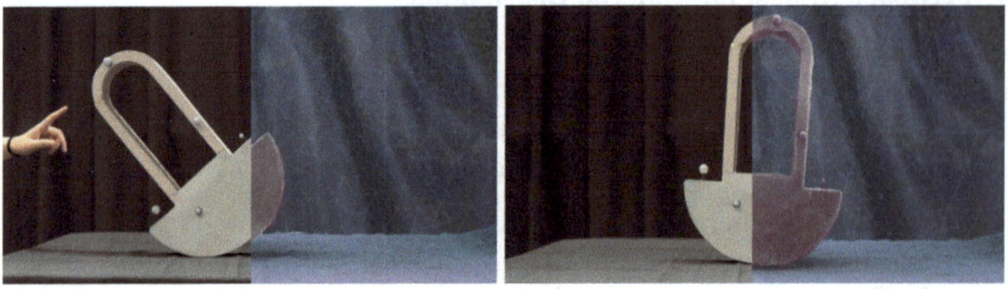

Fig. 8. Monitoring the desired motion and balance with the Opti-Track system and comparison to the rigid body simulation results. Left: the moment of applying initial motion force to the object in the first stable position, the object at the second stable position. This object is optimized to have equilibrium on any position along the curvature. Each image illustrates the physical object with Opti-Track markers on the left side, and the overlay motion tracking Data on the 3D scan mesh on the right side

Motion track of Assembly. This research conceives human-object interaction as a dynamic force for guided self-assembly. The hypothesis of this research focuses on programming the material for a specific motion behavior. The collaborative assembly of large-scale hollow elements is designed to use the human body weight as a part of the design for assembly. Therefore, two rest positions are targeted for the element: First, the rest position as the objects stand-alone (solo-equilibrium), and the second position as the

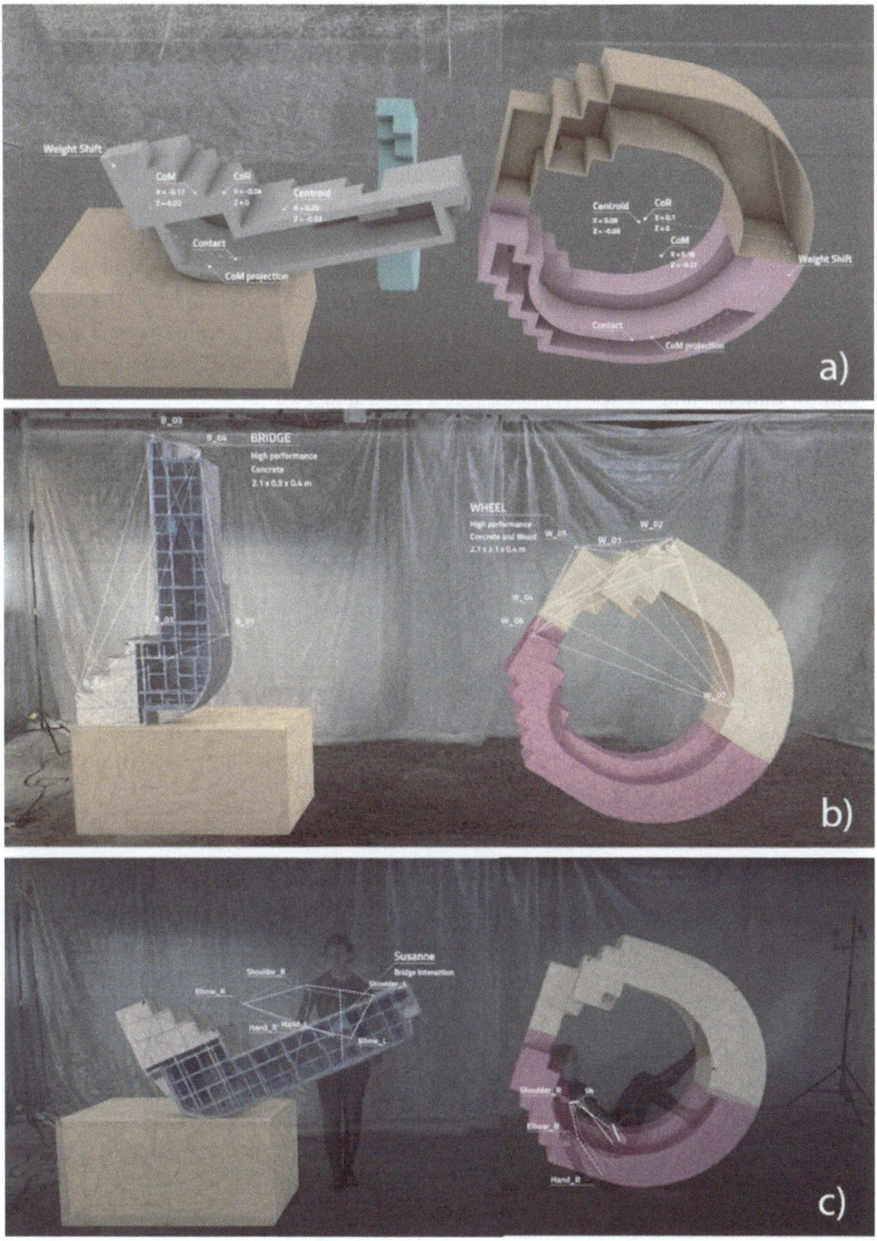

Fig. 9. The Motion-Tracking of hollow-concrete elements is designed and robotically fabricated for guided self-assembly. a) the simulation of assembly sequence, b) the stand-alone rest position (equilibrium) of the elements without additional human body weight. c) the collaborative rest position by the human body's weight as additional forces.

objects are forced to rest through the forces enacted by the human body (collaborative equilibrium) (see Fig. 9). The experiment with large-scale elements is designed with a team of choreographers, targeting the embodiment of the human body weight as a tool to force the elements to move for assembly (DDU Motion Tracking Animate Concrete, 2022). The two elements of this experiment gather to fulfill the different scenarios of applying the forces of humans. The L-shaped element (called a bridge) requires minimal initial force to bring it from vertical to horizontal. This force can be applied from the side or top of the element. The round-shaped element (called Wheel) requires larger moving forces, which should be applied by the human body's weight inside the object (see Fig. 10). The experiments with the large-scale element assembly and human body weight have been designed and simulated within the digital design framework. This target has been tested and evaluated with the motion-tracking of elements in collaboration with a team of dancers and choreographers. (see Figs. 8 and 11).

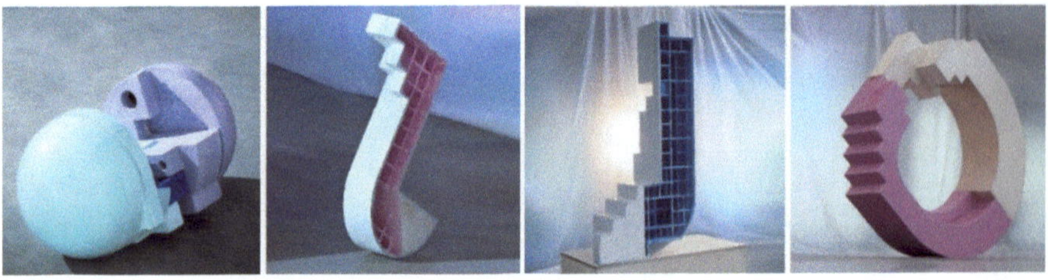

Fig. 10. Four Architectural Characters in 1:1 scale with the ability of self-guided Assembly through weight distribution, from Left to right: The Capsule; the Note-key; the Beam/Column; the Wheel.

Fig. 11. Architectural Choreography for guide self-assembly. Exploration on the human body weight in as the additional force for assembly in collaboration with choreographers and dancers.

4 Results

Digital simulation with Unity game engine in connection to CAD environment (in Rhino + Grasshopper) offers a robust framework for computation methods for geometry generation and physic simulations. Planning the motion through geometry is relatively new in the field of computational design. This research uses a simple geometry approach to calculate the interlocking, dry joint, and reversible connection principles and design the assembly through the motion of parts. Using rotational locks and cycloids allowed us to define geometric rules to fulfill the desires of reversible connections and kinetic self-assemblies and increase the system's complexity through generative models.

The physical results of this research are a series of hollow-concrete elements with programmed specific kinetic behavior. Due to the embedded motion behavior, we conceive these elements as characters. The large-scale characters are each focused on determining and illustrating different challenges. The L-shaped element (called a Beam/Column) demonstrates a counterintuitive self-erecting behavior despite the geometry magnitudes. The round shape (called Wheel) offers the two equilibrium positions, alone and after adding the human's body weight in a specific position. The S-shape (called Note/Key) is designed and calculated to keep the whole assembly locked together through the exact amount of weight and geometrical interlocking. The sphere shape (called Capsule) is designed so that two halves can roll- and lock through a single key element to stabilize the locking of the elements (see Fig. 10).

5 Discussion and Conclusion

The research "Animate Concrete" contributes to circular approaches in design and **"design for reuse"** by providing a framework for guided self-assembly and re-assembly. The robotic sequential casting method (RotoForm) enables counter-intuitive kinetic behavior of the concrete elements, which move unexpectedly due to their surprising (and invisible) mass distribution that prompts them to move. Moreover, the robotic fabrication method reduces material consumption significantly. The monitoring system of robotic sequential casting with thermal imaging accommodates the permanent adaption and self-correction of the robotic casting trajectory in each iteration. The feedback system with thermal imaging visualizes previously invisible internal material distribution during robotic sequential casting. These feedback loops help calibrate robotic fabrication with unpredictable material behavior. It shifts the complexity from preprogramming robotic trajectories towards a feedback-driven fabrication process. Moreover, the systematic investigation method gradually scales the system by observing the material behavior in the robotic casting procedure, as well as the motion behavior of elements (Animate Concrete DDU with TANZPOL, 2022). This system allows for iterative adjustment of the design-simulation parameters closer to the physical models in interaction with the human body and gravity.

6 Future Work and Outlook

The future developments target technical improvement of the design to a fabrication method for the future application of this technology to the construction industry. The motion tracking results with the large-scale elements illustrate the deviation between the target and the physical end position. This issue addresses the importance of calculating the surface friction parameter in the physical simulation. In addition, the material characterization and finite element simulations are essential next steps for this framework.

Adding a lightweight reinforcement system capable of resisting the tensile forces in the hollow elements is an important future step.

The dancers interacting with the Hollow-Characters served as proxies of inhabitants and users of reconfigurable spaces. Their knowledge and skills in balancing bodies allowed them to intuitively dance with building elements. In future work we seek to further explore this relationship of humans and reconfigurable building components of an ever-changing architecture.

Acknowledgement. The research project, developed at DDU, TUDa during S.2022, benefited significantly from the architecture students' contributions. Joshua Schäfer and Malcolm Unger played key roles in advancing the design simulation and motion tracking framework. Danial Ahmad, Bingqun Liu, Luke Schussler, and Daniel Schinkels provided crucial support in robotic fabrication. The motion tracking aspect was developed in collaboration with IAS TU Da, led by Dr. Boris Belousov and Timm Schneider, with additional support from Prof.Dr. Clemens Brünenberg.

References

Animate Concrete DDU with TANZPOL, 2022

Bächer, M., Whiting, E., Bickel, B., Sorkine-Hornung, O.: Spin-it: Optimizing moment of inertia for spinnable objects. ACM Trans. Graph. **33**, 96:1–96:10 (2014). https://doi.org/10.1145/2601097.2601157

Baines, T.: An integrated process for forming manufacturing technology acquisition decisions. Int. J. Oper. Prod. Manag. **24**, 447–467 (2004). https://doi.org/10.1108/01443570410532533

Barra, M.: The cycloid. Educ. Stud. Math. **6**, 93–98 (1975)

Bechthold, M., King, N.: Design robotics. In: Brell-Çokcan, S., Braumann, J. (eds.) Rob|Arch 2012, pp. 118–130. Springer, Vienna (2013). https://doi.org/10.1007/978-3-7091-1465-0_11

DDU Motion Tracking Animate Concrete, 2022

Dunaway, D., Rothbart, D., Gwinn, L.: Designing a process for robotically controlled rotational casting. Advanced Materials. (2022)

Fries-Briggs, G.: Mediating volumetric thresholds. In: McGee, W., Ponce de Leon, M. (eds.) Robotic Fabrication in Architecture, Art and Design 2014, pp. 211–221. Springer International Publishing, Cham (2014). https://doi.org/10.1007/978-3-319-04663-1_14

Gilibert, P., Mesnil, R., Baverel, O.: Rule-based generative design of translational and rotational interlocking assemblies. Autom. Constr. **135**, 104142 (2022). https://doi.org/10.1016/j.autcon.2022.104142

Griffith, S.T.: Growing Machines (Thesis). Massachusetts Institute of Technology (2004)

Kiss, R.M., Nagymáté, G., 2018 Application of OptiTrack motion capture systems in human movement analysis: A systematic literature review. Recent Innovations in Mechatronics (RIiM) Vol. 5(1). Accessed 15 Feb 23

Lendlein, A., Trask, R.S.: Multifunctional materials: Concepts, function-structure relationships, knowledge-based design, translational materials research. Multifunct. Mater. **1**, 010201 (2018). https://doi.org/10.1088/2399-7532/aada7b

Mehdizadeh, S., Tessmann, O.: Animate concrete: Materialization of concrete element kinetic assemblies. In: Yan, C., Chai, H., Sun, T., Yuan, P.F. (eds.) Phygital Intelligence. CDRF 2023. Computational Design and Robotic Fabrication. Springer, Singapore (2024). https://doi.org/10.1007/978-981-99-8405-3_33

Mehdizadeh, S., Zimmermann, A., Tessmann, O.: RotoColumn: A continuous digital fabrication framework for casting large-scale linear concrete hollow elements. In: Towards Radical Regeneration: Design Modelling Symposium Berlin 2022, pp. 476–486. Springer (2022)
Penrose, L.S.: Self-reproducing machines. Sci. Am. **200**, 105–117 (1959)
Retsin, G.: Discrete assembly and digital materials in architecture. In: Complexity & Simplicity-Proceedings of the 34th eCAADe Conference, pp. 143–151 (2016)
Shahab, A.R., Lloret Kristensen, E., Fischer, P., Gramazio, F., Kohler, M., Flatt, R.J., 2013. Smart dynamic casting or how to exploit the liquid to solid transition in cementitious materials. Presented at the 7th RILEM International Conference on Self-Compacting Concrete and 1st RILEM International Conference on Rheology and Processing of Construction Materials
Tessmann, O., Mehdizadeh, S.: Hollow-crete: Prestressed membranes as formwork for material efficient hollow concrete building elements. In: Impact: Design With All Senses: Proceedings of the Design Modelling Symposium, Berlin 2019, pp. 474–486. Springer (2020)
Tessmann, O., Mehdizadeh, S.: RotoForm: realization of hollow construction elements through roto-forming with hyper-elastic membrane formwork. In: Proceedings of the Symposium on Simulation for Architecture and Urban Design, pp. 1–7 (2019)
Tibbits, S.: Design to self-assembly. Archit. Des. **82**, 68–73 (2012)
Wangler, T., Roussel, N., Bos, F.P., Salet, T.A.M., Flatt, R.J.: Digital concrete: A review. Cem. Concr. Res. **123**, 105780 (2019). https://doi.org/10.1016/j.cemconres.2019.105780

Other Robot Bodies

Synergising Natural Material Complexity and Low-Impact Automation

Eszter Olah(✉), Mehrdad Zareian, and Moritz Dörstelmann

Professorship for Digital Design and Fabrication (DDF), Karlsruhe Institute of Technology (KIT), Karlsruhe 76131, Germany
{eszter.olah,mehrdad.zareian,moritz.dorstelmann}@kit.edu

Abstract. The construction industry is considered one of the world's largest raw materials consumers. It is contributing towards global resource depletion which is exacerbated by increasing demographic demands thus making the industrialisation of innovative sustainable technology crucial. Digital fabrication offers solutions by enabling higher material efficiency as well as the use of renewable bio-based materials. It is not limited to complex processes employing robotic arms to create finely graded building components but can also include facile, cost-effective and low-impact approaches. These can involve unique features to deal with materials less prominent in the construction repertoire due to their inhomogeneity. Customised low-cost and low impact solutions may aid the democratisation of specialised digital fabrication workflows by enabling easier transportability and accessibility, in addition to bypassing the dependence on special skills for operation. To showcase the application potential of these perspectives, this paper focuses on the evolution of a tailored Cartesian coordinate robot that - within the context of a semi-automated fabrication process - enabled the development of a novel construction system made of willow-earth hybrid building components. The evolution of this willow processing machine followed principles of low environmental impact and cost efficiency as its unique features were iteratively developed to deal with the intricate variations inherent to the raw material. The feasibility of the fabrication process was validated through multiple projects along its evolution, most recently the project 'ReGrow Willow' where 63 willow-earth components were prefabricated and installed to be exhibited at the Bundesgartenschau (Federal Gardening Show) in Mannheim in 2023.

Keywords: Low impact automation · Additive manufacturing · Plant-based material · Sustainable Construction · Customised Fabrication

1 Introduction

1.1 Contextualisation

The construction sector stands as a major contributor to the global depletion of resources [1] due to its significant consumption of raw materials [2]. Further exacerbated by the growing demand for new buildings [3], this sector is in dire need of considering alternative materials and methods to diversify the existing construction repertoire in order

M. Yablonina et al. (Eds.): Robotic Fabrication in Architecture, Art and Design 2024, pp. 116–128, 2026.
https://doi.org/10.1007/978-3-032-22607-5_8

to mitigate the effects of climate change and global resource depletion. Digital design and fabrication methods enable the mapping of material characteristics to adjust the fabrication process to reach greater material efficiency [4]. Nevertheless, for a scalable solution, output consistency must be considered. Bio-based materials - that are derived from materials of renewable biological origin [5] often require the homogenisation of an intrinsically anisotropic material in order to reach the required uniformity in output. This may necessitate a multi-step pre-processing stage that involves cost- and energy-intensive, complex fabrication workflows [6]. However, digital fabrication is not limited to complex systems involving robotic arms, it also enables solutions that embrace low-complexity and cost-effective automation. By applying these principles to the development of fabrication processes for bio-based - especially plant-based - materials, the emerging technologies can enable a reduction of dependence on specialised craft while allowing production scalability and automation feasibly, leading to the re-industrialisation of more sustainable and circular construction materials [7, 8] thereby lessening the environmental impacts of climate change [9, 10]. As a result, custom fabrication solutions can become accessible for communities that are normally excluded due to barriers of cost, availability or specialised knowledge [11].

This paper focuses on the evolution of a tailored Cartesian coordinate robot that - within the context of a semi-automated fabrication process- enabled the development of a willow-earth hybrid construction system [12, 13].

1.2 State-of-the-Art

Combining and adjusting existing manufacturing principles with new materials can accelerate the industrialisation of more sustainable construction materials. 3D printing allowed plant-based material fibres to be reintroduced into construction via additive manufacturing. From wood fibre reinforced PLA [14] to 'printing' with willow [15]: a common consideration - key in the 2.5D additive manufacturing processes - is the adhesion between the layers. Pre-processing enables a more uniform filament and therefore more control over the adhesion while taking away from the inherent anisotropic behaviour of the raw material

Key factors affecting the production scalability of a digital fabrication process are feasibility, reproducibility and automation [16]. Prefabrication can reduce and control some variables by ensuring a standard external environment [17], however, this results in a disadvantage with transportation. Reproducibility can be enhanced through standardisation. However, in the case of plant-based materials, this is inherently difficult and process-intensive due to the high degree of non-uniformity [16]. In the context of homogenisation of willow, an additive manufacturing process - with each strip treated with adhesives to form a cohesive filament [15] - may be feasible. Industrial wood manufacturing however is preceded by the homogenisation of natural wood by the application of steam, heat, ammonia or cold rolling as well as densification [18].

Developing automated robotic fabrication workflows for plant-based materials without standardisation or pre-processing encompasses challenges. Dealing with uncharacterisable variations might involve the implementation of a communication strategy forming a closed-loop feedback system which is able to update the virtual model and the manufacturing process [19]. 3D scanning or statistical analysis might be used to develop such a feedback system [20, 21]. Closed-loop systems could adapt to variations in pre-defined parameters without homogenising the material. There have been efforts to use reinforcement learning to characterise such variations and tackle challenges that bio-based materials pose from a fabrication perspective [22] however, the energy- and time-intensity may hinder the scalability potential.

Historical applications of plant-based materials in the vernacular building sector limited their pre-processing to soaking in the case of willow [23], which motivates the exploration of techniques that require minimal pre-processing. Such a technique is weaving - originally a characteristic of the textile industry - where a thin, bendable material is formed into a continuous woven 'sheet' of material either to be used as tension reinforcement for earth (historically) or concrete [24]. Since weaving techniques lend themselves to be used with long, slender materials, willow was proposed as a suitable plant-based material for the development of an automated weaving process.

1.3 Research Aim

The overarching research aim is to develop a digital fabrication system for a fast-regrowing plant-based material, willow (Salix). Starting from a material perspective, plant-based materials in their natural form are anisotropic with significant variance. By harnessing the construction potential of plant-based materials without introducing multiple levels of complexity in the pre-processing and fabrication stages a facile solution may emerge. The emergent fabrication system can be made more robust by incorporating readily available elements that can adapt to the variations of the material. Developing a process intended for efficient scalability - both in terms of building component scale as well as production rate - and automation could accelerate the reindustrialisation of sustainable materials. Furthermore, a fabrication system that consists of easily transportable and re-assemblable machinery, the application space - both geographically and socially - can be expanded as the fabrication process may utilise local materials, bypassing the need for specialised craftspeople. To drive the development of the envisioned custom fabrication system, the relationship between the required output quality - essential for structural requirements - and the principles of feasibility and low environmental impact have to be explored and compromises reached.

One of the main research lines at the Professorship of Digital Design and Fabrication at the Karlsruhe Institute of Technology has been focusing on reinterpreting vernacular practices of construction with bio-based, fast renewable materials through the digital perspective to explore their reintroduction into the current construction repertoire [8, 12, 13] by conceptualising and experimenting with low-impact, circular and localised solutions that could drive the industrialisation of more sustainable construction.

2 Methods

The overall fabrication process currently involves five steps as shown in Fig. 1 in order to produce the willow-earth hybrid components. The process utilises willow in its natural state with minimal pre-processing (watering or steaming). The weaving process is preceded by splicing and consists of willow extruded and laid weaving through a bed of metal poles (Fig. 2). Earth is then introduced into the cells of the woven component lending it significantly more compression strength [13]. The earth-filling process lends itself to the adoption of existing automated robotic earth extrusion tending towards an increased degree of automation. The following sections introduce the evolution of the splicing and weaving stages highlighting the main features that were implemented to enable the fabrication of the willow rebar.

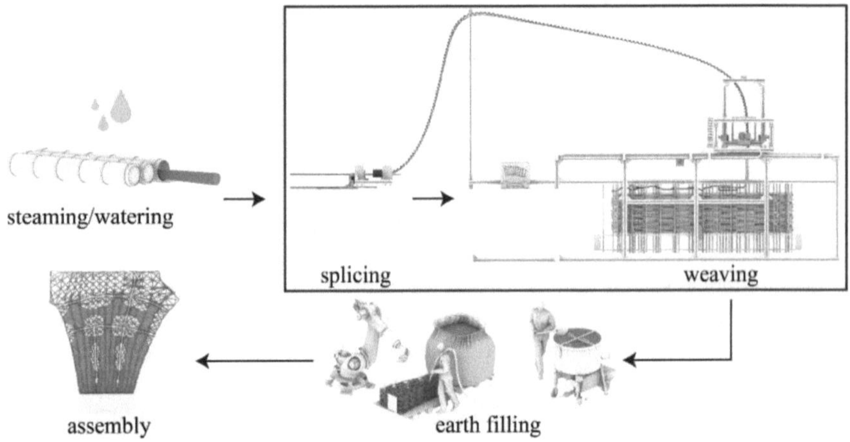

Fig. 1. Overview of the developed fabrication process.

2.1 Interdependent Process Designs for Custom Machinery and Architectural Components

Starting from a material perspective, manual material explorations were employed to gain familiarity with the raw material's inhomogeneities. Techniques traditionally used in vernacular architecture were also investigated to understand how such inherently slender and flexible materials can be made into building components by exploiting their features rather than necessitating homogenisation [8]. Taking inspiration from the textile industry [25], weaving and braiding logics were explored, for which the bending behaviour is key. Initial bending tests enabled an insight into this behaviour and a qualitative starting point for the design of the digital fabrication process could be established. This was an important step for the iterative development as - unlike for other bio-based materials, such as timber - there are no pre-existing standards that may be used to reliably estimate the variation in mechanical properties of willow. In addition to the bending radius, it was also verified that the bending location can seldom be predicted as it depends on the willow branches' internal structure in response to the external forces. Because of this unpredictability, established digital fabrication techniques - which rely on precise material behaviour and predetermined responses - are not applicable. Different strategies were explored for the manipulation of willow, for example, small-scale mobile robots and custom robot platforms on a rail guide. Both approaches offer scalability, allowing more robots to expand both system size and braiding pattern options. Yet, practical tests revealed a constraint between strength and size, and the number of robots needed. Using too many large robots in proximity is impractical, while small robots lack the necessary strength for bending willow, leading to an alternative approach. Robotic arms possess the versatility required for the manipulation of plant-based materials; research into willow-based additive manufacturing methods shows promising results for 3D printing willow stems [26] by standardisation. Motion planning with Cartesian robots is simple, fast and unlikely to fail due to singularity, angle limits or self-collision [27]. These types of digital fabrication machines enable a range of applications and are highly customisable for specific requirements [27]. Therefore, adopting a low-complexity approach, a Cartesian coordinate robot was developed.

Component-level tests revealed how thin willow branches enable structurally stable components to be utilised as integrated formwork and tension reinforcement for earth construction [12, 13]. Digital design tools were employed to explore scaled-up versions of the test components allowing quick design iterations and an insight into the requirements for the Cartesian robot to fabricate the conceptualised components. The parallel development of the fabrication system and the building component design allowed the two processes to inform each other and iteratively move their evolution forward. Small, circular cells showed promising structural behaviour (Fig. 3 Left), however, the complex weaving parameters – exact bending location, length, weaving tightness – proved challenging for digital weaving. The initial Cartesian weaving machine prototype highlighted the main obstacles regarding digital willow weaving that required the most development. Pulling individual willow branches (weft direction) through a bed of poles (warp direction) created friction, that – though necessary for structural integrity – hindered the fabrication process. It also limited weaving patterns

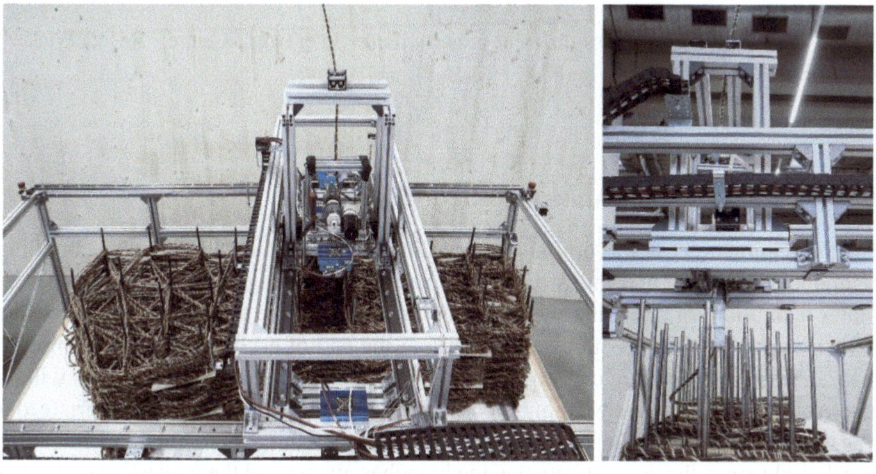

Fig. 2. Views of the Cartesian coordinate robot during the fabrication of willow components.

to straight lines (Fig. 3 Centre), which are less structurally favourable compared to interconnected patterns – as verified by structural simulations as well as by physical tests

[13]. Developing a continuous extrusion process that incorporates patterns developed and verified on the fabrication system means that an optimised cell distribution can be achieved (Fig. 3 Right) without friction hindering the fabrication and without the need to determine the length of each branch in the process.

Fig. 3. Pattern evolution across different projects. Left: 'InterTwig', Centre: 'RecodeWicker', Right: 'ReGrow Willow'.

2.2 Integrating Natural Material Behaviour

The fabrication process was optimised according to: inherent material features, building component evolution and industrialisation potential. Utilising willow branches necessitates a solution for encoding the length of each branch into the weaving process due to their variations. This requires either sensing the end of each branch or trimming it. Instead, a splicing solution can exploit varying lengths by creating a continuous willow macro-fibre. Preceding the weaving process, a splicer with a variable tension yarn feeder was thus devised (Fig. 4 Left), ensuring that the splicing thread held the branches sufficiently tight since the contact between the branches guaranteed the consistent quality of the macro-fibre. The resulting macro-fibre has a higher tensile strength compared to individual willow branches, lower deviation but a higher maximum in bending stiffness (as the willow branches are spliced superposed). Due to these variations, a standard extrusion process would not be suitable. An on-line measuring system could be adopted that detects

Fig. 4. Tailored fabrication features. Left: Willow macro-fibre splicer, Right: Custom extruder.

the instantaneous thickness variations and informs the extruder ahead to accommodate it. However, more robustly, spring-loaded, custom 3D-printed rollers were explored (Fig. 4 Right) as part of the extrusion mechanism as a hardware solution. The extruder followed a toolpath which was generated from the linear travel distance between the circular paths around the poles. This strategy minimised the travel distance while also maximising the distance from the obstacles (poles) thereby creating a safety margin to deal with the deflection of the poles which arise as the layers accumulate: the woven willow layers exert a force on the poles as tightness is essential for tension and friction within the willow rebar. Due to the non-standard nature of the willow macro-fibre, a straightforward coupling of the extruder speed with the machine translational speed is not suitable. Therefore, a synchronisation of the extruder speed with the XY motion is achieved by projecting each point on the toolpath curve onto a corresponding point on the intended willow path curve, thus dividing it into sub-curves which are then used to calculate the amount of extrusion necessary. Customising the script further enables the variation of the local tightness (such as the corners of the component), ensuring the willow macro-fibre is weaved tight thus lending structural stiffness to the components. Coupling these specialised features together resulted in a robust willow weaving Cartesian coordinate robot as depicted in Fig. 5 that enables the production of a range of differentiated building components with various adjustable parameters (e.g., rebar directionality, pole density, cell sizes, weaving tightness, macro-fibre thickness) thereby showcasing the feasibility of the iterative process that underpinned its development. Rapid prototyping techniques played a crucial role in expediting design iterations; especially in the context of the development of a custom fabrication solution, these techniques facilitate the quick testing of parts to optimise their interaction with existing elements and their sturdiness during the fabrication process.

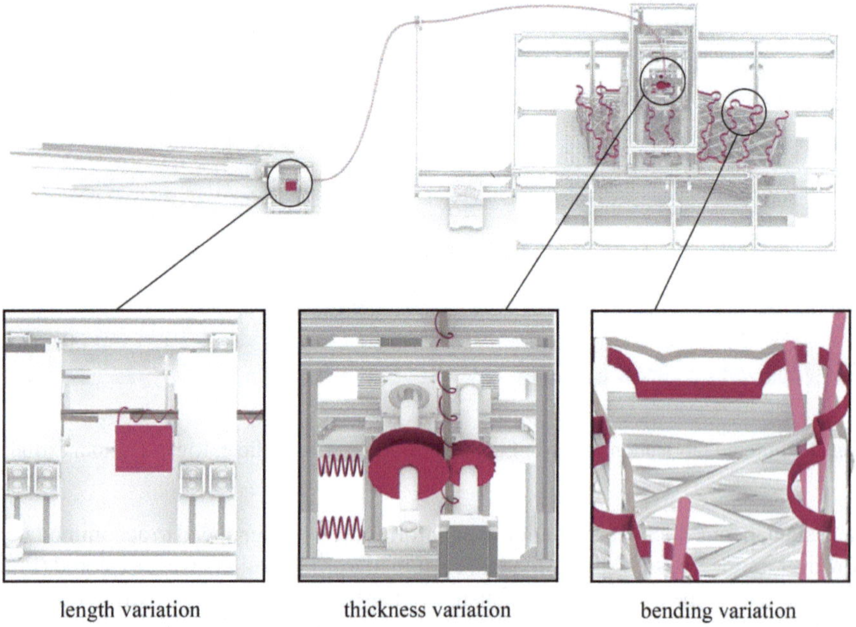

Fig. 5. Tailored solutions for dealing with natural material variations.

Fig. 6. Close-up details of the fabricated components for the project 'ReGrow Willow'.

3 Results and Reflection

The implementation of the Cartesian coordinate robot in the context of the whole fabrication system, enabled the production of 63 woven willow-earth hybrid components for the project 'ReGrow Willow' in the Bundesgartenschau (Federal Gardening Show) in Mannheim in 2023 (Figs. 6 and 8). This was a significant milestone in its evolution over three full-scale research demonstrators (Figs. 7 and 8) showcasing the viability of the fabrication system and thus meeting the overarching research aim.

Inhomogeneities on the material level are dealt with by the implementation of tailored features in fabrication system that adjust to such 'imperfections' without necessitating the need to exactly model every intricacy of the variations in length, bending and thickness of the material. The fabrication system is able to deal with changes that arise during fabrication itself by employing an optimised workflow for a custom machine toolpath for every component, highlighting the advantages of embracing the unpredictable, non-perfect nature of reality using facile solutions. Employing an iterative development strategy, adopting rapid prototyping techniques to build custom parts where necessary and to couple them with readily available elements to accelerate each iteration - with the first iteration informed by empirical assessments - means that each sub-part can be quickly tested and improved for durability. This way the development of several conceptualised features could be parallelised and thus their interaction and role in the whole fabrication process could be explored from the initial stages. Developing a system through increasingly more automation in each iteration meant that the empirical knowledge of the material was used the most effectively.

Fig. 7. Left: Full-scale research demonstrator 'InterTwig', Right: Prototype 'RecodeWicker'.

Without subjecting the input material to time- and energy-intensive or waste-generating standardising processes, the willow weaving robot showcased a cost-effective – it can be built from €5000 - and low-impact method for processing a plant-based material into formwork for earth filling. The efficacy of the fabrication process

Fig. 8. ReGrow Willow' demonstrator at the Bundesgartenschau (Federal Gardening Show) in Mannheim.

may be evaluated through physical labour and output quality. In the semi-automated proof-of-concept stage, one person was required to interact with the system to feed the 'splicer'. This is significantly lower in terms of physical labour input compared to manual weaving and splicing. Introduction for a new operator takes 10 min, thus making any machine operating expertise redundant. As the entire weaving path for a component is encoded into one file the operator does not need to understand weaving logic or follow the sequence. The output quality of the woven willow components has low variance as the range can be fine-tuned by different parameters in the process (spring tension, splicing speed, splicing yarn tension, weaving tension etc), this allows a greater control over the final product.

4 Conclusion and Outlook

Following the encompassing principles of minimising the embedded energy, cost and impact on the environment, a semi-automated fabrication system emerged. It demonstrates that material explorations as a starting point and an iterative development strategy catalysing process and fabrication integrated solutions - rather than complex simulation and modelling - can be an appropriate method for developing tailored fabrication processes. Taking the development further, a material feeding system has been conceptualised which instead of scanning and manipulating each branch, can robustly deal with a range of deviations through a material forwarding system. Employing off-the-shelf sensors around the extruder could set the willow tension automatically enabling a completely autonomous fabrication process. This is possible through the upgrade to a centralised, real-time controller (ROS) which can couple the currently decoupled splicing system with the Cartesian coordinate robot. Since the system was developed using widely available electrical and control components integrated with custom parts, deploying an increased level of automation is possible without large financial or time investment. Due to inherent features of the Cartesian coordinate robot, scalability of components in dimension as well as production rate not only proves feasible as the relative cost-effectiveness of the system means that reaching industrial-scale output levels could be achieved by parallel machines - with one supervisor - thus validating its potential for technology transfer towards industrial production. As the fabrication system contains a range of adjustable parameters, it is not inherently standardised to one plant-based material. Similarly long and slender materials, such as reed, poplar - local to the point of construction - can all be used in such a digital fabrication workflow, reinforcing its application potential in a global context. The underlying versatility and the fact that it consists of easily transportable and expandable modules – depending on component dimension – mean that the application space may be expanded both geographically and demographically. Thus proving that with minimal complexity a fabrication system that utilises sustainable – potentially local – materials can emerge. As the next stage of the research, the willow-earth hybrid component logic will be applied to arched ceiling slab components as part of a construction demonstrator to be exhibited at the Landesgartenschau (State Gardening Show) in Wangen im Allgäu in 2024. The principles of minimising energy, impact and complexity that underpinned the evolution of the system make it applicable in a range of different contexts and to different materials, which can foster a new perspective on the industrialisation of more sustainable fabrication solutions. Such solutions may circumvent the traditionally complex processes and instead enable robust solutions that adjust to the ever-changing intricacies of the real world without standardising them.

5 Acknowledgements

The authors would like to thank fellow researchers Erik Zanetti and Vincent Witt, as well as Atanaska Chausheva and all the students who contributed to the development and realisation of the project from the master courses ‘ReGrow Design’ and ‘ReGrow Build’. The research was funded by the Baden-Württemberg Ministry of Food, Rural Areas and Consumer Protection (MLR) as part of the Bioeconomy Innovation Programme for Rural Areas.

References

1. Guerra, B.C., Leite, F.: Circular economy in the construction industry: an overview of United States stakeholders’ awareness, major challenges, and enablers. Resour. Conserv. Recycl. **170**, 105617 (2021)
2. Zimmann, R., O’Brien, H., Hargrave, J., Morrell, M.: The Circular Economy in the Built Environment. Arup, London, UK (2016)
3. Freddie Mac: Housing Supply: A Growing Deficit. Accessed 16 Oct 2023 (2021). https://www.freddiemac.com/research/insight/20210507-housing-supply.
4. Kuzmenko, K., Feraille, A., Baverel, O.: Improving construction efficiency with digital fabrication: an environmental insight. IOP Conf. Ser.: Earth Environ. Sci. **323**(1), 012014 (2019)

5. Le, D.L., Salomone, R., Nguyen, Q.T.: Circular bio-based building materials: a literature review of case studies and sustainability assessment methods. Build. Environ. **244**(October), 110774 (2023)
6. MacDonald, K., Schumann, K., Hauptman, J.: Digital Fabrication of Standardless Materials, pp. 266–275, Austin (Texas), USA (2019)
7. Heisel, F., Hebel, D.E.: Building Better-Less-Different: Clean Energy Transition and Digital Transformation: Fundamentals-Case Studies-Strategies. Birkhäuser (2023)
8. Zanetti, E., Olah, E., Haußer, T., Casalnuovo, G., La Magna, R., Dörstelmann, M.: InterTwig—willow and earth composites for digital circular construction. In: Design for Rethinking Resources, pp. 491–511. Sustainable Development Goals Series (2023)
9. Agustí-Juan, I., Habert, G.: An environmental perspective on digital fabrication in architecture and construction. In: Proceedings of the 21st International Conference on Computer-Aided Architectural Design Research in Asia (Caadria 2016), pp. 797–806. CAADRIA (2016)
10. Agustí-Juan, I., Habert, G.: Environmental design guidelines for digital fabrication. J. Clean. Prod. **142**(January), 2780–2791 (2017)
11. Omer, M., et al.: Democratizing manufacturing–conceptualizing the potential of open source machine tools as drivers of sustainable industrial development in resource constrained contexts. In: Proceedings of the Conference on Production Systems and Logistics: CPSL 2022, pp. 256–266. Hannover: publishing (2022)
12. Zanetti, E., et al.: ReGrow willow - digital circular construction for earth-willow hybrid structures. In: Fabricate 2024. UCL Press (n.d.)
13. Casalnuovo, G., Zanetti, E., Haußer, T., Dörstelmann, M., La Magna, R.: Digital structural Design for Natural Composites: a case study of willow-earth hybrid construction. In: Habits of the Anthropocene: Scarcity and Abundance in a Post-Material Economy, Proceedings of the 43rd Annual Conference of the Association for Computer Aided Design in Architecture, II:282–92. Association for Computer Aided Design in Architecture (ACADIA) (2023)
14. Oksman, K., Skrifvars, M., Selin, J.-F.: Natural Fibres as reinforcement in Polylactic acid (PLA) composites. Compos. Sci. Technol. Eco-Compos. **63**(9) (2003)
15. Song, J., et al.: Processing bulk natural wood into a high-performance structural material. Nature. **554**(7691) (2018)
16. Bitting, S., Derme, T., Lee, J., Van Mele, T., Dillenburger, B., Block, P.: Challenges and opportunities in scaling up architectural applications of mycelium-based materials with digital fabrication. Biomimetics. **7**(2), 44 (2022)
17. Dörfler, K., Hack, N., Sandy, T.: Mobile robotic fabrication beyond factory conditions: case study mesh Mould wall of the DFAB HOUSE. Constr. Robot. **3**, 53–67 (2019)
18. Dawod, M., et al.: Continuous timber fibre placement: towards the design and robotic fabrication of high-resolution timber structures. In: Impact: Design with all Senses: Proceedings of the Design Modelling Symposium, pp. 460–473. Springer International Publishing, Berlin 2019 (2020)
19. Jeffers, M.: Autonomous robotic assembly with variable material properties. In: Robotic Fabrication in Architecture, Art and Design (2016)
20. Lharchi, A., et al.: Sensing and augmenting for adaptive assembly strategies. In: Digital Design Reconsidered: Proceedings of the 41st Conference on Education and Research in Computer Aided Architectural Design in Europe (eCAADe 2023), p. 269 (2023)
21. Rossi, G., et al.: Statistically modelling the curing of cellulose-based 3D printed components: methods for material dataset composition, augmentation and encoding. pp. 487–500 (2022)
22. Kalousdian, N.K., et al.: Learning robotic manipulation of natural materials with variable properties for construction tasks. IEEE Rob. Autom. Lett. **7**(2), 5749–5756 (2022)
23. Graham, A.: Wattle and Daub: Craft, Conservation and Wiltshire Case Study (2004). https://doi.org/10.13140/2.1.1875.2806.

24. Wu, C., Pan, Y., Yan, L.: Mechanical properties and durability of textile reinforced concrete (TRC)—a review. Polymers. **15**(18), 3826 (2023)
25. Brezine, C.: Algorithms and Automation: the Production of Mathematics and Textiles, Oxford (2009)
26. Eversmann, P., Ochs, J., Heise, J., Akbar, Z., Böhm, S.: Additive timber manufacturing: a novel, wood-based filament and its additive robotic fabrication techniques for large-scale, material-efficient construction. 3D Printing Add. Manufact. **9**(3), 161–176 (2022)
27. Morrison, D., et al.: Cartman: the low-cost cartesian manipulator that won the Amazon robotics challenge. In: 2018 IEEE International Conference on Robotics and Automation (ICRA), pp. 7757–7764. IEEE (2018)

Differential Boundary Deposition: Mobile Robotic Construction with Amorphous Materials

Matthew Johnson[1,2(✉)], Philipp Kragl[1], Michael Tucker[1,2], Nicolas Kubail Kalousdian[1,2], Samuel Leder[1,2], and Achim Menges[1,2]

[1] Institute for Computational Design and Construction, University of Stuttgart, Keplerstr. 11, 70174 Stuttgart, Germany
matthew.johnson@strabag.com, {philipp.kragl, mail}@icd.uni-stuttgart.de, mptucker22@gmail.com

[2] Cluster of Excellence Integrative Computational Design and Construction for Architecture (IntCDC), University of Stuttgart, Keplerstr. 11, 70174 Stuttgart, Germany

Abstract. This paper presents a robotic construction system enabled by a novel deposition technique for amorphous materials, differential boundary deposition (DBD). DBD involves the continuous addition of a fluid material around the perimeter of a moving robot, thereby creating a tube structure along the robot's path through space. The DBD system presented in this paper demonstrates the method using a mobile robot that can build along a freeform path through space. This system could extend the field of collective robotic construction (CRC) by allowing for the additive manufacturing of complex networked structures by a team of mobile constructor robots. This is possible due to not only the mobility of the robot, but also the ability of the tube structures to fuse with each other as they are built.

The core sections of this paper describe the conceptualization of DBD, the design of the demonstration system, and the validation of key performance criteria through physical experimentation. The implications of these results are discussed in the context of CRC applications. Further research is outlined that will explore the application of the system to design and construction.

Keywords: Collective robotic construction · Mobile robotics · Amorphous materials · On-site robotic construction

1 Introduction

The level of automation and digitization in the construction industry is low but rapidly rising [1]. Mechanized prefabrication of discrete building elements has become widespread, but onsite digital construction of entire building-scale systems remains rare. Industrially applied digital construction methods often resemble static manufacturing machines that

M. Johnson, P. Kragl, M. Tucker—Contributed equally.

M. Yablonina et al. (Eds.): Robotic Fabrication in Architecture, Art and Design 2024, pp. 129–141, 2026.
https://doi.org/10.1007/978-3-032-22607-5_9

have been scaled up and brought to the building site [2], bringing with them inherent limitations of scale or geometry. The introduction of more sophisticated robotic systems has proven challenging due to the dynamic, unpredictable nature of the construction site as compared to laboratories and factory floors. The emerging field of collective robotic construction (CRC) seeks to address these challenges through teams of robots that work together to build structures larger than themselves [3]. Many CRC projects also focus on the assembly of discrete building materials [4], but amorphous materials such as concrete and thermoplastics offer advantages in their ability to handle tolerances and construction inaccuracies. This paper presents a robotic construction technique that was developed for CRC with amorphous building materials: Differential Boundary Deposition (DBD).

Fig. 1 DBD construction process (experiment 4)

This paper focuses on the basic principles of the DBD system and their application to robotic construction. A deposition device was developed and validated through a series of experiments designed to showcase tube construction behavior. This device then became part of a mobile construction robot that was used for a second series of tests, which demonstrated the ability of the system to construct along an arbitrary path through space and fuse adjacent tubes into a larger overall structure. The resulting robotic construction system could extend the field of CRC by allowing for the construction of networked tube structures by teams of mobile construction robots.

2 Background

2.1 Deposition of Amorphous Materials

Continuous construction approaches generally involve the application of amorphous materials, which are defined as any building material that behaves compliantly and lack a predefined shape, therefore taking their form from their environment or the circumstances of their deposition [4]. These materials can fuse with themselves during deposition, obviating the need for an explicit joining or fastening action within a construction process.

One of the first mechanized construction techniques for amorphous materials was slip forming, a hybrid of casting and extrusion that emerged in the early 20th century

[5]. The Smart Dynamic Casting project from ETH Zürich digitized slip forming via a customized formwork end effector that was manipulated by a stationary industrial robot to produce bespoke column elements [6].

Mobile robotic projects dealing with amorphous materials often employ extrusion techniques analogous to desktop 3D printing [7–9]. The Minibuilders project from the Institute for Advanced Architecture of Catalonia is unique in showing a CRC approach to amorphous deposition, wherein a heterogeneous team of mobile robots each perform a single construction task [9].

2.2 Collective Robotic Construction

Peterson et al. define CRC as a subfield of robotic construction wherein teams of small, autonomous machines collaborate to build structures larger than themselves. These systems can be beneficial on the construction site due to their resilience towards environmental irregularities or mechanical breakdowns [10].

CRC systems are diverse, but can be categorized according to several key system attributes which affect the design and application of such robotic systems. One such attribute is the dichotomy between systems where construction robots add to a growing structure from outside and systems wherein robots use the growing structure as a locomotion substrate. In the first category, the wall-climbing robots presented by Yabloninina et al. [3] climb or drive across the ground or wall surfaces of a space in order to wind filament material onto a growing structure, while the Flight-Assembled Architecture project from Gramazio Koehler Research utilized a team of drones that fly through free space to deposit material [11]. Meanwhile TERMES utilized custom-manufactured building blocks that could be both grasped and traversed by wheeled robots [12]. In the Fiberbots project from the MIT Media Lab robots continuously wind material around themselves to form a tube structure that they can drive forwards through [13]. In general, CRC systems that use their building materials for locomotion automatically extend their construction range by adding material to the structure, rather than having it pre-determined by the features of the environment. For this project, a system that locomotes across its own structure was selected in order to further capitalize on the geometric flexibility of amorphous materials.

In some CRC systems, the presence of a robot collective is essential for the functionality of the overall system, while in others a single robot can operate alone, with collective behavior enhancing the speed, efficiency or build space of the construction system. In the robotic assembly system introduced by Leder et al. robots consisting of a single rotary joint grasp timber struts and form kinematic chains, moving each other and being moved in turn along arcing paths through space [14]. While a single machine would not be able to perform meaningful construction tasks, the robot collective is able to achieve complex 3-dimensional manipulation of material. In contrast, projects such as FiberBots and TERMES feature robots that are able to individually move through space and manipulate material [12, 13]. This project also proposes such a construction system, wherein a single robot is able to build on its own but a robotic collective can achieve greater efficiencies and unlock more complex geometries. Therefore, to validate the system principles presented in this paper it was only necessary to build one construction robot.

3 Methods

3.1 Differential Boundary Deposition

The core principle of DBD is the continuous addition of material around the perimeter of a robot body that moves forwards through space. Material is added around the entire perimeter simultaneously, but the rate of deposition can vary programmatically at different points on the perimeter (Fig. 1). Thus, while the overall cross section of the tube resembles the cross section of the depositor "swept" along its path through 3D space, the wall thickness and texture of the deposited structure can be variably controlled. Additionally, the wall thickness properties can be kept consistent along a curving path by synchronizing the deposition rates to the instantaneous speed of their respective points on the perimeter of the depositor.

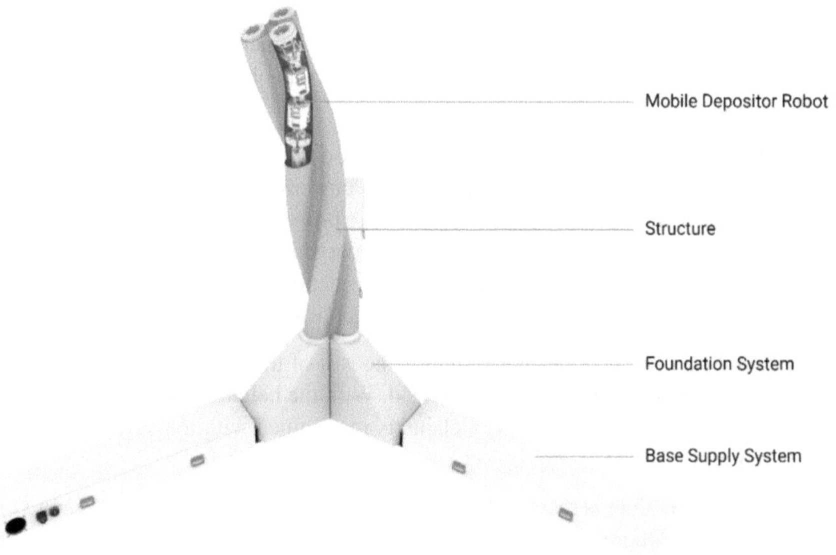

Fig. 2 DBD System Components

The DBD construction system presented here consists of four discrete elements: a mobile depositor robot, a foundation, a base support unit, and the built tube structure between the two (Fig. 2). The mobile depositor robot is the most technically complex part of the system. It was designed as a series of modules, which allows for the removal, replacement or reconfiguration of hardware components to meet different performance requirements. The base support unit simultaneously serves as a supply of material and power, and an initial locomotion substrate. The structure itself becomes the locomotion substrate once the mobile robot enters it. The structure also contains the tether of power, material and data cables, and guides the compliant robot body through space. The foundation serves as the interface between the base supply unit and the structure and can be permanently left behind at the end of construction.

ABS thermoplastic was selected as a demonstration material for the experiments presented in this paper. Compared to other amorphous materials, thermoplastics have relatively high tensile strength without additional reinforcing material [15], enabling a larger architectural design space. Additionally, a key material feature of thermoplastics is the reversibility of the curing process. Cooled thermoplastic can be reheated and returned to its workable state a practically unlimited number of times, meaning that two robots can construct a fused joint between their two tubes without needing to meet at a coordinated point in time.

3.2 Mobile Robotic Construction

Continuous deposition along a moving path through space requires the synchronous performance of several physical tasks, including forward motion of the depositor, excretion of material, heat management, self-sensing and calculation. The robotic system was developed through an iterative co-design process, wherein discrete hardware modules and their attendant control software were developed in parallel to each other and subjected to frequent testing. The final mobile constructor robot consisted of six hardware modules (Fig. 3):

Deposition. This module consists of the heated deposition head and a liquid-cooled curing ring. The deposition head consisted of a custom milled radial aluminum nozzle with embedded heating elements and temperature probes. Eight threads of ABS filament are fed into the heating chamber, where they are merged as they liquefy due to the internal geometry of the depositor. The perimeter of the depositor is divided into eight zones, each of which has an individually controllable deposition rate (Fig. 4a).

The cooling ring both thermally shields the rest of the robot and cures the material upon its emergence from the deposition head. With the help of active cooling a temperature difference of more than 200° Celsius is maintained within a vertical distance of less than 5 mm (Fig. 4b).

Steering. This module is responsible for controlling the instantaneous curvature of the tube structure, and therefore the robot's path through space. The only actively controlled linkage between modules takes the form of a six degree of freedom Stewart platform [16] able to orient the deposition head according to a target frame relative to the steering module's base.

Locomotion. This module consists of three sub-modules: two radial clamps and one linear actuator. While material is being deposited, the linear actuator simultaneously pushes the deposition head forward through the tube. When the linear actuator reaches the limit of its motion, deposition is halted and the upper radial clamp is engaged. The lower clamp can then be safely disengaged and the linear actuator retracted, pulling the tail section of the robot up through the tube.

Extrusion. This module hosts eight stepper motors fitted with standard 3D printer extruder gears. Each stepper motor is individually controllable and pushes one strand of filament forwards to the deposition head.

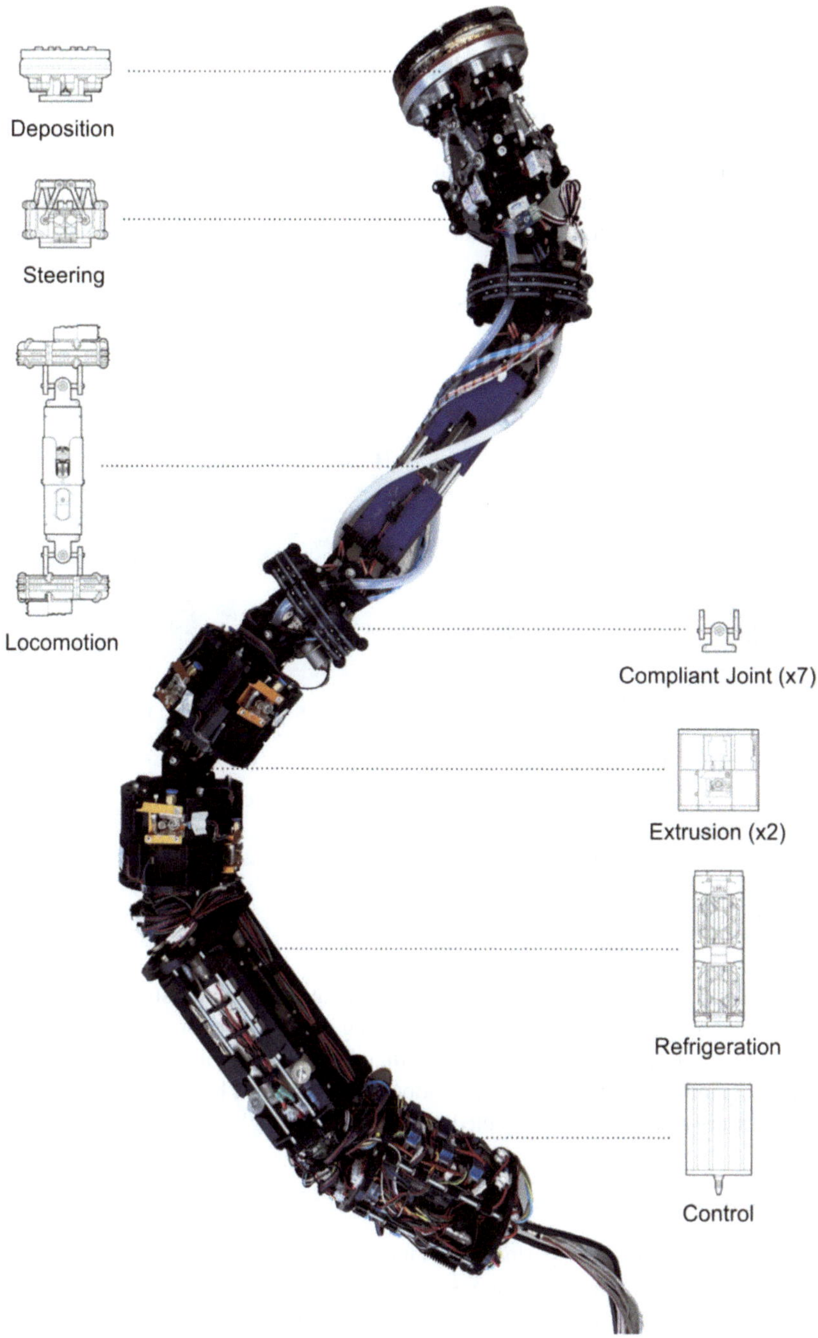

Fig. 3 Mobile depositor robot modules

Refrigeration. Peltier elements in this module chill coolant liquid while fans exhaust waste heat to the environment. A primary fluid loop circulates liquid through hosing to the deposition module and back to the Peltiers, while a secondary loop within the module moves heat from the Peltiers to the exhaust fans.

Control. This module houses an Arduino Mega fitted to an assembly of three custom-manufactured printed circuit boards. These boards host motor drivers, voltage converters, and logic units needed to drive the motors and sensors throughout the other modules. The onboard computer receives serial commands from an external broadcast, interprets them into robot control, and reports self-sensing data through the same serial channel.

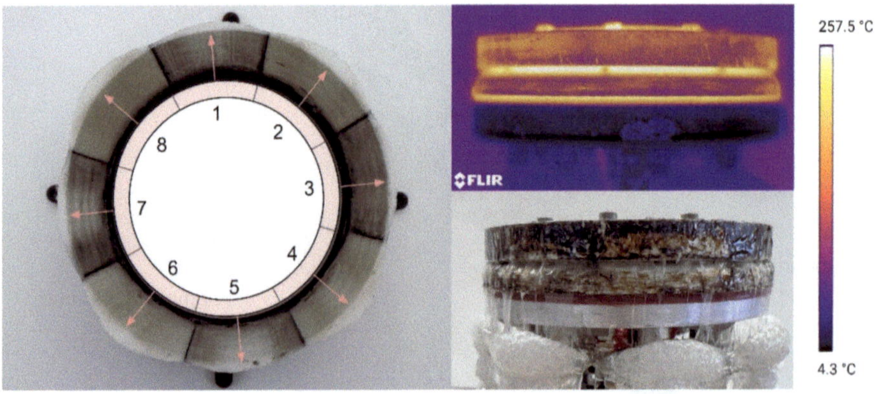

Fig. 4 a) deposition zones. b) Depositor and cooling ring.

4 Results

Two series of experiments were conducted to validate the DBD system. The first series tested the deposition method in a stationary setting, while the second tested the prototype mobile construction robot. Qualitative and quantitative data collected throughout these experiments influenced the iterative development of the robot hardware, and the samples produced were analyzed and subjected to structural testing.

4.1 Linear Deposition Experiments

Setup. The first series of experiments dealt only with the components directly required for the deposition of material. A deposition device consisting of the deposition module and the material supply modules was mounted on a linear actuator, while prototypes of the refrigeration and control modules remained stationary. Four tests were conducted with this setup in order to validate the basic principles of DBD.

Results. Test 1.1 was the first attempt at depositing material, and compared two varieties of ABS plastic from different manufacturers. Extrusion rate and linear actuation speed were manually adjusted until "strips" of a consistent thickness were produced by the better-performing material variant (Fig. 5a).

In test 1.2, the extrusion and actuation speeds from experiment 1 were confirmed in conjunction with the better-performing material variant. The result was a tube structure approximately 25 cm in length with a thickness ranging from 3 to 9 mm. The artifact from this experiment was the first to exhibit the "scalloped" cross section observed on all future samples produced with this hardware.

Test 1.3 leveraged the reversibility of thermoplastic curing to strengthen the bond between the deposited structure and a 3D printed base. The base was designed with radially spaced cavities in order to receive molten material from the deposition head, and 3D printed from ABS plastic with the same material properties as the main structure. At the beginning of the experiment, material was deposited in place until the base was filled. Once the base began to visibly deform, linear construction began. In the resulting artifact the base material and construction material formed a strong and consistent bond, effectively fusing into a single piece (Fig. 5b).

Test 1.4 aimed to validate the differential aspect of DBD by comparing the use of different extrusion rates within the same construction action. Three values were distributed across the eight deposition zones: a fast rate of 3 mm of filament per second, a slow rate of 1 mm per second, and an oscillating rate that moved between the two values over a period of 8 min. These different settings resulted in correspondingly varied cross-sectional properties, with a higher extrusion rate corresponding to a higher local wall thickness (Fig. 5c).

Fig. 5 a) Test 1.1 artefact. b) Test 1.2 foundation adhesion. c) Test 1.4 cross section variation.

4.2 Mobile Robotic Experiments

Setup. The second series of experiments implemented the full mobile robot prototype. The design specifications of the mobile robot include a 50-cm minimum turning radius, a three meter build length, and a 16 cm inner diameter. The base module mounted filament spools, transformed power from a standard wall socket, and provided a locomotion substrate for the robot before it entered the tube.

Control software was also developed that interprets design input, calculates motor speeds and other key data, and transmits commands to the computer onboard the mobile robot. Self-sensing data from the robot is incorporated into a feedback loop in order to automatically maintain a constant temperature and construction rate.

Three tests were conducted with the second hardware/software setup, with the aim of validating the DBD system as a digital construction process ready for CRC.

Results. In test 2.1, the mobile robot successfully steered along a curved path through space while constructing a homogeneous tube structure. The automatic modulation of deposition rates for each deposition zone was successful in producing a consistent wall thickness despite the zones covering different linear distances in the same amount of time (Fig. 6a). However, the radius of curvature of the built tube diverged from the design model. In other words, the mobile robot was able to steer in the correct direction but not to the correct degree, or at least not reliably.

In test 2.2, a second tube was constructed alongside the artifact from test 2.1. This time the robot was steered in the opposite direction, so that the tubes fused at their bases before diverging. Extra material was deposited in the regions adjacent to the fusing area in order to ensure a strong joint. The resulting artifact was essentially a single piece, with a solid bond where the cross sections of the two tubes overlapped (Fig. 6b).

Test 2.3 involved the extension of the tube from test 2.1 along a straight path, diverging 20° from vertical. The robot continued to climb until a tube approximately 2 m in length was produced. Despite the steadily increasing bending load on the structure, it exhibited no observable deflection under its own weight and that of the robot (20 kg). By the completion of the test the locomotion module had climbed fully within the structure, demonstrating that the tube structure can function as a locomotion substrate for the mobile robot.

Fig. 6 a) Interior of structure showing joint between test 2.1 and test 2.3. b) test 2.2, detail of joint.

5 Discussion and Conclusions

The experimental results validated the performance of the DBD system as a novel mobile robotic construction method. Because an individual tube structure can follow a 3-dimensional path curve, cross space unsupported, and fuse to other tubes (Fig. 7), the potential exists for teams of mobile robots to realize complex interconnected structures. CRC methods could be developed with an emphasis on the realization of structures that could not be achieved by a lone machine due to reasons of time, geometric interdependency, or structural integrity during construction.

Fig. 7 Mobile constructor robot, demonstration structure, and base supply unit

DBD compares favorably in several ways to the most widespread existing method for robotic thermoplastic deposition, fused deposition modeling (FDM). Because they are built up through cold-joined layers, traditional 3D prints always have a weak direction parallel to the flat deposition bed of the 3D printer. In contrast, DBD tubes are continuously deposited and have no layer joints. Moreover, FDM prints are inherently limited in size by the dimensions of the 3D printer, while DBD structures are theoretically unlimited in size because they are built from within by a mobile machine. These factors combine to make DBD a faster, stronger, and more flexible construction technique for the networked-tube morphologies envisioned for the system.

However, there are certain limitations that must be taken into consideration alongside the strengths of the system presented. The reliability of the robot system, the accuracy of steering, and the consistency of the structure all need improvement. Future iterations of the robot system will need to incorporate sensor-guided localization of the robots the ability of the mobile robot to sense its own state and its position in 3D space in order to collectively deposit 3D structures. Future work will evaluate the relative advantages and drawbacks of onboard local sensing and external "global" positioning.

Moreover, while DBD can achieve geometries that are impossible or impractical with other construction techniques, these geometries are currently described by a hollow cross-section that is swept along continuous curves in 3D space. While the range of parameters such as turning radius and cantilever distance can be improved by technical advances in the material-robot system, certain classes of structure are inherently unsuitable for DBD. For example solid structures are infeasible due to the location of the construction machine within the cross section of the structure, and sharp corners in the path curve are not achievable with the current hardware approach due to the turning radius of the mobile robot. Therefore, the research presented here should be considered as a specialized system for the production of hollow-tube morphologies, rather than as a universal means of producing any arbitrary geometry.

Fig. 8 Speculative application to spatial construction (after further research)

6 Outlook

Further work will investigate CRC applications of the DBD system. While the mobile constructor robot can operate individually, the mutual support of multiple tubes could enable morphologies that could not be stably constructed by a single machine, thereby extending the design space of the system (Fig. 8). Moreover, the next stage of experimentation will quantitatively evaluate the performance of the system in the context of contemporary construction practices in terms of cost, material efficiency, and construction speed in addition to morphological freedom. Metrics will be developed for comparison to state-of-the-art construction practices as well as alternative means of digital production, such as traditional 3D printing.

Several features of the construction system point towards interesting areas of application in the broader built environment. Because the hardware is fully contained within the cross section of the structure, it is possible to build within constrained and geometrically irregular spaces such as caves, cliffs or the interior of existing buildings. Conversely, the security of the robot within its structure could be beneficial for underwater or extraterrestrial construction, where gravitational conditions differ from the normal construction environment. The installation of hollow plastic tubes could be useful for the retrofitting of building services, such as cable channels and air ducts, into existing structures. DBD could also be applied to new construction either as on-site recyclable formwork for casting or as permanent structure for small-scale constructions. Moreover, the principles of DBD have the potential to be applied to other materials in a manner analogous to the extension of 3D printing technology to concrete and metals. This material transfer would require new hardware and imply different design parameters, but could enable areas of application that are not feasible for plastics. The exploration of these different applications could eventually lead to a family of robotic "species" that are suited to CRC with different materials and in different environments.

Acknowledgements. This research project was realized as a Master's thesis in the Integrative Technologies and Architectural Design Research M.Sc. Program (ITECH) at the University of Stuttgart, Germany, led by the Institute for Computational Design and Construction (ICD) and the Institute of Building Structures and Structural Design (itke).

The authors would like to thank Professor Thomas Wortmann (ICD) and Professor Jan Knippers (itke) for their support, insight and critical feedback during the development of this project.

This research was partially funded by the Deutsche Forschungsgemeinschaft (DFG, German Research Foundation) under Germany's Excellence Strategy – EXC 2120/1 – 390831618.

References

1. Arntz, M., Gregory, T., Zierahn, U.: Digitalisierung und die Zukunft der Arbeit: Makroökonomische Auswirkungen auf Beschäftigung Arbeitslosigkeit und Löhne von morgen. Zentrum für Europäische Wirtschaftsforschung, Mannheim (2018)
2. El-Sayegh, S., Romdhane, L., Manjikian, S.: A critical review of 3D printing in construction: Benefits, challenges, and risks. Archiv. Civ. Mech.Eng. **20**(34), pp. 1–25 (2020)
3. Yablonina, M., Menges, A.: Towards the development of fabrication machine species for filament materials. Willmann, J., Block, P., Hutter, M., Byrne, K., Schork, T. (eds.) In Robotic Fabrication in Architecture, Art and Design 2018, ROBARCH 2018. Springer (2019).
4. Napp, N., Rappoli, O.R., Wu, J.M., Nagpal, R.: Materials and mechanisms for amorphous robotic construction. In: 2012 IEEE/RSJ International Conference on Intelligent Robots and Systems, pp. 4879–4885. Vilamoura-Algarve, Portugal (2012). https://doi.org/10.1109/IROS.2012.6385718
5. Ketchum, M.S.: The Design of Walls, Bins, and Grain Elevators. The Engineering News Publishing Co, Chicago (1907)
6. Lloret-Fritschi, E., Reiter, L., Wangler, T., Gramazio, F., Kohler, M., Flatt, R.J.: Smart Dynamic Casting. In: Proceedings of the Conference on Smart Materials and Digital Fabrication (HPC/CICT), Tromsø, Norway (2017)

7. Sadeghi, A., Mondini, A., Mazzolai, B.: Towards self-growing soft robots inspired by plant roots and based on additive manufacturing technologies. Soft Robotics. **4**(3), 211–223 (2017)
8. Zhang, X., et al.: Large-scale 3D printing by a team of mobile robots. Autom. Constr. **95**, 98–106 (2018) ISSN 0926-5805.
9. Jokic, S., Novikov, P., Maggs, S., Sadan, D., Jin, S., Nan, C.: Minibuilders. In: Architectural Principles, Tools, and Processes, edited by Digital Vernacular, pp. 259-265 (2015).
10. Petersen, K.H., Napp, N., Stuart-Smith, R., Rus, D., Kovac, M.: A Review Of Collective Robotic Construction. Sci. Robot. **4**(28) (2019). https://doi.org/10.1126/scirobotics.aau8479
11. Augugliaro, F., et al.: The flight assembled architecture installation: cooperative contruction with flying machines. Control Systems, IEEE. **34**, 46–64 (2014). https://doi.org/10.1109/MCS.2014.2320359
12. Petersen, K., Nagpal, R., Werfel, J.K.: TERMES: An autonomous robotic system for three-dimensional collective construction. In: Durrant-Whyte, H., Roy, N., Abbeel, P. (eds.) Robotics: Science and Systems VII. MIT Press, Cambridge, MA (2011)
13. Kayser, M., et al.: FIBERBOTS: Design and Digital Fabrication of Tubular Structures Using Robot Swarms. Springer, Cham (2018)
14. Leder, S., et al.: Leveraging building material as part of the in-plane robotic kinematic system for collective construction. Adv. Sci. **9**, 2201524 (2022)
15. Prajapati, A., Rajpurohit, S., Patadiya, N., Dave, H.: Analysis of Compressive Strength of 3D Printed PLA Part. In: Advances in Additive Manufacturing and Joining. Springer, Singapore, pp. 259–268 (2021). https://doi.org/10.1007/978-981-15-9117-4_22
16. Stewart, D.: A platform with six degrees of freedom. Aircr. Eng. Aerosp. Technol. **38**(4), 30–35 (1966). https://doi.org/10.1108/eb034141

A Suspended Cable-Driven Parallel Robot (CDPR) and Reorientation Mechanism for the Semi-Automated Positioning of Prefab Assemblies

Steven Beites[1(✉)], Ethan McDonald[2], and Marc Arsenault[2]

[1] McEwen School of Architecture, Laurentian University, 85 Elm Street, Sudbury, ON P3C1T3, Canada
sbeites@laurentian.ca

[2] School of Engineering and Computer Science, Laurentian University, 935 Ramsey Lake Rd, Sudbury, ON P3E 2C6, Canada

Abstract. In North America, the dependency of the Architecture, Engineering, and Construction (AEC) industry on traditional building techniques and materials significantly threatens to escalate housing prices to unsustainable heights, further aggravated by labor shortages and a declining number of young adults pursuing skilled trade positions. This research proposes an alternative approach to construction by developing a semi-automated mobile platform, aiming to accelerate construction through technological adoption and transform the way we design and construct our built environment. It presents two years of interdisciplinary research conducted in the realms of cable robotics, specifically focusing on the design and development of a cable-driven parallel robot (CDPR). This research introduces three key developments in the area of large-scale architectural assembly using CDPRs. It provides insight into the construction of the developed mechatronic system including mechanical design, cable actuation, motor and control systems. It presents the development of a software tool for the control, simulation and verification of the CDPR, and introduces an early prototype of a novel reorientation end-effector allowing for the pick-and-place assembly of prefabricated panels. The ongoing research culminates in a working prototype, characterized by flexibility, strength, portability, and modularity. This work contributes to the evolving field of CDPRs, and although in the early stages of development, it provides valuable insight for a broader audience interested in CDPRs as an alternative method for in-situ construction.

Keywords: Cable robotics · Automated construction · In-situ construction · Housing · Prefabricated assemblies

1 Introduction

Housing supply is one of the greatest challenges we are facing today. The World Economic Forum has estimated that two billion homes will be needed in the next eighty years in order to keep up with the world's rising population [1]. Housing shortages have

M. Yablonina et al. (Eds.): Robotic Fabrication in Architecture, Art and Design 2024, pp. 142–158, 2026.
https://doi.org/10.1007/978-3-032-22607-5_10

a detrimental effect on communities, leading to overcrowding, the rapid deterioration of housing stock, unaffordable market rents and potentially homelessness. Accordingly, governments are prioritizing housing development in order to meet current needs of its citizens, yet solutions to this pressing issue will not arise from policy strategy alone. To meet such ambitious targets, a fundamental shift is required across all stages of design and construction. In North America, The Architecture, Engineering and Construction (AEC) industry, which continues to be bound by traditional methods and materials of construction, will only continue to drive the cost of housing to unreachable heights. Intensified by labor shortages in the skilled trades and short construction seasons (in northern regions), this research proposes an alternative method of construction through the development of a semi-automated construction platform as a means to build faster and more economically.

According to a Construction Cost Survey conducted by the National Association of Home Builders (NAHB), it was found that in 2022, construction expenses accounted for 60.8% of the typical home sale price [2]. Given that labor represents 50–70% of these construction costs, the potential for automated processes within a prefabricated framework, as presented through this research, could present a substantial opportunity for cost reduction, significantly reducing construction timelines and minimizing traditional labor needed on site. Additionally, as the cable-driven parallel robot (CDPR) semi-automates the positioning and placement of prefabricated assemblies, the human operator remains and continues to be an important actor in the process. In a society increasingly reliant on technology and amidst the declining interest and supply in skilled trade positions, leveraging on-site technology enhancements and fostering human-robot interaction (HRI) methods could provide a unique opportunity in drawing young adults into the construction industry.

2 Background

CDPRs consist of a mobile platform (MP) attached to a fixed frame by several cables that are controlled through actuating winches. Compared to traditional industrial robotic arms, they offer new possibilities and advances in their application for construction due to their ability to span large distances, their increased mobility, scalability and lower associated costs. Employed historically in material/cargo handling applications, high-speed tracking photography and live broadcasting (e.g. Sky- Cam) [3–5], CDPRs have been seldom explored within architectural practice yet have immense potential to transform current methods of construction. Notable examples such as robotic bricklaying systems [6], the automated installation of curtain wall modules [7], the in-situ construction of designed granular materials [8], and the additive manufacturing of cementitious material for large scale construction [9] are all promising applications that have the potential to transform the AEC industry.

Additional research projects such as the COGIRO cable robot [10], developed by TECNALIA and LIRMM, have demonstrated the potential in industrial applications for warehouse storage and retrieval systems, while the research conducted by the IAAC, using this same aforementioned CDPR, focused its efforts on construction tasks, and more specifically large scale adobe 3D printing [11].

Although the aforementioned precedents exhibit numerous similarities in the mechanical design of the CDPRs, they offer minimal information regarding their design and assembly and significantly diverge in their intended applications. This research attempts to offer insight into the design and development of a CDPR platform, while presenting a pick-and-place framework for the semi-automated positioning of prefabricated assemblies. Adapting this technology for the precise needs of handling prefabricated building components requires a specialized approach in developing the CDPR's architecture and its associated components. It requires an understanding of the interdependencies among architectural design, material assemblies, and robotic processes to ensure the provision of a system capable of managing the required payload, scalable to the application, integrated with software for design and simulation, and equipped with the necessary rotational and translational motion capabilities. This customized approach brings its own unique opportunities and paves the way for innovative solutions that will be further presented.

3 Methods

This paper presents research conducted in collaboration between engineering and architectural teams over a two-year period focusing on the design and development of a CDPR as an alternative method for in-situ assembly. It presents three significant advancements in the field of large-scale architectural assembly using cable robots. First, it provides information on the technical implementation of the platform including the mechanical design of the CDPR, which is developed based on kinematic sensitivity and payload requirements. Details are provided for the design of the frame including the cable actuation and routing system, motor selection, and control systems leading to the design and development of a working prototype. (Figure 1) Its target application is for small structures such as the construction of single family homes, row houses and coach houses. Secondly, the research introduces the development of a software tool for the control, simulation and verification of the CDPR in order to facilitate architectural design-fabrication exploration. Lastly, in order to achieve the pick-and-place operations of prefabricated assemblies, the research presents the early development of novel end-effector allowing for the multi-rotational positioning of elements while maintaining a center mass to prevent external disturbance moments in the system.

The work presented herein contributes to the continued development of CDPRs for automated construction applications. Although early in its development, it offers insight into the advancement of a CDPR as an alternative method for large scale assembly, with emphasis placed on the mechanical system, end-effector and robotic control.

3.1 Material Assembly

To initiate the research and facilitate the testing of the CDPR's pick-and-place capabilities, a prefabricated panel was produced. The WikiHouse modular building system [12] was chosen for its open-source platform, which simplified the design and manufacturing processes and effectively met the testing needs of the CDPR. A modified Wikihouse wall panel, made from 18 mm plywood and measuring 1500 mm × 600 mm × 250 mm

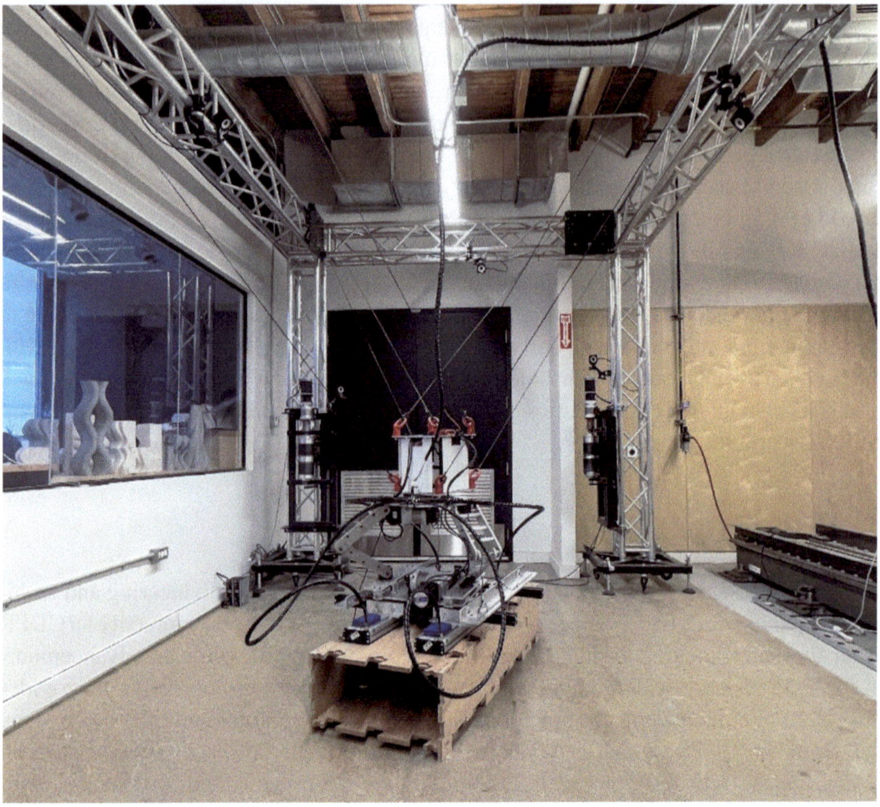

Fig. 1 CDPR prototype including frame and winch assembly, cable routing system, mobile platform, reorientation mechanism and material assembly.

was fabricated (Fig. 1). This wall panel did not include any additional architectural subassemblies such as insulation or cladding, as it was deemed sufficient for the initial testing phases.

3.2 CDPR Selection

CDPRs differ from serial robots by relying on flexible cables rather than motor-driven joints. These flexible cables are manipulated through multiple winches, enabling the MP to navigate in three-dimensional space by adjusting the cable lengths. Maintaining tension within a specified range for all cables is crucial as any slack could result in the MP becoming under-constrained.

The COGIRO cable robot, a robot well known for its large workspace and high payload capabilities and developed by TECNALIA & CNRS-LIRMM, uses 8 winches allowing for independent control of all cables. As such, it has the capacity to rotate the MP at desired angles however within certain limits. The largest rotation it can achieve with an optimized robot geometry is 105° about a vertical axis, and (+/−) 40° about each horizontal axis [13]. Although the COGIRO's mechanical design is interesting, it implied limitations in terms of its rotational capacity. For our intended application, which necessitates the placement of prefab assemblies and the ability to rotate a panel from a horizontal to a vertical position, a translation of 360° about a vertical axis, and 90° about a horizontal axis was required. In order to achieve this motion, an auxiliary mechanism was developed that eliminated the need for the MP's 2-Degrees of Freedom (DoFs) of rotation on the vertical axis. Considering the CDPR's cables were arranged pairwise to form parallelograms (Fig. 2a, b), this provided the added benefit of reducing the required quantity of actuators to 4 (as opposed to 8 for the COGIRO) since the equal length of cables within a parallelogram may be controlled by a single motor-driven winch unit.

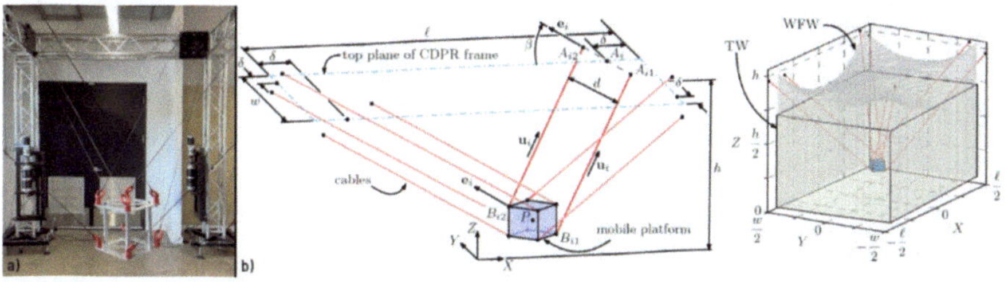

Fig. 2 (a) Current prototype: 8 cables & 4 winches (b) Illustration of the current prototype's wrench feasible workspace and task workspace.

3.3 Frame Design

Truss members of CDPRs vary depending on the loading requirements. A truss made of 6082-T6 aluminum with a cross-section of 290 mm was selected for this research as it represented a cost-effective option capable of supporting the required loads.

The targeted CDPR frame (Fig. 3) is a rectangular prism with dimensions of 10 m in length, 6.67 m in width, and 6.67 m in height. Certain requirements were identified in the early onset of the research, such as the ability of the MP to position assemblies measuring 1.2 m x 2.4 m in any orientation while avoiding collisions. To achieve this, the robot had to operate at a height of 4.5 m. Moreover, the robot was required to support a maximum expected payload of 100 kg which includes the weight of the assembly, the MP and the end-effector. Due to space limitations within the lab, the WIP prototype was limited to a workspace measuring 6 m × 3.5 m × 3 m, however all associated components were designed and fabricated to meet the requirements stated above.

3.4 Winch Design

The cable actuation system, commonly known as a winch, comprises a motor-powered drum and a level winding mechanism. The winding mechanism chosen was a spooling-helper mechanism [15], a system composed of pulleys or guides that follow the cable exit point on the fixed drum to ensure even distribution of the cable.

The spooling-assist mechanism, as depicted in Fig. 4, was custom fabricated, incorporating two pulleys designed for guiding each cable onto the drum. These pulleys are

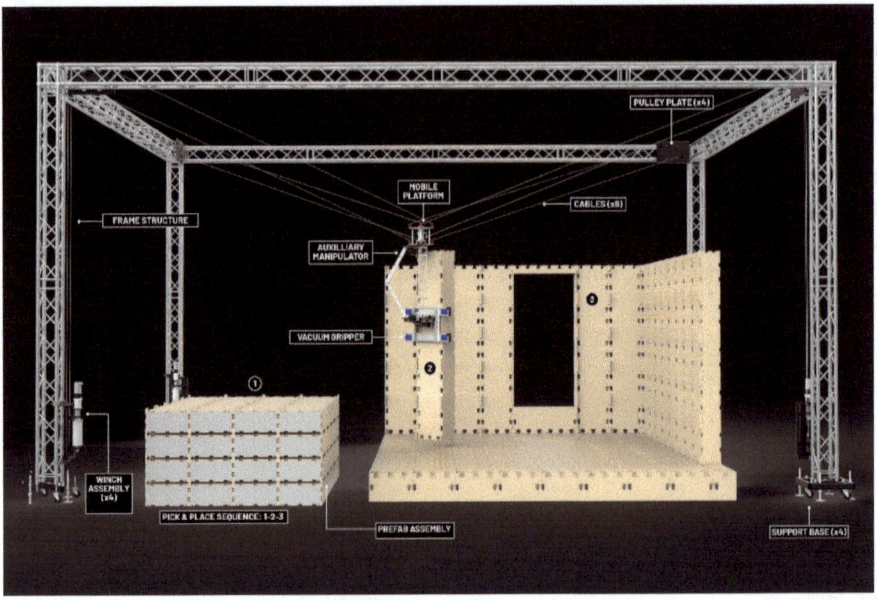

Fig. 3 Targeted CDPR architecture and application.

affixed to a linear guide and are actuated by a lead screw. The cable routing ensures that the lead screw always bears tension, thus preventing buckling and enabling the screw to support greater thrust loads.

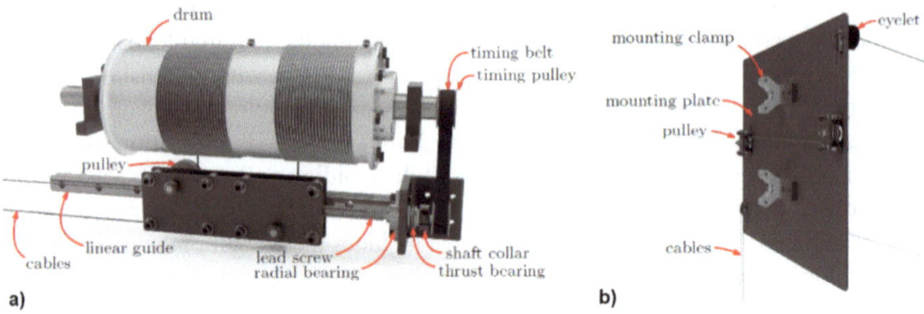

Fig. 4 Renderings of (a) the cable actuation system and (b) the cable routing system.

3.5 Motor, Gearbox, Brake Selection and Controller Integration

The controller employed in the CDPR is the Teknic ClearCore Controller. It receives cable trajectory data from a computer via a serial connection, then translates this data into step and direction signals, which are sent to the Teknic servo motors. They are equipped with an integrated driver that processes these signals, determining the necessary motor positions, velocities, and accelerations to achieve the desired cable path. The brakes, positioned between the motors and gearboxes, serve to halt any unintended movement of the MP during a power outage by counteracting drum rotation.

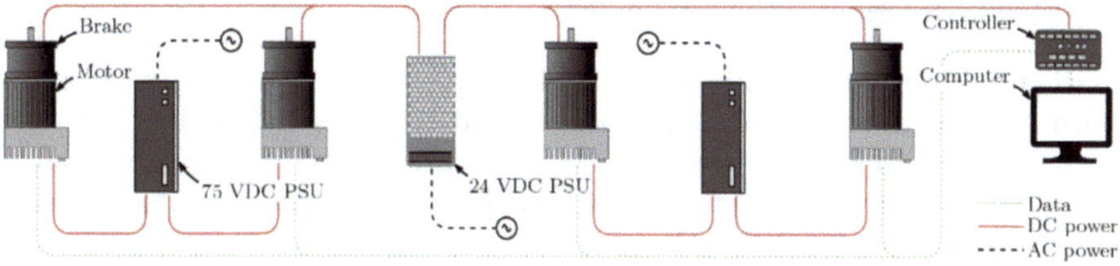

Fig. 5 Integration of the controller with the motors, brakes, and the necessary power supply units (PSUs).

4 CDPR Modeling Methods

CDPRs feature a straightforward mechanical structure, characterized by their limited number of components. Despite this simplicity, the task of accurately modeling these systems introduces complex challenges. Engineers looking to analyze and design CDPRs typically depend on numerical computing applications such as MATLAB, which offer robust capabilities for data examination, equation solving, and optimization processes, albeit without specific features for CDPR design or analysis. Alongside MATLAB, a variety of software tools have been developed specifically for the detailed examination and simulation of cable robots, including kinematics, dynamics, control, and workspace analysis, for example:

WireX [16] is an open-source platform designed to facilitate the comprehensive analysis and development of cable-driven parallel robots. It provides functionalities for users to evaluate or craft new CDPR geometries, conduct forward and inverse kinematics studies, calculate structural and stiffness matrices for particular poses, assess collision risks, define workspaces, and generate path plans to G-code files [16].

CASPR [17] is a software platform designed for the modeling, analysis, and simulation of cable-driven parallel robots. It is an open-source tool that integrates with MATLAB, providing users with capabilities to perform a wide range of computational analyses specific to cable robots. CASPR facilitates the modeling of both forward and inverse kinematics, CDPR control and determination of workspace feasibility and overall is very useful for modeling and analyzing CDPRs.

Despite the aforementioned platforms, there remains a significant gap for a tool that seamlessly integrates CDPR modeling and analysis with the practical considerations for design and fabrication, through integrated CAD functionality. The ability to simulate the CDPR within a design environment prior to physical making is critical to its early adoption within an architectural design framework. Design fabrication tools such as Kukalprc [18], which allows for the control and simulation of serial robots, have provided designers (with little to no previous experience in robotics) with unlimited design exploration capabilities. Accordingly, in order to facilitate architectural fabrication innovation while addressing the complexities associated with parallel robots, the development of a design tool was initiated for the control and simulation of the CDPR.

4.1 MATLAB/Grasshopper Integration

Initial efforts were aimed at incorporating MATLAB into Rhinoceros' Grasshopper to enhance design experimentation while preserving the precision in modeling, analyzing and simulating CDPRs. Python was utilized within Grasshopper for MATLAB integration. MATLAB scripts were then developed to evaluate wrench feasibility, forward kinematics (FK), inverse kinematics (IK), and cable tensions among other aspects. Although these simulations were accurate, the linkage between MATLAB and Grasshopper fell short in efficiently visualizing motion paths. This inefficiency was primarily due to the prolonged durations needed to execute simulations, as calculations occurred at every segment along a curve or path. While MATLAB excels in the modeling and analysis of CDPRs, its incorporation with Rhinoceros 3D faced challenges, hindering swift design iterations and the quick simulation of assembly processes.

4.2 CDPR Studio

CDPR Studio was developed as a plug-in for Grasshopper (GH) within McNeel's Rhinoceros software. CDPR Studio was developed in C#, consisting of a suite of custom GH components capable of generating trajectories for a variety of fabrication applications. The plug-in calculates the forward kinematics (FK) and inverse kinematics (IK), the kinematic sensitivity of a given pose, cable tensions outside the acceptable range, and provides users with the ability of creating a user defined CDPR frame, MP and end-effector. It also provides users with a method to visualize Cartesian space and joint-space trajectories, as well as out of range cable tensions while calculating the cable lengths and cable length change at every point of the Cartesian space trajectory. CDPR Studio 0.2 consists of the following core components: the robot builder, the solver/simulation previewer, the trajectory planner and the trajectory sender. Every component takes in distinct inputs and generates the necessary data types for analyzing, verifying, and simulating the intended fabrication applications, such as pick-and-place operations or 3D printing.

4.3 Robot Builder Component

The Robot Builder component (Fig. 6) requires the user to select a predetermined frame configuration, MP and end-effector, with the added functionality of creating a custom CDPR configuration as needed. Inputs such as DoFs, anchor points and selection of mesh geometry are used to generate a preview of the CDPR and simulate the robot before executing the trajectory. This component assembles the robot system using the robot, MP, and end-effector specified by the user and outputs the robot system parameters as well as the CDPR components as mesh geometries.

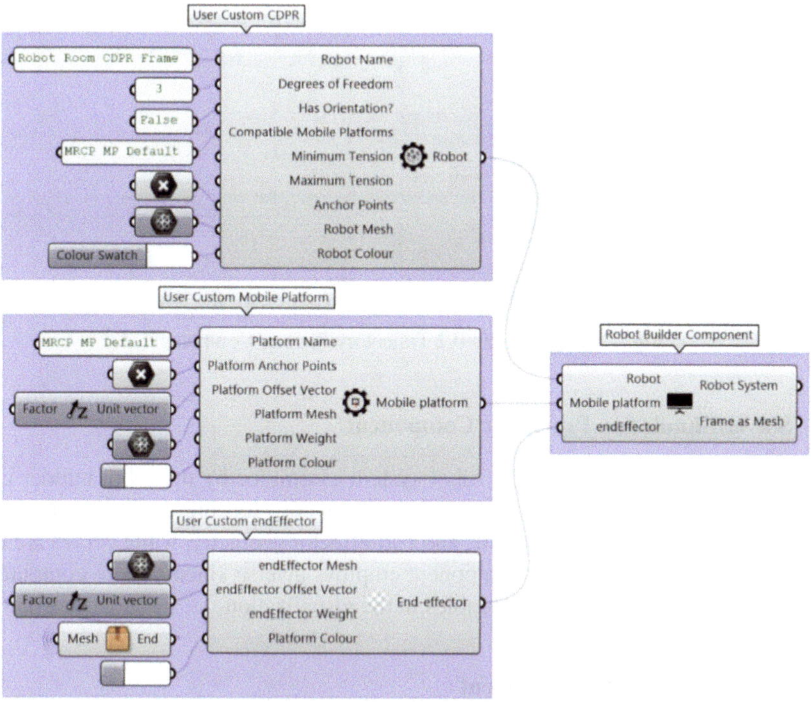

Fig. 6 CDPR Studio 0.2 Robot Builder Component

4.4 Trajectory Planner Component

CDPR Studio offers two trajectory planning methods. The first method, known as the Curve Trapezoidal Constant-Orientation Method, involves moving the MP along a user-defined curve using a trapezoidal velocity profile. Users select a curve for traversal and input parameters such as acceleration of the move and coast velocity. Position calculation along the path occurs at each timestep using a trapezoidal time scaling function, ensuring controlled motion by smoothly transitioning between the start and end points of the path, while inverse kinematics calculates the cable lengths at each point along the trajectory.

The second method, known as the Linear Trapezoidal Method, is used to move the MP between frames using a trapezoidal velocity profile. Users provide a list of frames to traverse, along with acceleration and coast velocity parameters. At each timestep, both position and orientation are determined using a trapezoidal time scaling function for smooth motion and seamless transition between frame orientations. Similarly, inverse kinematics is utilized to calculate the cable lengths at each point along the trajectory.

These methods offer distinct approaches to trajectory planning, catering to different user requirements and scenarios. Both methods output lists consisting of a series of states (Fig. 7). These state lists include the position, orientation and cable lengths at each timestep required for the CDPR to traverse the trajectory.

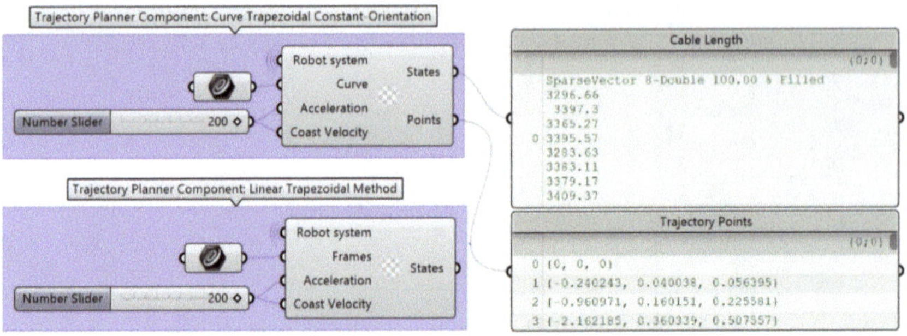

Fig. 7 CDPR Studio 0.2 Trajectory Planner Component.

4.5 Solver/Simulation Previewer Component

The solver component utilizes the robot system assembled by the robot builder component to visualize the robot within the Rhino viewport. It achieves this by employing meshes to represent the robot, its MP, and end-effector geometry, while curves are used to depict the cables. The solver component employs inverse kinematics to compute the necessary curves for rendering the cables in the visualization.

4.6 Trajectory Sender Component

The trajectory sender component facilitates the transmission of a trajectory to the robot's controller via a serial connection. Users initiate the process by selecting the COM port and baud rate. Subsequently, a list generated by one of the trajectory planners is transmitted state-by-state to the robot's controller.

The transmission process operates iteratively, with the trajectory sender component fulfilling requests for new states from the robot's controller at each timestep. These states are then transmitted until all states within the trajectory have been received by the robot's controller and used to execute the motion. This systematic approach ensures seamless communication between the trajectory planner and the robot's controller, enabling precise execution of the planned trajectory.

In summary, CDPR Studio 0.2 (Fig. 8) incorporates methods for defining the CDPR, the MP, and associated tools. While still in alpha development, the plug-in aims to allow users to effectively employ CDPR Studio without needing extensive knowledge of robotic cable systems. Moreover, CDPR Studio integrates path-planning methods that utilize geometry from Rhino, providing users with crucial insights into the feasibility of planned paths and allows them to simulate the CDPR within a design environment before implementation. This feature gives users the confidence to execute their intended tasks in the software, aligning their digital plans with real-world outcomes.

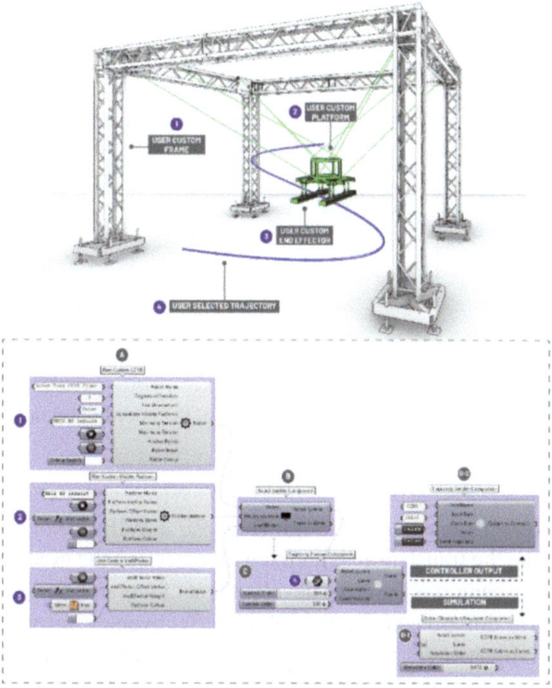

Fig. 8 CDPR Studio 0.2 Workflow (A-D2).

5 Reorientation Mechanism

The CDPR's cables are arranged pairwise to form parallelograms, each controlled by a single motor-driven winch unit, thus eliminating its capacity to provide rotational motion. Due to the limited ability to generate large rotations, an auxiliary mechanism attached to the MP was needed in order to reorient panels for assembly. While the current CDPR prototype can withstand external disturbance moments, their occurrence is undesirable due to the potential for unequal cable tension within a given parallelogram. To prevent these disturbances, it is important to keep the horizontal position of the combined center of mass (COM) of the gripper and assembly constrained to a vertical line passing through the center of the MP while the panel is being manipulated. In doing so, this prevents any disturbance moments from acting on the MP and thus this became the primary goal for the design of the auxiliary mechanism.

As an attachment to the MP, commercially available manipulators, such as collaborative robots and small industrial arms were explored as potential reorientation mechanisms however they were deemed unsuitable for this application. Despite providing the DoFs required for rotation, they would be unable to maintain the center of mass constant during the pick-and-place operation. Furthermore, the size of the panel would be constrained to the reach of the manipulator when rotating a panel from a horizontal to a vertical position, thus limiting the size of the assembly and overall design possibilities.

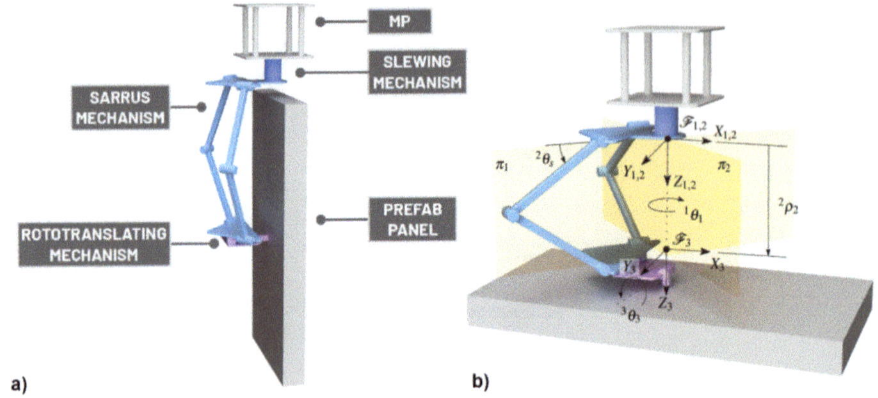

Fig. 9 (a) Illustration of the proposed mechanism and (b) the coordinates defining the pose of the prefab panel.

Accordingly, based on an early design (Fig. 9), a WIP prototype was developed (Fig. 10), to facilitate the translational motion while meeting the goals mentioned above. The subsequent stage centers around the development of this auxiliary mechanism, which serves as an integral component for the robot's effectiveness in panel assembly. The selection of the auxiliary mechanism's topology hinges on the requirement to provide the necessary DoFs essential for completing designated tasks. To enable the CDPR to effectively carry out its intended task, the auxiliary mechanism requires 2-DoFs for rotation of the panel using a vacuum gripper, with one axis oriented vertically and the other horizontally. Additionally, it should provide 1-DoF for translation along a vertical axis, allowing the panel to be lowered to avoid potential collisions with the CDPR's cables.

This mechanism integrates both rotational and translational motions, incorporating a revolute joint for controlling orientation around the horizontal axis and a prismatic joint to maintain the horizontal position of the combined COM of the panel and the mechanism. This positioning ensures that the COM consistently aligns with a vertical line passing through the center of the MP. The coordination of these translation and rotation movements allows for the attainment of both objectives using a single actuator. This synchronization is achieved by employing a wrapped cam mechanism linked to the panel's rotation, and the cam profile is chosen to ensure that, irrespective of the panel's horizontal orientation, the COM of the mechanism and the panel aligns with the vertical line running through the center of the MP.

As the panel rotates from a horizontal to vertical orientation, the common COM undergoes both horizontal and vertical shifts. While the vertical shift has no impact on the moment applied to the MP, the horizontal shift demands attention due to its potential to induce a significant disturbance moment, which could compromise the robot's accuracy.

The horizontal shift of the common COM is mathematically expressed as:

$$\Delta x_r = \frac{m_p e_p \sin(\theta)}{m_p + m_r}$$

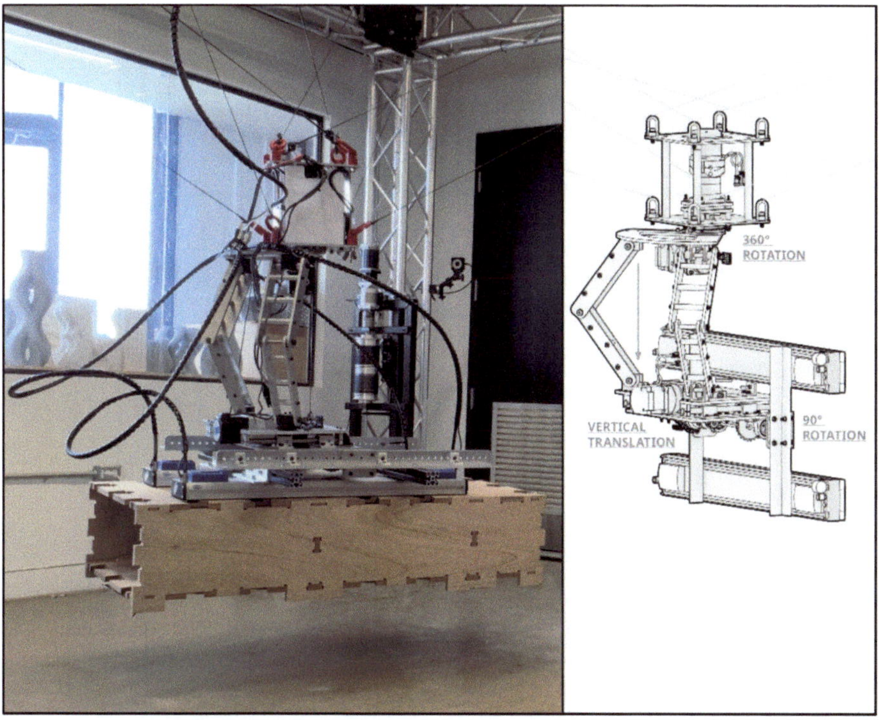

Fig. 10 Prototype of reorientation mechanism and illustration displaying rotational and translational motion capacity (right).

Here, m_p and m_r represent the masses of the prefab and the rototranslating mechanism, respectively, e_p denotes the distance between the COM and the rotation axis, and θ represents the panel angle.

To counteract the horizontal motion of the common COM, the prismatic joint is actuated with the function $x_p = -\Delta x_r$ using the wrapped cam mechanism, effectively nullifying the common COM's horizontal motion. This process is depicted in Fig. 11b, where the horizontal position of the common COM is plotted against the panel angle. The plot also includes the horizontal position of the prismatic joint. The combination of these adjustments yields no horizontal motion of the common COM, as evidenced by the corrected COM curve. This was further simulated in Autodesk Inventor (Fig. 11c) where the COM remains fixed during a translation of 90° about a horizontal axis.

A subsequent validation is yet to be conducted with the physical prototype. As the panel transitions from a horizontal to vertical orientation, the mechanism will need to be monitored through position plots to ensure it maintains its alignment relative to the ground. The observation of this phenomenon would confirm that the common COM is capable of maintaining its horizontal position throughout the rotation process.

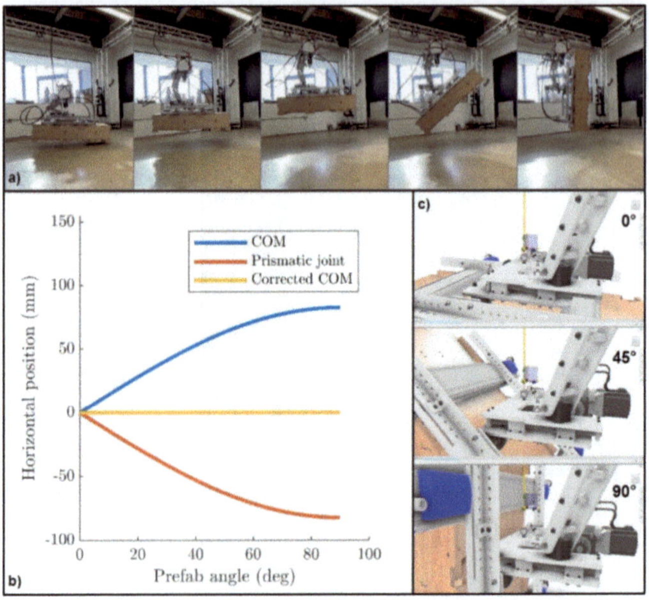

Fig. 11 (a) Image sequence of a pick-and-place procedure (b) Common COM and prismatic joint horizontal positions as a panel is rotated. (c) Simulation depicting Autodesk Inventor's center of gravity based on assembly weight and volume.

6 Calibration

Given that a pair of CDPR cables share one actuator and must be of equal length to form parallelograms, calibration could be challenging as this configuration removes the ability to adjust each cable independently in order to achieve the necessary tension. A digital twin of the CDPR was developed through a LIDAR scan of the actual environment to confirm the MP's location and cable lengths. However, inaccuracies in manufacturing the CDPR components, coupled with the system's inability to adapt and provide immediate feedback, rendered this approach somewhat ineffective. As a result, Optitrack's motion capture and 3D tracking system was used to identify and locate specific home positions and calculate cable lengths in real-time with the use of eight (8) Prime13 cameras placed on the truss frame and sixteen (16) passive markers positioned in specific locations across the MP and cable plates. Together with Motive's NAtNet SDK and an adapted grasshopper plugin [21], real-time streaming was captured facilitating the integration of Optitrack motion tracking data into Rhino's 3D environment with sub-millimeter precision. As such, the home position, both orientation and position was identified and calibrated in both digital and physical environments thus maintaining the required cable lengths and ultimately cable tension to ensure proper deployment.

7 Limitations and Future Developments

As the research is on-going, there are limitations with the selected components. The current motors limit the maximum payload (100 kg), overall workspace, as well as velocity and acceleration (250 mm/s). However, these outputs can be increased without the need to substitute the motors through a variety of methods such as trajectory optimization, optimized gear ratios, etc.

Furthermore, the current reorientation mechanism's design, reliant on a wrapped cam to drive the prismatic joint, cannot adjust to mass changes, limiting its ability to counteract the COM's horizontal motion unless the panel's weight is constant. Future work will seek to address this by considering alternative designs for dynamic adjustment, possibly removing the need for a wrapped cam mechanism altogether. This improvement aims to neutralize COM movement across different sized and weighted panels that are typically found in prefab residential construction, such as floors, walls, and roof assemblies. Additionally, since the mechanism's translation along the vertical axis currently relies on a single cable, improvements in rigidity are anticipated through the addition of extra cables. This modification is expected to enhance the stability of the mechanism, particularly during vertical axis translations.

Future work also includes the on-going development of CDPR Studio, as well as the integration of the OptiTrack motion capture with the Robot Operating System (ROS) allowing for two-way communication, additional robot control, navigation, and object manipulation, thus improving the CDPR's ability to interact with its environment. By integrating OptiTrack with ROS, the CDPR will gain improved control and navigation capabilities. As OptiTrack's precise motion capture system provides real-time data about the robot's position and orientation, this data can be used to enhance the robot's ability to move accurately within its environment, avoiding obstacles and following designated paths. Furthermore, this cross-platform integration will allow the CDPR to detect and recognize objects and prefab panels in its environment. This is particularly valuable in the assembly tasks, where the CDPR will be used to place and manipulate prefabricated components, as well as precisely identify different prefab panels and determine their optimal placement sequence. This recognition capability can significantly streamline the assembly process and minimizes errors.

8 Conclusion

In conclusion, the pressing global challenge of housing supply necessitates innovative solutions that go beyond traditional methods of construction, highlighting the need for a fundamental shift in the way we design and construct buildings to meet ambitious housing targets. The research presented in this paper addresses this need by exploring the potential of cable-driven parallel robots as an alternative method for in-situ construction.

While robotics in construction is not a new concept, this research takes a unique approach by focusing on the design and development of a CDPR for in-situ assembly. The development of a working prototype, characterized by flexibility, portability, and modularity, is a key milestone in this ongoing research. This prototype is tailored for the construction of small structures such as single-family homes, row houses, and coach houses, offering pick-and-place operations of prefabricated assemblies and innovative end-effector capabilities.

Additionally, the paper introduces a novel software tool designed for the control, simulation, and verification of the CDPR, enhancing architectural design-fabrication exploration. While the research presented is still in its early stages, it represents a significant contribution to the advancement of CDPRs for automated construction applications.

In sharing this work, the paper aims to reach a wider audience interested in the potential of CDPRs as a promising and alternative approach to in-situ construction. As the demand for housing continues to grow, innovative solutions like the one presented here are essential for addressing the housing supply challenge and reshaping the AEC industry. This research underscores the importance of interdisciplinary collaboration and the pursuit of novel technologies to meet the housing needs of the future.

References

1. World Economic Forum. The world needs to build 2 billion new homes over the next 80 years. Accessed October 10, 2023. https://www.weforum.org/agenda/2018/03/the-world-needs-to-build-more-than-two-billion-new-homes-over-the-next-80-years.
2. National Association of Home Builders. 60% of a Home's Sales Price Goes to Construction Costs." Accessed Jan 4, 2024. https://www.nahb.org/blog/2023/03/60-percent-of-home-sales-price-goes-to-construction-costs.
3. Brown, G.W.: Suspension System for Supporting and Conveying Equipment, Such as a Camera, 1986.
4. Qian, S., Zi, B., Shang, W.W., Xu, Q.S.: A review on cable-driven parallel robots. Chin. J. Mech. Eng. **31**(1), 66 (2018)
5. Cone, L.L.: SKYCAM: An aerial robotic camera system. Byte. **10**(10) (1985)
6. Bruckmann, T., Boumann, R.: Simulation and optimization of automated masonry construction using cable robots. Adv. Eng. Inform. **50**, 101388 (2021)
7. Iturralde, K., et al.: A cable driven parallel robot with a modular end effector for the installation of curtain wall modules. In: ISARC. Proceedings of the International Symposium on Automation and Robotics in Construction, vol. 37, pp. 1472–1479. IAARC Publications (2020)
8. Dierichs, K., Kyjánek, O., Loučka, M., Menges, A.: Construction robotics for designed granular materials: In situ construction with designed granular materials at full architectural scale using a cable-driven parallel robot. Constr. Robot. **3**, 41–52 (2019)
9. Hahlbrock, D., et al.: Cable Robotic 3D-printing: additive manufacturing on the construction site. Constr. Robot. **6**(3), 305–318 (2022)
10. Tecnalia: Cable Driven Parallel Robotics for industrial applications. Accessed October 10, 2023. https://youtu.be/px8vwNerkuo.
11. Dubor, A., Izard, J.-B., Cabay, E., Sollazzo, A., Markopoulou, A., Rodriguez, M.: On-site robotics for sustainable construction. In: Willmann, J., Block, P., Hutter, M., Byrne, K., Schork, T. (eds.) Robotic Fabrication in Architecture, Art and Design 2018: Foreword by Sigrid Brell-Çokcan and Johannes Braumann, Association for Robots in Architecture, pp. 390–401. Springer International Publishing (2018)
12. WikiHouse: Accessed March 04, 2024. https://www.wikihouse.cc/.
13. Gouttefarde, M., Collard, J.-F., Riehl, N., Baradat, C.: Geometry selection of a redundantly actuated cable-suspended parallel robot. IEEE Trans. Robot. **31**(2), 501–510 (2015)
14. COGIRO: Accessed March 4, 2024. https://www.lirmm.fr/cogiro/.

15. Idà, E., Mattioni, V.: Cable-driven parallel robot actuators: State of the art and novel servo-winch concept. Actuators. **11**(10), 290 (2022)
16. Pott, A.: WireX–An Open Source Initiative Scientific Software for Analysis and Design of Cable-Driven Parallel Robots. IFTOMM World (2019)
17. Lau, D., Eden, J., Tan, Y., Oetomo, D.: CASPR: A comprehensive cable-robot analysis and simulation platform for the research of cable-driven parallel robots. In: 2016 IEEE/RSJ International Conference on Intelligent Robots and Systems (IROS), pp. 3004–3011. IEEE (2016)
18. Braumann, J., Brell-Cokcan, S.: Parametric robot control. Integr. CAD/CAM Archit. Des., ACADIA. **11**, 242–251 (2011)
19. McDonald, E., Beites, S., Arsenault, M.: CDPR studio: A parametric design tool for simulating cable-suspended parallel robots. In: International Conference on Computer-Aided Architectural Design Futures, pp. 344–359. Springer Singapore, Singapore (2021)
20. Harris, C.R.: Array programming with NumPy. Nature. **585**, 357–362 (2020)
21. Hughes, R. Povl-Sonne, F.: Tracker: A Grasshopper plugin for NaturalPoint's NatNet. https://github.com/rhughes42/optitrack-gh.

Out of Plane

CHANDELIER'1244: Towards Cyber-Physical Experimentation with Structural Adaptation in Architecture

Mathias Maierhofer(✉) and Achim Menges

Institute for Computational Design and Construction, University of Stuttgart, Stuttgart, Germany
mathias.maierhofer@icd.uni-stuttgart.de

Abstract. This research investigates methods for combining lightweight construction and sensor-based actuation for structural adaptation of architectural elements. If successful, structurally adaptive elements have the potential to achieve considerable resource savings. Given that research on this topic has recently expanded to include applications in architecture, this paper investigates ways to make its underlying methods and principles accessible to the architectural community. This is approached through the development of a design-to-operation method that augments architect-driven experimentation and speculation with strategies for structural design, actuator placement and control. In addition, the paper applies this method to the development of an interactive installation that showcases the concept of structural adaptation through a furniture-scale object conceived to engage non-experts with the intrinsic behaviors of adaptive structures in a safe and playful manner.

Keywords: Adaptive structures · Robotic architecture · Active structural control · Cyber-physical systems

1 Introduction

1.1 Motivation and Aim

In this research, structural adaptation refers to a notion of lightweight construction that proposes the enabling of structures to not only withstand, but to actively adapt their load-bearing behavior in response to adverse influences (Sobek, 2016). Despite its background in various branches of engineering, research into structural adaptation has recently expanded to explore its application in architecture (Blandini et al., 2022). Similar development can be observed in the architectural discourse itself, where robotic principles are increasingly explored as a means to actively operate, e.g., adapt or reconfigure, architectural spaces (Bier et al., 2018). The research presented in this paper seeks synergies across these two discourses. Specifically, it aims to make the concept of structural adaptation accessible to the community of robotic architecture, and in doing so, to complement the further investigation of structural adaptation with the perspective of architectural design.

M. Yablonina et al. (Eds.): Robotic Fabrication in Architecture, Art and Design 2024, pp. 162–173, 2026.
https://doi.org/10.1007/978-3-032-22607-5_11

1.2 Background

1.2.1 Structural Adaptation

Research on structural adaptation, or active structural control, dates back to the 1960s (Zuk and Clark, 1970). Over the past decades, this concept has been investigated with respect to various application domains, structural systems and adaptation strategies (Korkmaz, 2011). The strategy underlying this paper was first articulated in 2001 (Sobek and Teuffel). Recently, it was more formally introduced under then notion of 'Ultra-lightweight Construction' (Sobek, 2016), given the aim to unlock hitherto unachievable resource savings through the combination of traditional lightweight construction with principles of active structural control. It proposes the augmentation of structures with sensor-actuator mechanisms, to actively manipulate, i.e., homogenize, forces and deformations resulting from adverse influences. In doing so, safety and serviceability limits can be met with significantly smaller cross-sections, leading to potential material savings of up to 70% together with a doubling in service life (Ostertag et al., 2020). Notably, stiffness-governed systems, such as long-span or high-rise structures, were found to be an application domain where adaptive solutions can outperform their passive counterparts both ecologically and economically (Senatore et al., 2018).

This strategy was first applied to the ultra-lightweight 'Stuttgart Smart Shell' (Neuhaeuser et al., 2013), and more recently to the 'D1244' (Fig. 1), the world's first adaptive high-rise building (Blandini et al., 2022). The latter presents a 12-story spatial truss structure, equipped with sensors and hydraulic actuators to manipulate forces and deformations through minimal and virtually imperceptible length changes.

Fig. 1 Exterior and interior view of the D1244

1.2.2 Robotic Architecture

The field of robotic architecture explores the utilization of robotic principles and entities for advanced fabrication, and increasingly, for the operation, e.g., adaptation, of inhabited architectural structures (Bier et al., 2018). Work related to the latter is based on a shift in the conception of buildings from static and deterministic artifacts towards complex, dynamic and open-ended systems. Such systems are inherently cybernetic, a notion used across disciplines and first introduced to the architectural discourse in 1969 (Pask). Besides laying the groundwork for contemporary design computation (Werner, 2019), cybernetics stimulated the exploration of feedback-based interactions between spatial systems across various scales, their environments and human occupants (Steenson, 2022).

With increasing accessibility to software and hardware for digital and physical computing, cybernetic principles experienced a renaissance in the early 2000s. This has stimulated a plethora of proposals for architecture that is proactive (Oosterhuis, 2003), adaptive (Schnädelbach, 2010), interactive (Fox, 2018), or more generally, robotic (Kilian, 2018). Among others, this includes adaptive façades (Attia et al., 2020), reconfigurable partition walls (Yablonina et al., 2021), or transformable furniture objects (Bier et al., 2018).

1.3 Relevance and Scope

With respect to full-scale, i.e., multi-story buildings, most instances of robotic architecture have so far been conceived of, or implemented as, secondary systems embedded within passive load-bearing structures. This paper investigates the means required to expand explorations of robotic operation to the domain of load-bearing structures.

In addition to the potential for low-emission construction, the architectural relevance of structural adaptation lies in the opportunity to unlock new structural morphologies beyond the constraints of established typologies. This is due to the fact that the latter exhibit only limited adaptability, which in the case of truss structures, describes the extent to which a given structural layout allows for the adaptation of its load-bearing behavior (Steffen et al., 2022). However, the relationship between structural topology, adaptability as well as actuator layout is by no means trivial and presents a considerable obstacle to the accessibility of structural adaptation (Maierhofer and Menges, 2022).

This challenge is addressed based on two research objectives: The first objective is to establish an accessible design-to-operation method to support and inform architect-driven exploration and speculation. The second objective is to apply this method to a human-scale prototype through which the imperceptible and often counter-intuitive behaviors of adaptive structures can be explored in a safe and playful manner. Both objectives are addressed through the development and implementation of the cyber-physical installation CHANDELIER'1244, which translates the concept of structural adaptation to an interactive furniture-scale object.

Moreover, the installation also served a case study to evaluate its underlying methods in terms of their capacity (1) to support the identification of adaptable topologies and corresponding actuator layouts, (2) to adequately detect and adapt to adverse element load cases, (3) and finally, to demonstrate this process in an engaging and comprehensible way.

2 Methods

2.1 Overview

CHANDELIER'1244, hereafter referred to as C1244, presents a composition of two structures: (1) the 'bench', an adaptive truss structure that serves as a seating element for up to eight people, and (2) the 'lamp', a suspended light structure identical in dimensions and topology to the bench located beneath it (Fig. 2). The installation is conceived around the following scenario: Visitors occupy the bench, which, in turn, detects resulting live loads in real-time. If necessary, it would initiate the actuation of specific members such that stresses are redistributed form over- to underutilized members. The lamp, on the other hand, simultaneously displays the changing levels of utilization through different colors and hence, quite literally, sheds light on the otherwise imperceptible process of stress-homogenization.

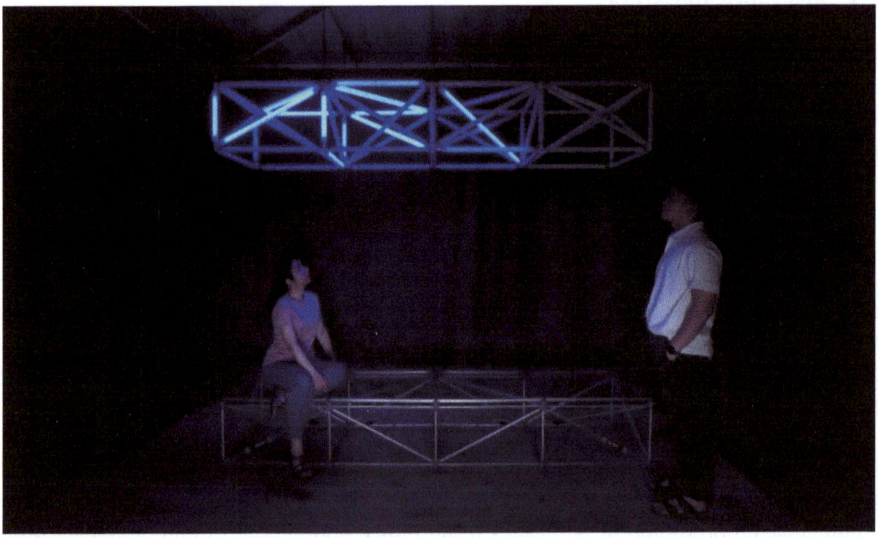

Fig. 2 C1244 visualizing the distribution of element utilization due to the live load of one occupant. Dark blue = low utilization, Turquoise = medium utilization, Red = high utilization

The following sections examines and evaluates the design-to-operation methods that were established throughout the development of this installation. The proposed methods are implemented using Rhino, Grasshopper and Arduino, i.e., technologies that are commonly used in, and hence accessible to, the architectural community.

2.2 Structural Design and Adaptability

The adaptability of a given truss topology relates to its internal distribution of statical indeterminacy (Geiger et al., 2020). This information can be derived from Redundancy Matrices (RM), which specify - on the level of each structural member - the extent to which its potential actuation, i.e., minimal change of length, influences all other members. The calculation of the RM was implemented in C# according to von Scheven et al. (2021), and deployed as a plug-in to Grasshopper. By means of this plug-in, the topology of the C1244 was designed such that all members exhibited some degree adaptability, defined by a corresponding diagonal entry in the RM larger than 0, on a scale from 0 to 1.

Starting from the configuration shown in Fig. 3a, this was achieved by iteratively adding diagonal elements, and thus by increasing statical indeterminacy, in the proximity of existing elements with a redundancy contribution of 0 (illustrated in yellow). Fig. 3b shows the final layout that satisfies the defined adaptability requirement using six additional diagonal members. Support conditions were defined at the outset of this process. However, rather than supporting each lower node, three out of ten potential support points were removed to stimulate a larger variety of stress distributions for different loading scenarios. For example, a visitor's live load at unsupported conditions would cause significantly higher and more challenging stresses compared to the same load applied at supported areas.

The proposed structure measures 2.4 m in length, 0.6 m in width and 0.4 m in height. It consists of 20 joints (including seven supports) and 59 structural elements. be implemented instead. Six diagonal members with the highest redundancy contribution, and thus highest influence on the rest of the structure, were selected for actuation. Through the simulation of virtual load cases, the ability of this actuator layout to sufficiently manipulate forces was validated using the structural model described in Sect. 2.4. While this was procedure proved to be sufficient for the scope of this investigation, more sophisticated actuator-placement algorithms (Wagner et al., 2018) could

2.3 Materialization and Hardware

2.3.1 Bench Element

The structural members of the bench element are realized as aluminum profiles ('Makerbeams') with a cross-section of 10 × 10 mm. Vertical elements are directly attached to 3d-printed joints (tough PLA) through threaded rods. Other elements are equipped with brackets at both ends that attach to joints with a metal bolt and allow for 1-DOF in rotation. Polycarbonate sheets mounted on 15 × 15 aluminum profiles form a seating surface that rests on the load-bearing structure and distributes the occupant's load to its upper nodes. Excluding this layer, the self-weight of the load-bearing structure is 4.6 kg, which equals to a weight of 1.7 kg per square meter.

The hardware installed for adaptation includes: (1) ten sensors that detect and measure live loads; (2) six linear actuators that redistribute resulting forces by shortening and extending the length of structural members by few millimeters; and (3) two microcontrollers for data processing and communication.

Ordinary load cells rated for 50 kg, such as those found in bathroom scales, were chosen as sensors. Sensors are placed within the interface between the top layer and the upper truss nodes. Given that these sensors are typically used in groups of four – forming one complete Wheatstone bridge – bridge completion was performed for each sensor. In addition, HX711 signal amplifiers were implemented.

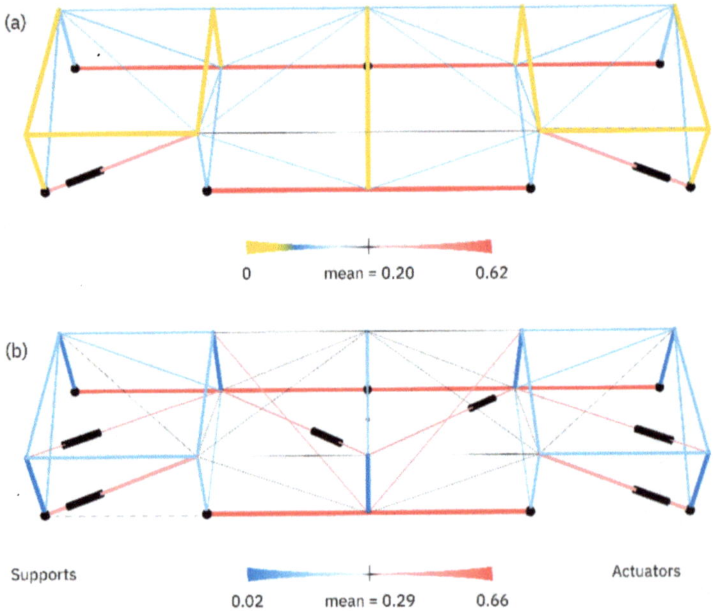

Fig. 3 Adaptability, i.e., redundancy distribution, of the structure (a) before and (b) after the addition of six diagonal elements

For actuation, 'Actuonix P16-P' linear actuators with position-feedback and corresponding motor shields were selected, due to their compact form-factor and a comparatively high payload of 300 N. Actuators were installed in series, meaning that they interrupt the structural element into which they are integrated (Blandini et al., 2022). To avoid damage due to potential element buckling during element-elongation, each actuator is supported by two metal rods that facilitate in-axis movements through sliding connections. A similar detail was used to resolve intersecting cross-bracings in the center of the structure. Both sensors and actuators are controlled using two WiFi-enabled 'Arduino Rev 2' microcontrollers. Figure 4 provides an overview of the hardware used. All cross-sections, dimensions and actuation forces were validated using the digital structural model discussed in Sect. 2.4.

2.3.2 Lamp Element

In line with the topology of the bench element, the lamp consist of 3d-printed joints and linear elements composed of several parts: A flat 12 × 3 mm aluminum profile with addressable 'WS2812B' LED-strips on both sides, and 3d-printed spacer parts at each end, onto which a translucent acrylic tubes are attached (Fig. 5). The resulting LED components are clustered into 17 internally wired units, which, in turn, are powered by three external 5 V power supplies. A single ESP32 microcontroller is used to control the total of 2x1972 pixels, which on average equates to around 2x33 pixels per element.

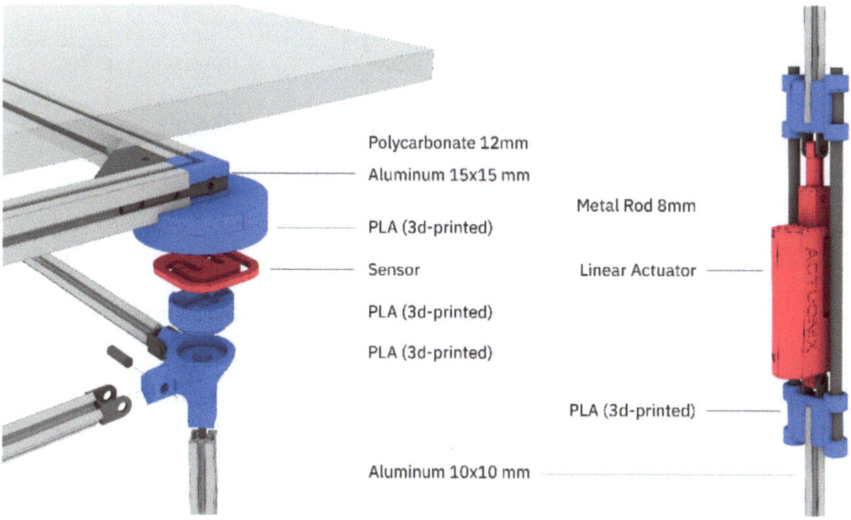

Fig. 4 Bench element hardware overview

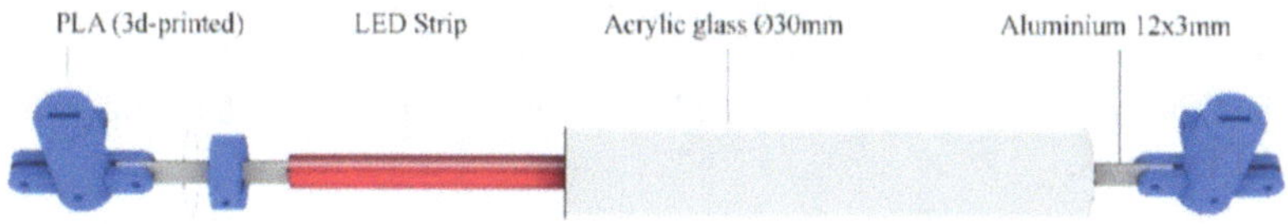

Fig. 5 Lamp element hardware overview

2.4 Control and Communication

Contrary to the D1244, where stress distributions are monitored using strain sensors in the majority of structural members, this research proposes an alternative approach. Rather than relying on exact strain measurements for each element, this work utilizes sensor data solely to quantify live loads at each upper node. Element stresses, in turn, are computed in a digital structural model and assumed to be enough of an approximation to compute required actuator strokes.

The digital structural model is developed using 'Karamba3d' to compute element utilizations for present occupant loads and 'Opossum' (Wortmann, 2017) for their optimization, i.e., adaptation. At the highest level, the digital model is based on one input and two outputs. (Fig. 6).

The input constitutes a list of ten load values (0-50 kg) resulting from the load sensors. These loads are applied as point loads to the structural model. Resulting element utilizations (0–100% per element) are calculated using Karamba3d and converted to a color scheme from dark blue to bright red, which is dispatched to the corresponding LED pixels (RGB). As soon as at least one structural member exceeds the specified utilization limit of 70% the adaptation loop is triggered. The latter represents a single-objective optimization (RBFOpt), the objective being the minimization of the largest

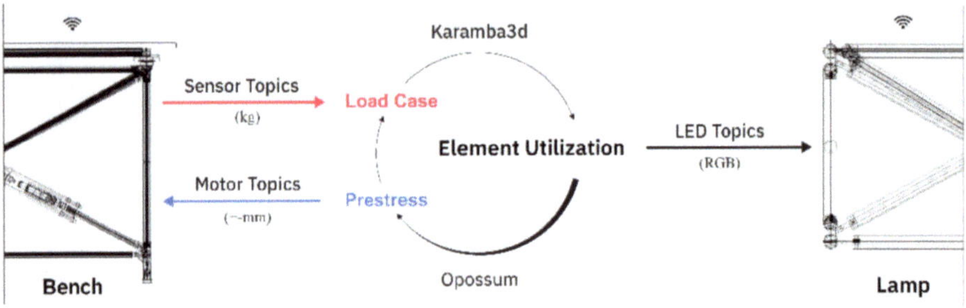

Fig. 6 Control and communication loop underlying the adaptation process

utilization value present, with the variables being the actuator strokes from −0,5 to 0.5 mm, which are added as pre-stress loads to the structural model. By means of this process, a reduction of element utilizations by up to 60% could be achieved.

The exchange of information between the digital and the physical setup was established using the internet-based 'MQTT' protocol, which is commonly used for IoT applications. MQTT operates on the publish-subscribe model, where devices or applications publish and/or subscribe to 'topics' associated to certain types of information. In the case of C1244, the structural model subscribed to the sensor topics (visitor loads) and published both to the motor (actuator strokes) and LED topics (color values); the bench published to sensor topics and subscribed to the motor topics; and the lamp subscribed to the LED topics.

3 Results and Discussion

C1244 was installed within the D1244 during a one-month event dedicated to introducing the general public to new developments in the fields of architecture, engineering and urban planning. Therefore, the majority of visitors were non-experts and had only received a brief introduction to concept of structural adaptation upon arrival. They were then guided to the installation in groups of around 10–15 people and invited to take a seat.

Depending on the amount of occupants and their position on the bench, C1244 would unfold different behaviors, here categorized into three behavioral patterns:

1. In the event of insignificant loading, e.g., only one person, a faint color scheme visualizing low element utilizations appeared and remained static since a redistribution of stresses was not required (Fig. 2).
2. In the event of medium loading, e.g., due to a concentrated load of two to four people in unsupported areas, the initial color scheme would update and visualize the adapted state, allowing visitors to observe how stresses were relocated to underutilized elements (Figs. 6 and 7). This observation was additionally enhanced by audible yet hardly visible actuator movements.
3. In the event of extreme loading, e.g., five or more occupants, initial colors would remain static or change only slightly, since a redistribution of stresses was not possible due to overall high utilization.

Fig. 7 Element utilization before actuation. Dark blue = low utilization, Turquoise = medium utilization, Red = high utilization

According to visitors, all three patterns contributed to the their familiarization with the present concept of adaptation - albeit in different ways. Pattern #2 was best suited to convey the key principle of stress re-distribution in the event of localized stress peaks, pattern #1 and #3 showed that such structures are not dependent on actuation at all times, and retain minimum serviceability in the event of extreme influences (Fig. 8).

Although visitor feedback was generally positive, at times, their experience was affected by the optimization loop that requires around 5 s to converge. For example, with (excited) children moving across the bench swiftly, this delay would result in the display of color schemes for an already outdated loading scenario. Future work will address this shortcoming by substituting the current optimization algorithm with the potentially more appropriate simplex algorithm (Adeli and Kamal, 1986). However, the otherwise successful implementation of the C1244 showcases the general potential of the established workflow to support the development of a meaningful adaptive structure using software and hardware that is readily-available to the architectural community.

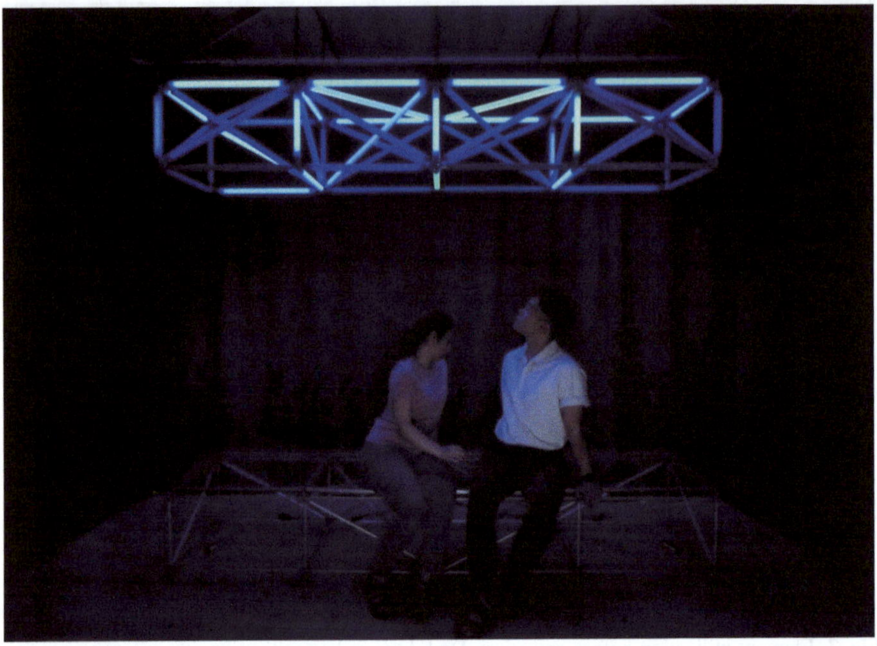

Fig. 8 Element utilization after actuation. Dark blue = low utilization, Turquoise = medium utilization, Red = high utilization.

4 Conclusion

The research presented in this paper marks a first attempt to increase the accessibility of structural adaptation and its potentials in architecture. At a methodological level, this is addressed through the development of a design-to-operation method that supports architect-driven exploration and speculation; and at a conceptual level, through the application of this method in the development of a furniture-scale installation designed to engage non-experts with the intrinsic behaviors of structural adaptation in a safe and playful manner. In doing so, this work contributes to the field of structural adaptation in the sense that it allows to complement previous achievements with the perspective of architectural design. In turn, it contributes to the field of robotic architecture, by introducing potential avenues for future research and collaboration. With respect to the theme of this conference, this work hopes to further showcase the role of robotics beyond fabrication and construction processes. However, this is not to argue for a shift in focus but for a broader conception of robotics in architecture, where machine-intelligence not necessarily "remains external to what is made but becomes fully embedded and absorbed in the system to be constructed" (Menges, 2015).

Acknowledgements. This work was supported by the German Research Foundation (DFG, Deutsche Forschungsgemeinschaft, Project-ID 279064222), as a part of the collaborative research center CRC 1244 (SFB1244) "Adaptive Skins and Structures for the Built Environment of Tomorrow" project A02. In addition, the authors would like to thank their student assistants Shirin

Shevidi and Donghwi Kang; Fabian Eidner and Oliver Moldow for their support in setting up the installation; their colleague Max Zorn for the support with Opossum, Samuel Leder for the support with MQTT communication; as well as Florian Geiger, Jan Gade and Prof. Manfred Bischoff for their contributions to the implementation and interpretation of the redundancy matrix.

References

Adeli, H., Kamal, O.: Efficient optimization of space trusses. Comput. Struct. **24**(3), 501–511 (1986)

Attia, S., Lioure, R., Declaude, Q.: Future trends and main concepts of adaptive facade systems. Energy Sci. Eng. **8**(9), 3255–3272 (2020)

Bier, H., Liu Cheng, A., Mostafavi, S., Anton, A., Bodea, S.: Robotic building as integration of design-to-robotic-production and -operation. Robotic Building, 97–119 (2018)

Blandini, L., et al.: D1244: Design and construction of the first adaptive high-rise experimental building. Front. Built Environ. **8**, 814911 (2022)

Fox, M.: Interactive architecture: adaptive world. Chronicle Books (2016)

Geiger, F., Gade, J., Von Scheven, M., Bischoff, M.: A case study on design and optimization of adaptive civil structures. Front. Built Environ. **6**, 94 (2020)

Kilian, A.: The Flexing Room Architectural Robot: An Actuated Active Bending Robotic Structure using Human Feedback, In: ACADIA - Re/calibration: on imprecision and infidelity (2018)

Korkmaz, S.: A review of active structural control: challenges for engineering informatics. Comput. Struct. **89**, 2113–2132 (2011)

Maierhofer, M., Menges, A.: Methods for integrating architectural design intent into the agent-based design of (adaptive) truss structures. In: Proc. of the 42nd Annual Conference of the Association for Computer Aided Design in Architecture, pp. 318–325 (2022).

Menges, A.: The new cyber-physical making in architecture: Computational construction. Archit. Des. **85**(5), 28–33 (2015)

Neuhaeuser, S., Weickgenannt, M., Witte, C., Haase, W., Sawodny, O., Sobek, W.: Stuttgart SmartShell - A full scale prototype of an adaptive shell structure. J. Int. Assoc. Shell Spat. Struct. **54**(4), 259–270 (2013)

Oosterhuis, K.: Hyperbodies. Springer Science & Business Media (2003)

Ostertag, A., et al.: Reliable design of adaptive load-bearing structures with focus on sustainability. In: Proc. of the 30th European Safety and Reliability Conference and 15th Probabilistic Safety Assessment and Management Conference, pp. 1–5 (2020)

Pask, G.: The Architectural Relevance of Cybernetics. Archit. Des. **7/6** (1969)

Schnädelbach, H.: Adaptive architecture - A conceptual framework. In: Proc. of MediaCity: Interaction of Architecture, Media and Social Phenomena. pp. 522–38 (2010)

Senatore, G., Duffour, P., Winslow, P.: Exploring the application domain of adaptive structures. Eng. Struct. **167**, 608–628 (2018)

Sobek, W.: Ultra-lightweight construction. Int. J. Space Struct. **31**(1), 74–80 (2016)

Sobek, W., Teuffel, P.: Adaptive systems in architecture and structural engineering. Smart Struct. Mater. **4330**, 36–45 (2001)

Steenson, M.W.: Architectural intelligence: How designers and architects created the digital landscape. MIT Press (2022)

Steffen, S., Blandini, L., Sobek, W.: Analysis of the inherent adaptability of basic truss and frame modules by means of an extended method of influence matrices. Eng. Struct. **266**, 114588–114588 (2022)

von Scheven, M., Ramm, E., Bischoff, M.: Quantification of the redundancy distribution in truss and beam structures. Int. J. Solids Struct. **213**, 41–49 (2021)

Wagner, J.L., et al.: On steady-state disturbance compensability for actuator placement in adaptive structures. Automatisierungstechnik. **66**(8), 591–603 (2018)
Werner, L. C.: Gordon Pask and the Origins of Design Cybernetics. Design Cybernetics – Navigating the New, pp. 65–84 (2019)
Wortmann, T.: Opossum - Introducing and Evaluating a Model-based Optimization Tool for Grasshopper. In: Proc. of the 22nd CAADRIA Conference, pp. 283–292 (2017)
Yablonina, M., Ringley, B., Brugnaro, G., Menges, A.: Soft office: A human–robot collaborative system for adaptive spatial configuration. Constr. Robot. **5**(1), 23–33 (2021)
Zuk, W., Clark, R.: Kinetic Architecture. Van Nostrand Reinhold (1970)

Non-Planar 3D Printing of Double Shells

Ioanna Mitropoulou[1(✉)], Amir Vaxman[2], Olga Diamanti[3], and Benjamin Dillenburger[1]

[1] Digital Building Technologies, ETH, Zurich, Switzerland
mitropoulou@arch.ethz.ch
[2] Institute of Perception, Action and Behaviour, The University of Edinburgh, Edinburgh, UK
[3] Institute for Geometry of the Department of Mathematics, TU Graz, Graz, Austria

Abstract. We present a method to fabricate double shell structures printed in transversal directions using multi-axis fused-deposition-modeling (FDM) robotic 3D printing. Shell structures, characterized by lightweight, thin walls, fast buildup, and minimal material usage, find diverse applications in prototyping and architecture for uses such as façade panels, molds for concrete casting, or full-scale pavilions. We leverage an underlying representation of transversal strip networks generated using existing methods and propose a methodology for converting them into printable partitions. Each partition is printed separately and assembled into a double-shell structure. We outline the specifications and workflow that make the printing of each piece and the subsequent assembly process feasible. The versatility and robustness of our method are demonstrated with both digital and fabricated results on surfaces of different scales and geometric complexity.

Keywords: Robotic 3D Printing · Non-planar · Strips · Digital Fabrication · Double Shell · Print Path Design · FDM · Digital Fabrication

1 Introduction

Recent developments in Additive Manufacturing (AM) have opened new possibilities for 3D printing objects with unprecedented geometric complexity. We focus on fused deposition modeling (FDM), whereby a linear molten thermoplastic material is extruded from a hot nozzle into layers that accumulate to produce an object. FDM is relatively inexpensive and flexible (Gao et al. 2015) and has gained significant popularity among industrial and hobbyist practitioners.

The introduction of robotic arms in FDM printing enables the scaling up of the printing process. In larger-scale printing, material deposition often takes the form of a hollow shell, where each layer is realized by a single path without multiple offsets or infill. Shells have the advantage of less material usage and faster buildup, and thus are used for various architectural applications such as façade panels (Sarakinioti et al. 2018; Taseva et al. 2020), bespoke floors (Aectual 2019), and molds for concrete casting (Jipa et al. 2017; Leschok and Dillenburger 2019; Burger et al. 2020).

M. Yablonina et al. (Eds.): Robotic Fabrication in Architecture, Art and Design 2024, pp. 175–191, 2026.
https://doi.org/10.1007/978-3-032-22607-5_12

However, the realization of single-layer FDM shells has some inherent disadvantages. First, the thickness of the shell is constrained by the size of the extrusion nozzle. For FDM printing using filament extrusion, this is typically between 0.5 and 2.5 mm, which can cause instabilities when scaling up. Second, due to the one-dimensional nature of the paths, there is an asymmetry in the quality of the approximation of the surface: the direction parallel to the paths is smoothly approximated, while the direction orthogonal to the paths is only approximated with a staircase of subsequent sections (Fig. 2a). This adversely affects both aesthetics, with a staircase effect in the direction orthogonal to the paths, and structural strength, as the resulting object is anisotropic, with considerably higher strength along the print direction (Ulu et al. 2015). While both the staircase effect and anisotropy are hard to avoid since they are inherent to the printing process, they are properties that can be controlled to improve the strength and quality of the printed object.

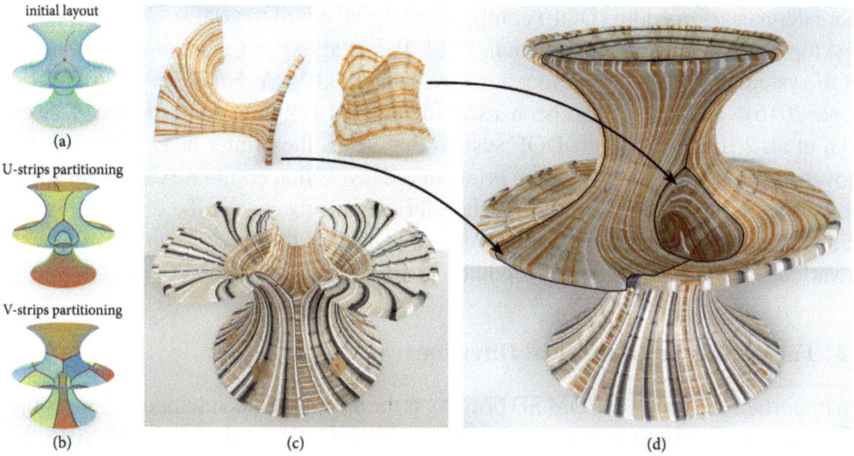

Fig. 1. FDM robotic non-planar 3D printing of double shells along transversal directions. Fabrication of the Costa minimal surface using paths aligned with principal curvature directions. (a) Input Strip-Decomposable Quad (SDQ) mesh. (b) Fabrication-aware partitioning for the *U* (top) and the *V* (bottom) strip networks, each partition is shown with a different color. (c) Printing of individual pieces and assembly. (d) Final prototype.

To that end, we propose the design of double (Figs. 1 and 2c) instead of single (Fig. 2b) shells, with two outer walls printed using non-planar paths in transversal directions and connected with sparse links (ribs) along a hollow interior. This adds structural depth to the resulting object, thus improving stability. It also allows the distribution of imperfections more evenly along different directions on each side of the shell.

2 Related Work

2.1 Non-Planar Slicing and Multi-Axis Printing

Traditional planar 2.5D printing (Gibson et al. 2009) is based on the accumulation of planar layers along a single build direction. Multiple approaches for non-planar printing still follow the layered approach but allowing for curved layers. In perhaps the first work in that direction (Chakraborty et al. 2008), the printer can dynamically change -values within individual layers. Since then, multiple works have pursued the direction of decomposing a shell or volume into curved layers (Ahlers et al. 2019; Allen and Trask 2015; Chen et al. 2019; Etienne et al. 2019; B. Huang and Singamneni 2015; Lim et al. 2016; Llewellyn-Jones et al. 2016; Pelzer and Hopmann 2021). Curved slicing produces non-planar print paths that can greatly reduce support, improve surface quality (in particular, reduce staircase artifacts), and improve mechanical properties.

To facilitate the fabrication of non-planar paths, one can use a robotic arm with more degrees of freedom (DOF) compared to the standard triaxial system of traditional desktop printers (Keating and Oxman 2013; Pan et al. 2014; C. Wu et al. 2017). Multi-DOF systems have been used to print wireframe models (Y. Huang et al. 2016; R. Wu et al. 2016). They have also been used for layered printing; for example, Dai et al. (Dai et al. 2018) use a multi-DOF system that keeps the printer head fixed and freely moves the object during printing to print curved layers that eliminate sacrificial support. However, moving the object during the fabrication process can be challenging when printing larger objects. Moving the print tool instead of the object is an alternative used in various research projects (Mitropoulou et al. 2020; Zhang et al. 2022).

2.2 Fabrication in Transversal Directions

An important challenge of FDM 3D printing is the anisotropic behavior of printed objects due to the weak adhesion between layers along the vertical axis (Ulu et al. 2015; Fang et al. 2020). In response to this challenge, researchers have explored alternative printing paths using transversal directions, such as aligning print directions with stress lines (Tam and Mueller 2017), also integrating carbon fiber (Kwon et al. 2019). Utilizing transversal directions to enhance fabrication isn't confined to just 3D printing. This approach is prevalent in fabrication processes that involve linear elongated segments, where the discontinuous direction has weaker bonding than the continuous one. Examples include crafting surfaces with paper strips (Takezawa et al. 2016) or metal rods (Ma et al. 2020). The key advantage is that when directions intersect, the strengths in one direction compensate for the vulnerabilities in the other, resulting in a more robust final product.

2.3 Strips

Utilizing strips to represent the underlying geometry is common in such fabrication processes because the geometric characteristics of strips match the elongated fabrication units. Strips have been used to design single-curved façade panels (Pottmann et al. 2008), textile ribbons (Schüller et al. 2018) and developable surfaces (Verhoeven et al. 2022).

In the same spirit, (Mitropoulou et al. 2024) design transversal strip layouts overlayed into a Strip-Decomposable Quad (SDQ) mesh (Fig. 2d), considering fabrication-related properties and user-controlled alignment. They propose using SDQ meshes as an intermediate representation to describe non-planar 3D printing paths. Our work follows that paradigm; however, we take advantage of both transversal strip networks to fabricate double instead of single shells printed along transversal directions. Further, we extend this scheme with a partitioning method that guarantees both topological and geometric feasibility for printing.

3 Overview

We begin with a set of two transversal strip networks, U (depicted in blue) and V (depicted in green), combined in an SDQ mesh (Fig. 2d). The novelty of this work consists of considering each strip network to represent the non-planar print paths of each side of a double shell, printed in transversal directions. The challenges we tackle to enable this fabrication mode are the following.

- Partitioning strip networks into printable patches that can be fabricated with our print setup (Sects. 4.1, 4.2).
- Addition of fabrication-enabling details and assembly into one connected double shell object (Sect. 4.3).

3.1 Path Representation

Paths are generated by subdividing N times each strip S_i. Unlike planar printing, here, the print direction T (i.e., the orientation of the robot's print head while printing) and the layer height h are variable. Fig. 2e shows how paths are generated on a strip S_i and the resulting T and h for two points a and b. T is given by the direction of the transversal (green) edge, and h is given by the distance from the neighboring subdivision.

3.2 Printing Standing Shells

Next, we describe our non-planar 3D printing setup of standing shells (Fig. 3, top), i.e., shells where only the first path lies on sacrificial support, and all subsequent paths lie on previously deposited paths. The robotic arm orients the extruder (i.e., the print tool) towards the print direction T at each position and prints sequentially on a sacrificial base, from first to last path, in an uninterrupted motion. This sequential logic requires a print sequence, where each path being printed lies fully on previously deposited paths. Since the layer height h—i.e., the distance between consecutive layers—varies, the extruder updates the material flow rate at each print position to match the quantity of material necessary for filling the current layer height. On the other hand, the layer width w is considered constant throughout the process.

The robot's degrees of freedom provide considerable flexibility for reorienting the tool around the object and depositing material in various orientations. However, the range of possible orientations for material deposition is limited. During the buildup of paths, print angles that are unfavorable for material deposition can appear (Fig. 3, bottom

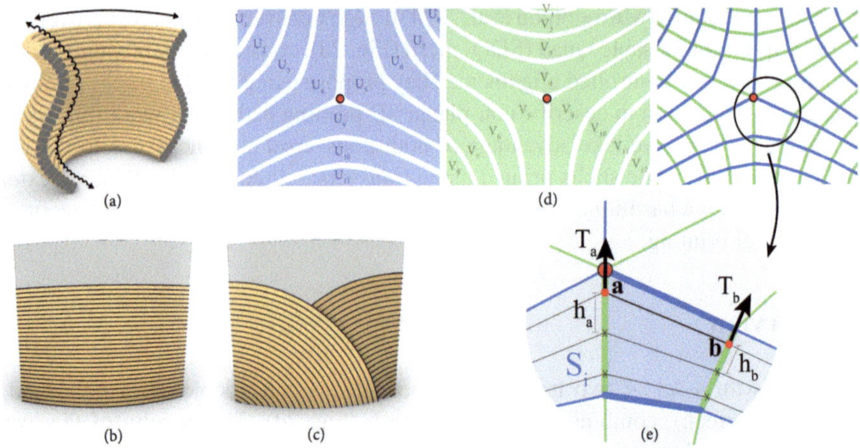

Fig. 2. (a) In 3D printed surfaces, one direction is smoothly approximated, while the other is approximated with a staircase of subsequent sections. (b, c) Diagram of a single and a double shell with each side printed in transversal directions. (d) Two transversal strip networks, U (left) and V (middle), overlayed into one SDQ mesh (right), illustraton from (Mitropoulou et al. 2024). (e) Paths generation by subdividing a strip S_i showing the layer height h and print direction T for two points a, b.

right). For that, we set a maximum print angle variation γ that should not be exceeded (for our setup $\gamma = \pi/2$). In addition, the robot cannot print outside of its reachable space; we define a bounding box of reachable space (for our setup, it has dimensions 50 x 50 x 50 cm), and all print geometry must lie within it.

Partitioning Constraints

Considering the characteristics of the process outlined above, we propose a partitioning scheme of strips that considers the following constraints.

1. *Topologically feasible.* All patches must have a topologically simple sequence of paths so that there is an unambiguous print sequence. This should consist of a single non-branching sequence of curves that layer upon each other so that each path being printed lies on a previously deposited path.
2. *Geometrically feasible.* For each patch, the dimensions must fit within the bounding box, and the angle variation should not exceed the maximum print angle variation.
3. *Non-overlapping seams.* The partitioning cuts on the two strip networks U and V must not overlap so that upon assembly, the two shells form one connected object.

3.3 Workflow

Our workflow proceeds as follows.

- Begin with an SDQ mesh.
- Carry out **topological partitioning** that separates the strip networks into simply-connected patches (Sect. 4.1).

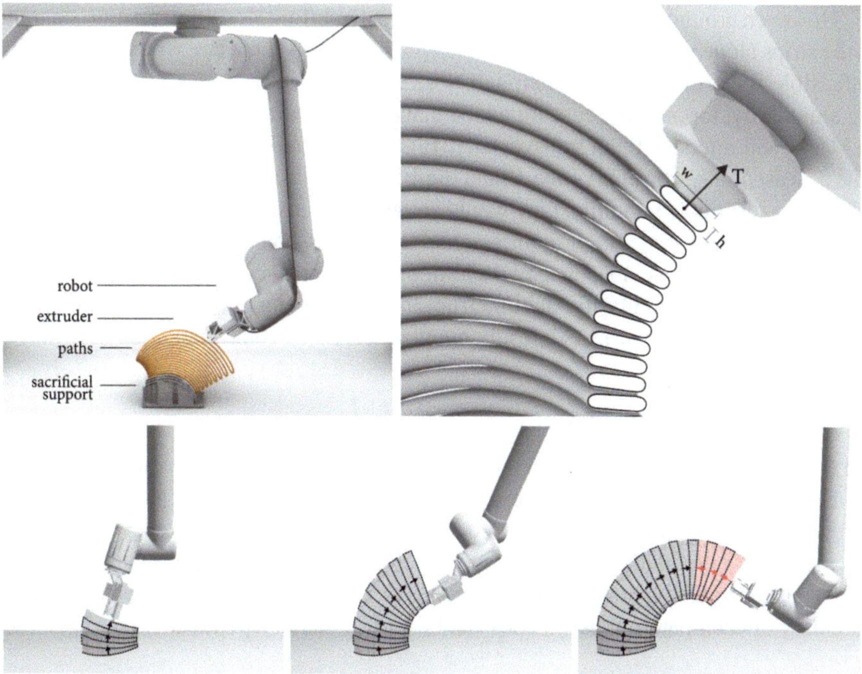

Fig. 3. Top: Overview of a standing shell's robotic 3D printing process. Bottom: Print angles during the buildup of paths can become unfeasible (right, in red) if the print direction varies a lot.

- Carry out **geometric partitioning** into patches that respect the angle and size bounds (Sect. 4.2).
- Incorporate **assembly-related details** and compute fabrication data (Sect. 4.3).
- Print and assemble the individual pieces to obtain the final output (Sect. 5).

4 Method

4.1 Topological Partitioning

Topological partitioning (Fig. 4) aims to separate the strip networks U and V into simply-connected patches so that each patch has an unambiguous ordering (up to orientation) and can be printed as a standing shell.

In (Mitropoulou et al. 2024), a partitioning scheme is detailed that cuts the strip network along the separatrices of the singularities and separates handles, thus splitting a strip network into simply-connected parts, as shown in Fig. 4a and 4d. These cuts continue until they form closed loops or terminate on open boundaries. The difference here is that we aim to partition both U and V (i.e., both Fig. 4a and 4b, similarly both Fig. 4d and 4e), and we have the additional constraint that the cuts of the two networks should not overlap. As long as the partitioning is parallel to the direction of strips (i.e., no strip is cut perpendicularly), then the cuts of U and V are non-overlapping, as they are applied in transversal directions, for example, see the partitioning of the D6 singularity (Fig. 4a, b, c).

However, around the D2 singularity (Fig. 4d, e, f), cutting along the separatrix parallel to the strips alone would not separate the self-incident strip at the cut, which creates paths unfeasible for printing due to their self-obstructing geometry. As a result, around the D2 singularity, topological cuts are added that are both parallel to the strips (Fig. 4d, e, in black) and intersect them transversally (Fig. 4d, e, in red). This causes an overlap of cuts as, essentially, the cuts of the U and the V network are identical. If those cuts were allowed to terminate naturally, they would create an extensive sequence of edges where both sides share a cut. Then, upon assembly, the final object would be disconnected there. To avoid this, we introduce two additional transversal cuts C_U and C_V in distance d_q edges from the singularity, that interrupt the overlapping cut (the resulting reduced overlap is marked in cyan). As a result, the pieces are still connected upon assembly.

4.2 Geometric Partitioning

We next partition the patches further to adhere to the sizing and angle constraints (Fig. 5). To avoid overlapping cuts in this step, we keep track of the edges that participate in cuts on each network and disallow adding overlapping cuts. Since every edge sequence of the mesh can potentially become a cut, there are always more than enough edges to use as cuts despite this restriction. The geometric partitioning cuts are interrupted at the boundary of each patch; thus, patches can be partitioned independently.

Partitioning to Fit Size Bounds.
In each patch, we find the two transversal strips S_U and S_V with the largest area. In addition, we compute the maximum extent of each patch using Principal Component Analysis (PCA). Each resulting principal direction is best aligned with either S_U or S_V. If a direction exceeds the maximum size, we partition the according strip by adding the appropriate number of equidistant cuts (in number of quads), ignoring edges that would cause overlap. The resulting partitions are checked again, and the process is repeated until no piece exceeds the set bounds. Figure 5 illustrates the partitioning of the U and V networks of a doubly-curved surface that is, by design, larger than the bounding box.

Partitioning to Fit Angle Bounds.
While the robot is printing with the extruder oriented in the print direction T, its orientation change should not exceed a certain threshold γ. Note that the angle with which the robot is printing depends on the piece's orientation on the build platform. This is why we are not examining angles with the gravity direction (these are defined later when the piece's print orientation has been decided) but rather angle variations, which are independent of the piece's orientation on the print platform.

Assume we are working on a patch of U strips; its U-strips (blue) lie along the paths' direction, and its V-strips (green) are transversal to it. As a result, the total angle variation $A = \sum \mid \alpha_{V_i} \mid$ (the sum of absolute angle differences of neighboring T) within V-strips measures the variation during the buildup of the piece. We consider the V-strip with the greatest angle variation A_{max}. If $A_{max} > \gamma$, then we cut the strip with $k = ceil(A_{max}/\gamma)$ equidistant (in number of quads) cuts, again ignoring edges that would cause overlap. All resulting patches are rechecked, and the process is repeated until no patches have V strips that exceed the A_{max}. We show the process in Fig. 5, using different angle thresholds γ.

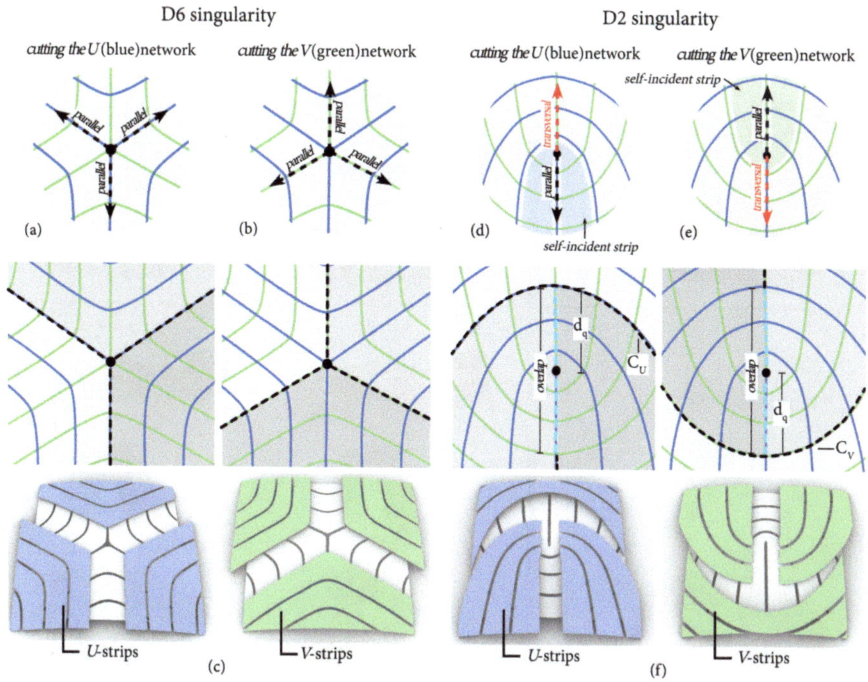

Fig. 4. Topological partitioning of the two singularities most commonly found in SDQ meshes for both U and V strips. The strips' outline is displayed as smooth for easier comprehension. (a, b, c) D6 singularity partitioning along separatrices. (d, e, f) D2 singularity partitioning along separatrices. The cyan line illustrates the overlapping cut on the two sides, bounded by the transversal cuts C_U, C_V in distance d_q edges from the singularity.

4.3 From Digital to Physical

In the following, we provide the steps for preparing the digital double shell model to be ready for printing.

Surface Thickening – Offsetting

The double-shell surface consists of two shells printed along transversal directions and a hollow interior with sparse ribs connecting the shells. The total thickness t of the shell can be determined by the user, where in most prototypes, we use the default value $t = 4n$ (Fig. 6a), where n is the nozzle diameter (2.5 mm in our setup). To account for the thickness of the shells, we offset the original surface on each side with default distance $t/2 - 0.5n = 1.5n$ in the direction of the normals of the mesh, which we compute for each quad by triangulating it and finding the average normal of its triangles. This offsetting scheme is adequate for our application since our shapes have low surface detail, and the quad mesh has sufficiently high resolution.

Managing Inaccuracies During Assembly

The assembly of the printed pieces might be hampered by small inaccuracies present in

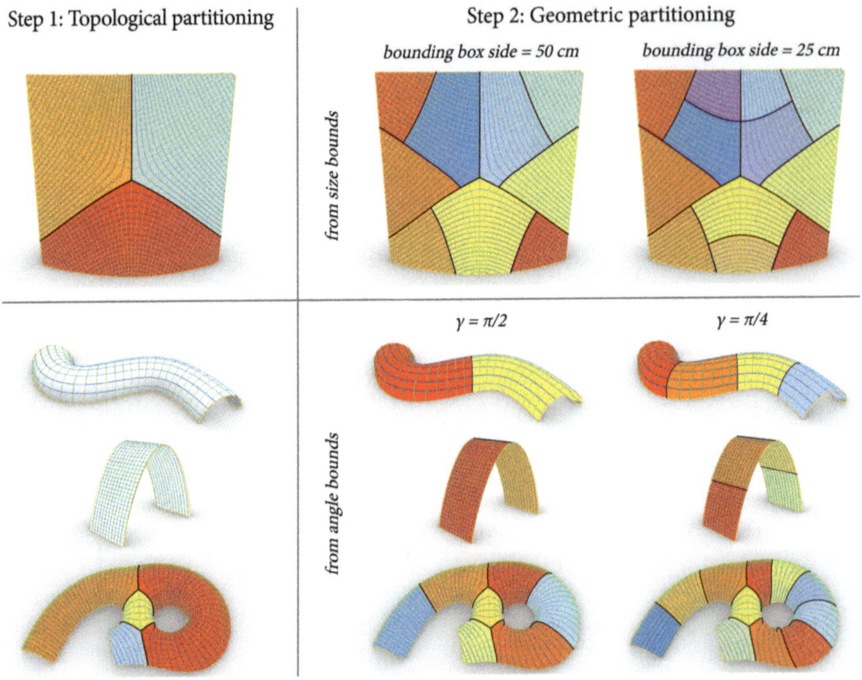

Fig. 5. Partitioning steps: first, topological partitioning (left), and second, geometric partitioning (right). Top: Partitioning a doubly curved surface from size bounds using two different bounding box sizes. Bottom: Partitioning various surfaces from angle bounds using two different γ thresholds. In both cases, only the U-strips partitioning is displayed.

the prints. They originate either from inevitable process imperfections or minor deformations of the pieces during and after the print due to uneven shrinkage while cooling. To account for them, we strategically introduce tolerances by removing a single strip on each boundary between patches. We do this only on one side (by default, the V side, Fig. 6b) to avoid creating holes that would otherwise appear at the intersection of gaps. For small inaccuracies, this strategy is sufficient to compensate for errors on both sides of the shell.

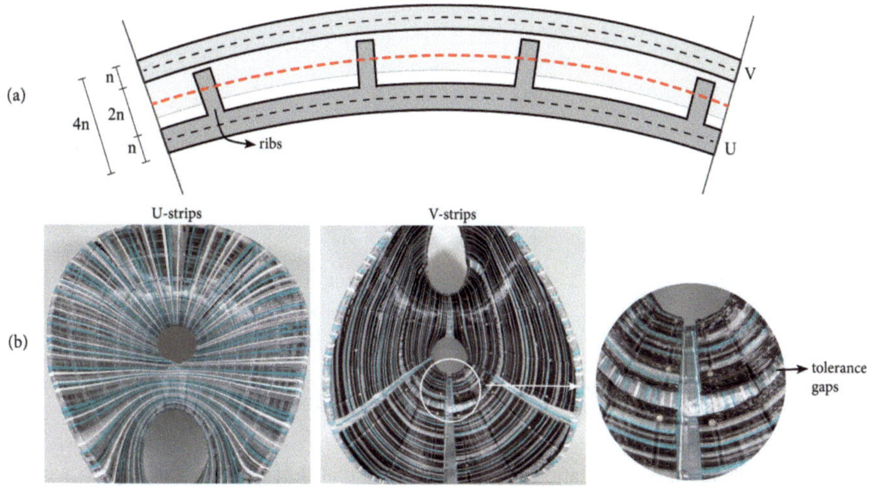

Fig. 6. (a) Offsetting the original surface (in red dotted line) on both sides to create a double shell of thickness $4n$. (b) Accounting for tolerances by creating one side with gaps (right, V-strips).

Orienting for Fabrication

To decide on the global print orientation of a piece, we calculate the average $\vec{h}$ of all directions T (Fig. 7a, left, in black). We also calculate the average $\vec{m}$ of the first strip's T directions (Fig. 7a, left, in red). Then, the piece's global orientation is calculated so that the vector $0.5\left(\vec{h} + \vec{m}\right)$ is aligned with the vertical $\vec{z}$ direction (Fig. 7a, right).

Sacrificial Support

Once the pieces have been oriented for fabrication, a supportive base is created upon which the piece can be printed. It consists of a scaffolding, a lighter lower part produced by a liner hatch to save time and material, and a platform, an upper part, which connects to the printed object and consists of layered offsets of the first path of the patch (Fig. 7b).

Ribs and Connections

Along each wall, we print rigidifying ribs, which help with the overall stability and link the two sides of the shell upon assembly. The ribs follow the direction transversal to the one we are printing and have a default distance of 4 strips (chosen for aesthetics). Similarly to the print paths, the ribs intersect transversely on the two sides of the shell. To create an interlocking connection between the two sides, we interrupt the ribs of one side on the intersections with the ribs from the other side (Fig. 7 c, d). This provides a guide for positioning the pieces during assembly while allowing minor adjustments by sliding, as we leave an additional tolerance gap between the ribs at their intersections. The pieces are kept in place using screws on rib intersections (visible in Fig. 6b right), which are drilled manually. The assembly process (displayed in Fig. 9 for prototypes d and e) consists of assembling one piece from each side in an alternating pattern at every step so that each new piece is half attached to the existing structure.

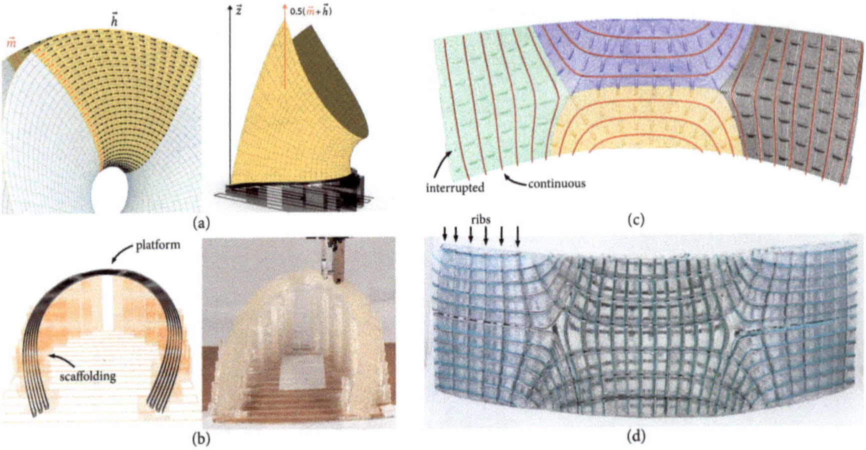

Fig. 7. (a) Calculation of orientation for fabrication so that $0.5\left(\overrightarrow{m} + \overrightarrow{h}\right)$ is aligned with the vertical direction $\overrightarrow{z}$, where $\overrightarrow{h}$ is the average of all directions T (vectors in black), and $\overrightarrow{m}$ is the average of the first strip's T directions (vectors in red). (b) Sacrificial base diagram (left) and printing (right). (c) Rib interruptions for interlocking the two sides of the shell. The continuous ribs of one wall are denoted with red lines. The ribs on the other side (colored per patch) are interrupted to create an interlocking. (d) Printed prototype.

Printing Process

Our printing setup consists of a universal robot UR10 mounted overhead to gain better access to all sides of the printed object (Fig. 8). The plastic extruder tool has a nozzle with a diameter of 2.5 mm and is controlled by an Arduino Mega board. The printing process is carried out at a constant linear robot speed at the extrusion point (15 mm/s while printing the object and 23 mm/s while printing support). We vary the extrusion rate to match the volume of material required to achieve the desired path thickness at every printing position. The change of filament color is carried out manually without stopping the printing process, and its purpose is solely aesthetic, to emphasize the path orientations and differentiate the two shells visually.

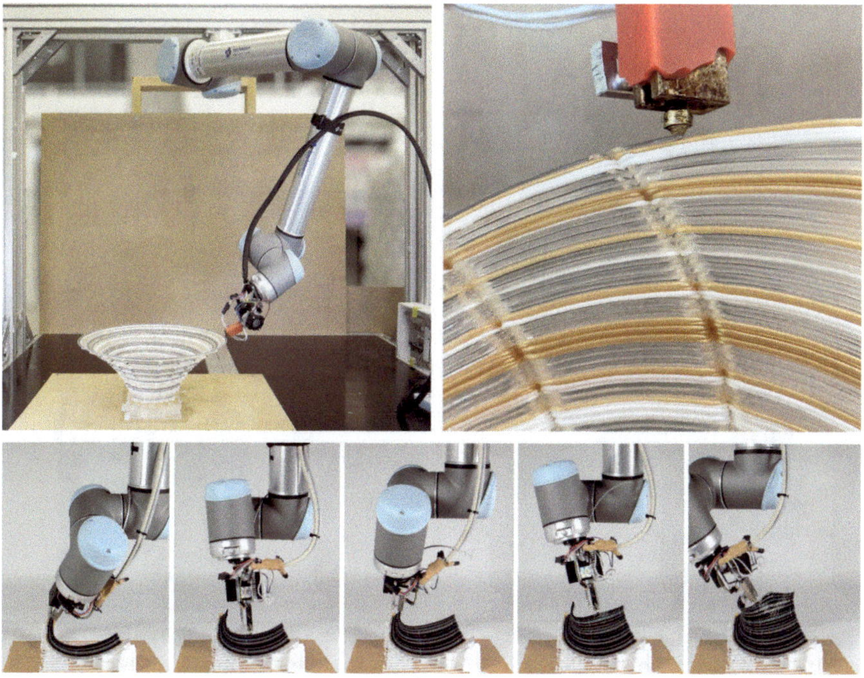

Fig. 8. Top: Robotic setup for fabrication. Bottom: timelapse of printing, total time 35 min.

5 Results

We present a variety of printed prototypes created with the described methods. Figure 9 presents the Costa minimal surface with the full pipeline from the initial input SDQ mesh to the final fabricated result. The shape is partitioned into 7 pieces on one side and 11 on the other. The prototypes shown in Fig. 9a, b, c are created with an SDQ mesh aligned to user-drawn directional constraints. The curved surfaces (Fig. 9a, b) are fabricated with 3 pieces on each side, while the elongated curved surface (Fig. 9c) is partitioned into 5 and 6 pieces. The Chen Gackstatter minimal surface (Fig. 9d) is generated using boundary-aligned constraints; the paths of one side are constrained to be parallel, and the paths of the other side to be orthogonal to the boundary. It is segmented into 5 and 16 pieces. Finally, the Batwing minimal surface (Fig. 9e) is generated with curvature-aligned directional constraints and is segmented into 9 and 12 pieces. Table 1 presents fabrication details for all the prototypes.

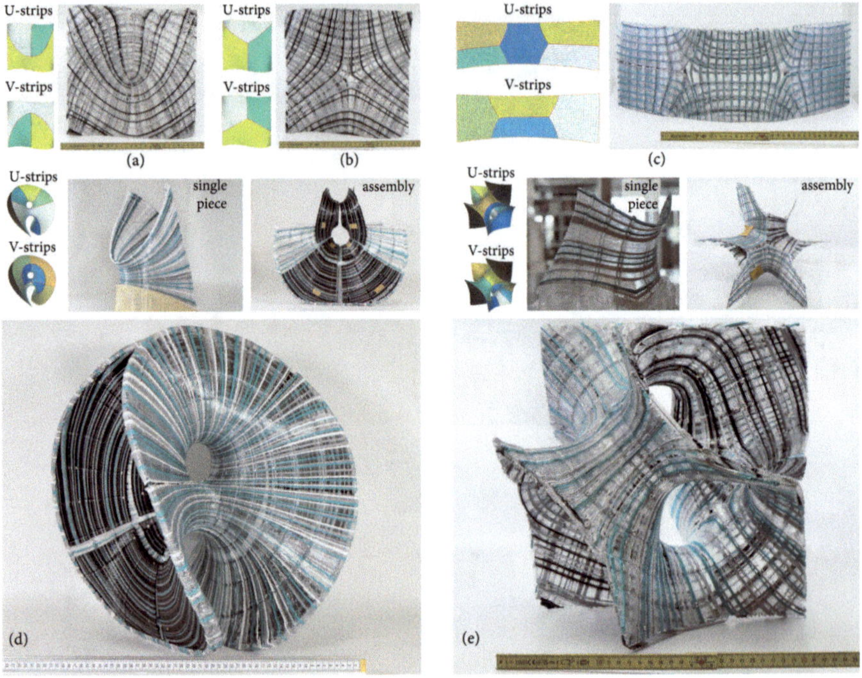

Fig. 9. (a, b) Curved surfaces with paths aligned to two different user-drawn directions. (c) Elongated curved surface with paths aligned to user-drawn directions. (d) Chen Gackstatter minimal surface with paths aligned to boundary directions. (e) Batwing minimal surface with paths aligned to curvature directions. More details on the prototypes are displayed in Table 1.

Table 1. Details of prototypes presented in Figs. 1 and 9.

model	Dimensions (cm)	#sings	#geometric cuts	# pieces ***U***	# pieces ***V***	Total print time (hrs)	% sacrificial support
Fig. 1	60 x 60 x 51	4	12	7	11	51.5	49%
Fig. 9a	30 x 6 x 30	1	0	3	3	9.6	52%
Fig. 9b	30 x 6 x 30	1	0	3	3	8.9	48%
Fig. 9c	55 x 17 x 10	1	0	5	6	8.1	54%
Fig. 9d	50 x 50 x 60	4	8	5	16	46	42%
Fig. 9e	30 x 30 x 30	7	4	9	12	16.8	28%

6 Conclusion

Our pipeline successfully produces quality print paths for various shapes. We note some limitations and challenges.

Challenges of double shells. Despite the merits of printing double shells in transversal directions, this fabrication mode also has notable challenges. Printing a double shell implies that the surface is essentially printed twice, which leads to more pieces, more sacrificial support for their base, and thus longer fabrication times. Further advancing the partitioning method to consider the material efficiency of support structures can help reduce the waste material. Also, the printing in transversal directions takes place in separate pieces that are held together with sparse point links of small surface area (screws). As a result, the connection of transversal directions is significantly weaker compared to other approaches that print both directions directly on each other (Tam and Mueller 2017; Kwon et al. 2019).

Assembly. A well-known challenge in FDM, which our work is also susceptible to, occurs in the assembly of the printed paths. Minor inaccuracies caused by imperfections of printing, such as over-extruding or deformations of the pieces during cool-down, may cause them not to fit well together. When printing double shells, these can be especially problematic as the interfaces between pieces have a very large area; essentially, the entire pieces are interfaces, meaning that any imperfection can impair the assembly process. Future work into designing smarter part connections, such as puzzle joints or joints that take advantage of the double-shell thickness of the components, could prove very beneficial to the final quality of the assembled 3D-printed object.

Offsetting. A topic worth further study is how to offset shells optimally; our normal-based offsetting only offers an approximate solution and may need to be adapted for more intricate shapes.

6.1 Outlook

This work opens the door to a variety of promising future applications.

From multi-color to multi-material printing. We use colors to differentiate the curved paths visually. Such changes in the extrusion's content can have significant future applications. Printing different materials, instead of just different colors, can create functional material gradients that serve specific performance goals. For instance, fiber-reinforced filament can be used strategically to improve the resulting piece's structural properties. The development of specialized hardware that feeds the correct material at the correct times during the print would contribute to such controlled functional material changes.

Applications in architecture. 3D-printed shells in architecture find various applications but are often limited by their low thickness and anisotropy, which can be effectively mitigated through the use of double shells printed using non-planar layers. Printing each shell side separately and assembling them afterward is beneficial for embedding functional equipment like tubes or wires and easily accessing them for repairs. Also, it facilitates the incorporation of reinforcement between the two shells, enhancing the overall functionality and durability of the shell structure. Further, add-on printing is a key application of non-planar shell printing, enabling direct material deposition onto existing elements to improve efficiency, promote reuse, and facilitate repair or reinforcement. Structural optimization is another important application. By exploiting the anisotropic properties of 3D printed materials, objects can be optimized for specific load cases. Printing each side of the shell with a different orientation allows further control over the anisotropic properties, leading to additional mechanical improvements.

6.2 Reflection

Our research on the non-planar 3D printing of double shells using multi-axis FDM resonates with the conference's theme, "Beyond Optimization," challenging traditional approaches in robotic fabrication. We address this theme through two key aspects:

Embracing real-world complexity in assembly tasks. An important part of our assembly method is the acknowledgment and proactive management of inaccuracies. Unlike traditional approaches that strive for hyper-accurate results—often a challenging and resource-intensive endeavor—our method anticipates and accommodates inaccuracies by incorporating gaps in the seams for handling tolerances. In the real-world conditions of fabrication, absolute precision is often unattainable. This acknowledgment of fabrication imperfections demonstrates a shift from pursuing unfeasible precision to developing robust and adaptable systems that do not fail when unavoidable imperfections occur.

Broadening the scope beyond the technical. Our research presents a novel way for printing double shells using non-planar paths. Beyond functional efficiency, this control opens avenues for artistic expression and aesthetic enhancement. By manipulating print paths, we unlock new design possibilities, creating textures and forms that enhance both the visual and tactile qualities of the final product, which can be particularly valuable in artistic applications.

References

Aectual. 2019. Pattern Terrazzo Floors. Accessed February 8, 2024. https://www.aectual.com/architectural-products/flooring/variants/pattern-terrazzo/intro.

Ahlers, D., Wasserfall, F., Hendrich, N., Zhang, J.: 3D printing of non-planar layers for smooth surface generation. In: In 2019 IEEE 15th International Conference on Automation Science and Engineering (CASE), pp. 1737–1743. IEEE Press (2019). https://doi.org/10.1109/COASE.2019.8843116

Allen, R.J.A., Trask, R.S.: An experimental demonstration of effective curved layer fused filament fabrication utilising a parallel deposition robot. Addit. Manuf. **8**, 78–87 (2015). https://doi.org/10.1016/j.addma.2015.09.001

Burger, J., et al.: Eggshell: Ultra-thin three-dimensional printed formwork for concrete structures. 3D Print. Addit. Manuf. **7**(2), 48–59 (2020). https://doi.org/10.1089/3dp.2019.0197

Chakraborty, D., Reddy, B.A., Choudhury, A.R.: Extruder path generation for curved layer fused deposition modeling. Comput. Aided Des. **40**(2), 235–243 (2008). https://doi.org/10.1016/j.cad.2007.10.014

Chen, L., Chung, M.-F., Tian, Y., Joneja, A., Tang, K.: Variable-depth curved layer fused deposition modeling of thin-shells. Robot. Comput. Integr. Manuf. **57**, 422–434 (2019). https://doi.org/10.1016/j.rcim.2018.12.016

Dai, C., Wang, C.C.L., Wu, C., Lefebvre, S., Fang, G., Liu, Y.-J.: Support-free volume printing by multi-axis motion. ACM Trans. Graph. **37**(4), 1–14 (2018). https://doi.org/10.1145/3197517.3201342

Etienne, J., et al.: CurviSlicer: Slightly curved slicing for 3-Axis printers. ACM Trans. Graph. **38**(4), 1–11 (2019). https://doi.org/10.1145/3306346.3323022

Fang, G., Zhang, T., Zhong, S., Chen, X., Zhong, Z., Wang, C.C.L.: Reinforced FDM: Multi-axis filament alignment with controlled anisotropic strength. ACM Trans. Graph. **39**(6), 1–15 (2020). https://doi.org/10.1145/3414685.3417834

Gao, W., et al.: The status, challenges, and future of additive manufacturing in engineering. Comput. Aided Des. **69**(August), 65–89 (2015). https://doi.org/10.1016/j.cad.2015.04.001

Gibson, I., Rosen, D.W., Stucker, B.: Additive manufacturing technologies: Rapid prototyping to direct digital manufacturing, 1st edn. Springer Publishing Company, Incorporated (2009)

Huang, B., Singamneni, S.: A mixed-layer approach combining both flat and curved layer slicing for fused deposition modelling. Proc. Inst. Mech. Eng. B J. Eng. Manuf. **229**(12), 2238–2249 (2015). https://doi.org/10.1177/0954405414551076

Huang, Y., et al.: FrameFab: Robotic fabrication of frame shapes. ACM Trans. Graph. **35**(6), 1–11 (2016). https://doi.org/10.1145/2980179.2982401

Jipa, A., Bernhard, M., Dillenburger, B., Ruffray, N., Wangler, T., Flatt, R.: SkelETHon formwork 3D printed plastic formwork for load-bearing concrete structures. Blucher Des. Proc. **3**(12), 345–352 (2017). https://doi.org/10.5151/sigradi2017-054

Keating, S., Oxman, N.: Compound fabrication: A multi-functional robotic platform for digital design and fabrication. Robot. Comput. Integr. Manuf. **29**(6), 439–448 (2013). https://doi.org/10.1016/j.rcim.2013.05.001

Kwon, H., Eichenhofer, M., Kyttas, T., Dillenburger, B.: Digital composites: robotic 3d printing of continuous carbon fiber-reinforced plastics for functionally-graded building components. In: Willmann, J., Block, P., Hutter, M., Byrne, K., Schork, T. (eds.) Robotic Fabrication in Architecture, Art and Design 2018, pp. 363–376. Springer International Publishing, Cham (2019)

Leschok, M., Dillenburger, B.: Dissolvable 3DP formwork. Water-dissolvable 3D printed thin-shell formwork for complex concrete components. In: Bieg, K., Briscoe, D., Odom, C. (eds.) ACADIA 19: Ubiquity and Autonomy. Proceedings of the 39th Annual Conference of the Association for Computer Aided Design in Architecture, pp. 188–197. Association for Computer Aided Design in Architecture (ACADIA) (2019)

Lim, S., Buswell, R.A., Valentine, P.J., Piker, D., Austin, S.A., De Kestelier, X.: Modelling curved-layered printing paths for fabricating large-scale construction components. Addit. Manuf. **12**, 216–230 (2016). https://doi.org/10.1016/j.addma.2016.06.004

Llewellyn-Jones, T., Allen, R., Trask, R.: Curved layer fused filament fabrication using automated toolpath generation. 3D Print. Addit. Manuf. **3**(4), 236–243 (2016). https://doi.org/10.1089/3dp.2016.0033

Ma, Z., et al.: Designing robotically-constructed metal frame structures. Comput. Graph. Forum. **39**(2), 411–422 (2020). https://doi.org/10.1111/cgf.13940

Mitropoulou, I., Bernhard, M., Dillenburger, B.: Print paths key-framing: Design for non-planar layered robotic FDM printing. In: Symposium on Computational Fabrication. SCF '20. Association for Computing Machinery, New York, NY, USA (2020). https://doi.org/10.1145/3424630.3425408

Mitropoulou, I., Vaxman, A., Diamanti, O., Dillenburger, B.: Fabrication-Aware Strip-Decomposable Quadrilateral Meshes. Computer-Aided Design. **168**, 103666 (2024). https://doi.org/10.1016/j.cad.2023.103666

Pan, Y., Zhou, C., Chen, Y., Partanen, J.: Multitool and multi-axis computer numerically controlled accumulation for fabricating conformal features on curved surfaces. J. Manuf. Sci. Eng. **136**(3), 031007 (2014). https://doi.org/10.1115/1.4026898

Pelzer, L., Hopmann, C.: Additive manufacturing of non-planar layers with variable layer height. Addit. Manuf. **37**, 101697 (2021). https://doi.org/10.1016/j.addma.2020.101697

Pottmann, H., et al.: Freeform surfaces from single curved panels. ACM Trans. Graph. **27**(3), 1–10 (2008). https://doi.org/10.1145/1360612.1360675

Sarakinioti, M.-V., et al.: Development and prototyping of an integrated 3D-printed façade for thermal regulation in complex geometries. J. Facade Des. Eng. **6**(2), 29–40 (2018). https://doi.org/10.7480/jfde.2018.2.2081

Schüller, C., Poranne, R., Sorkine-Hornung, O.: Shape representation by zippables. ACM Trans. Graph. **37**(4), 1–13 (2018). https://doi.org/10.1145/3197517.3201347

Takezawa, M., Imai, T., Shida, K., Maekawa, T.: Fabrication of freeform objects by principal strips. ACM Trans. Graph. **35**(6), 1–12 (2016). https://doi.org/10.1145/2980179.2982406

Tam, K.-M.M., Mueller, C.T.: Additive manufacturing along principal stress lines. 3D Print. Addit. Manuf. **4**(2), 63–81 (2017). https://doi.org/10.1089/3dp.2017.0001

Taseva, Y., Eftekhar, N., Kwon, H., Leschok, M., Dillenburger, B.: Large-scale 3D printing for functionally-graded facade. In: Anthropocene, Proceedings of the 25th International Conference of the Association for Computer-Aided Architectural Design Research in Asia (CAADRIA) 2020, vol. 1, pp. 183–192 (2020)

Ulu, E., Korkmaz, E., Kubilay Yay, O., Ozdoganlar, B., Kara, L.B.: Enhancing the structural performance of additively manufactured objects through build orientation optimization. J. Mech. Des. **137**(11), 111410 (2015). https://doi.org/10.1115/1.4030998

Verhoeven, F., Vaxman, A., Hoffmann, T., Sorkine-Hornung, O.: Dev2PQ: Planar quadrilateral strip remeshing of developable surfaces. ACM Trans. Graph. **41**(3), 1–18 (2022). https://doi.org/10.1145/3510002

Wu, C., Dai, C., Fang, G., Liu, Y.-J., Wang, C.C.L.: RoboFDM: A robotic system for support-free fabrication using FDM. In: In 2017 IEEE International Conference on Robotics and Automation (ICRA), pp. 1175–1180 (2017). https://doi.org/10.1109/ICRA.2017.7989140

Wu, R., Peng, H., Guimbretière, F., Marschner, S.: Printing arbitrary meshes with a 5DOF wire-frame printer. ACM Trans. Graph. **35**(4), 1–9 (2016). https://doi.org/10.1145/2897824.2925966

Zhang, T., et al.: S3-Slicer: A general slicing framework for multi-axis 3D printing. ACM Trans. Graph. **41**(6), 1–15 (2022). https://doi.org/10.1145/3550454.3555516

Polymers, Fabric and Steel: A Hybrid Robotic Fabrication System for Bespoke Doubly-Curved Structural Concrete

Jon Krähling Engholt[1(✉)] and Dave Pigram[2]

[1] Aarhus School of Architecture, Aarhus, Denmark
jonengholt@gmail.com
[2] University of Technology Sydney, Sydney, Australia

Abstract. This paper describes an interwoven material, fabrication, and computational design system for bespoke structural concrete. The system combines robotically 3D-printed polymers, fabric-formwork, and robotically bent steel rebar reinforcement. The ecological imperative to reduce the embodied carbon associated with concrete use is well known. This system contributes towards this imperative in two ways. First, by enabling the cost-efficient production of doubly-curved forms with variable thickness and the optional inclusion of holes, the system opens the possibility of significant structural optimization and, therefore, minimization of the concrete itself. Second, by drastically reducing the material investment and waste associated with traditional methods of producing doubly-curved formwork. These qualities are achieved by employing the inherent structural capacity of bespoke steel reinforcement to support and define the form of fabric shuttering — reducing or eliminating the need for alternate shuttering support. Beyond optimization, the fabrication technique allows for the production of architectural elements with unique, compelling, and diverse characters, expanding the formal possibilities of concrete in architecture. Further, the method exemplifies an approach to design that celebrates the imprints of making, the artifacts of production, and outcomes with qualities that exceed those in the digital models that preceded them. The paper follows an inquiry into a robot-based design- and fabrication system that renders aesthetically pleasing reinforced concrete forms with less material use - both in formwork and cast structure. The project follows an explorative trajectory through iterations of physical experiments that produce a range of small prototypes and techniques. As the system matures, the project focuses on realizing a 1:1 three-component structural wall prototype demonstrating the intertwined techniques' potential.

Keywords: Fabric formwork · Digital fabrication · Robotics · Concrete · Steel

1 Introduction

Despite being the most consumed material globally (U.S. Geological Survey 2018), concrete is relatively rarely articulated as anything more than a structural means. Whether cast in situ or erected from precast elements, concrete structures often find themselves

M. Yablonina et al. (Eds.): Robotic Fabrication in Architecture, Art and Design 2024, pp. 193–209, 2026.
https://doi.org/10.1007/978-3-032-22607-5_13

hidden behind facades of more noble materials - or delegated to serve as infrastructure. The constraints of formwork construction additionally lead the vast majority of structures to contain excessive material. However, as the scarcity and origins of natural resources become an ever-growing part of the architectural agenda, it is pertinent to question concrete-based building practices.

Flexible formwork is a cost-efficient technique with immense formal potential; however, this latent potential simultaneously addresses challenges in - and notions of - control. These challenges can unfold by the employment of digital fabrication. Simultaneously, the precision and delicacy of digital fabrication techniques might bring forth material qualities and elevate the experience of a material often perceived as primitive and rough.

Flexible formwork approaches typically introduce a range of complexities, including reliance on patterning and creating joints via sewing. While double-curvature is readily achievable, significant limitations remain on the specific forms that are practically achievable. The introduction of additive manufacturing via robotic plastic extrusion alleviates many complications associated with fabric formwork while increasing the variety of possible shapes and surface qualities.

The system, described in detail below, combines five elements: flexible fabric shuttering; a polymer network that is 3D-printed onto the shuttering to control its deformation; bespoke bent and welded steel rods that define local and global form, support the fabric shuttering and serve as permanent reinforcement for the final element; a rigid structure to control and support the flexible fabric shuttering along its perimeter; the concrete itself. This combination of elements enables the production of (ideally structurally optimized) doubly-curved forms with non-uniform thicknesses and the ability to incorporate holes without the usual economic and material costs and waste such constructions conventionally imply.

2 Background

2.1 Fabric Formwork

Due to its liquid beginnings, concrete offers immense formal flexibility. However, this flexibility is limited in practice by the cost, complexity, and waste of shuttering and formwork production. These constraints have led to the dominance of flat forms created from rigid - often timber - formwork and an avoidance of curved geometries. Flexible formwork has formed a wide range of concrete structures and has offered exciting new structural and architectural possibilities. Replacing rigid molds with flexible materials presents many practical advantages and opportunities for improved structural efficiency (Hawkins et al. 2016).

Formwork stiffness can be adjusted by pre-stressing or orientating the fabric to articulate warp and weft direction (Hawkins et al. 2016). However, leaving intricate control of fabric micro-behavior aside, clamping the flexible formwork to manage concrete thickness in double-sided casts is a predominant means of control - and thus formal expression. The combination of, on one side, digital fabrication and analysis and, on the other, fabric formwork presents an appropriate match that in itself is not novel: managing the contingent nature and strategic control of deformable membranes calls for the level of apprehension and precision that such tools offer. Whether combining digital form-finding methods and flexible casting to optimize architectural components structurally (Veenendaal et al. 2011a), designing and engineering the local behavior of weaved membranes (Ahlquist and Menges, 2013), or utilizing the precision of 6-axis robots to control the flexibility of fabric formwork (Culver et al. 2016), concrete and flexible formwork certainly supports exploitation of the inherent capacities between material agency and digital control. The study and subsequent realization of a full-scale complex concrete shell structure at ETH Zürich further suggest a promising combination of a steel cable net and fabric formwork, utilizing both flexibility and structural logic in one coherent design (Méndez Echenagucia et al. 2018) (Fig. 1).

Fig. 1. Facade by Miguel Fisac and Kenzo Unno's wall system during casting

Locally tailoring the fabric's performance allows for negotiating global curvature and local features, adding character to the cast form. Alternately, adding a layer of (semi-open) rigid formwork outside the flexible membrane enables an aesthetic articulation of the concrete pressure while retaining a degree of control. This feature is exemplary in the built works of Kenzo Unno (since 1997) and Miguel Fisac (between 1969 and 2000), both of whom employ methods of 'frame restraints' in facade castings (West 2006; Veenendaal et al. 2011b). Such exterior restraints on otherwise flexible formwork benefit from the low complexity and effort in locally tuning the fabric by adding a rigid component. Fisac achieved this effect primarily on horizontally cast prefabricated facade elements, while Unno went further and developed a minimal-waste process of attaching exteriorly restrained fabric formwork to a steel reinforcement structure, removing any need for further formwork components. Unno employs only orthogonal and standard reinforcement and formwork components in his work. The system described here builds upon Unno's precedent to realize more complex three-dimensional forms by combining bespoke form-generation processes in both fabric formwork and steel reinforcement. Tailored rigidity thus promises a materially rich and complex output from an input of comparably simple manufacturing.

The P-wall (2009) by Matsys Design has demonstrated how strategically distributed rigid supports can transform a flat fabric-based formwork principle into a rich material articulation and simultaneously manifest a digitally designed intention of cast thicknesses (Kudless 2011). However, this technique, akin to Fisac's frame restraints, produces discrete horizontally cast elements with one curved and one flat surface that are not well suited to serve as reinforced structural elements. More recent implementations of fabric formwork include the roof of the HiLo research and innovation unit for NEST (1,1 Prototype 2017; Final construction 2021) that deployed a cable net-supported fabric shuttering system and KnitNervi (2022) (Popescu 2024). In both examples, the formwork was tension-active, introducing tight formal constraints and single-sided, necessitating sprayed-on mortar with its own associated complications (Block et al. 2017). A related approach to semi-flexible formwork material minimization (in this case for bespoke precast concrete elements) used folded laser-cut PETg to achieve the PreVault pavilion (Larsen et al. 2012) later refined for the Utzon/40 Pavilion (Maxwell et al. 2013).

2.2 Bespoke Steel Reinforcement

The second component in Unno's concrete wall system - the orthogonal rebar structure - supports the clamped geotextile during casting and reinforces the finished wall. A non-orthogonal revision of this steel support structure suggests a design and fabrication system based on Computer Numerically Controlled (CNC) fabrication. In particular, robotically assisted CNC production techniques offer highly flexible platforms for bespoke steel forming.

CNC cold-forming (bending) of metal rods using purpose-built equipment is common in the building industry, including forming ribbed bars for concrete reinforcement applications. These machines offer high formal flexibility; however, reinforcement designs rarely exploit this flexibility, predominantly employing 2-dimensional 90- and 45° bends. Recent innovations in cold-forming have focused on integrating 6-axis industrial robots within computational design-to-production systems that more fully exploit 3-dimensional bending on expanded applications. Such applications are often developed in conjunction with additional robotic fabrication processes, and the system described here fits into this category. These industrial robots can bend via custom end-effectors (Smigielska 2018), providing collision avoidance benefits or floor-mounted external axes (Maxwell et al. 2013; Pigram and McGee 2011), providing increased force and rigidity. Smigielska (2018) also contributed a method to increase the accuracy of adjustments for springback via machine learning.

3 Methods

We developed the 3D-print-tailored rebar-supported fabric shuttering technique through a series of generative design experiments, combining traditional concrete casting (pouring into formwork) and robotic manufacturing. These experiments initially focused on combining two formwork materials: Fabric and thermoplastic polymers. A 7-axis robot with a large-scale Fused Deposition Modelling (FDM) extruder printed patterns onto a fabric membrane held flat and in place with a vacuum table. During the concrete pour, the printed polymer acted as a lower-elasticity network, in between which the fabric membrane stretched and deformed in response to the fluid concrete's hydrostatic pressure. The printed pattern thus created a tailored variable behavior across the casting surface.

The project progressed in several stages. The first stage focused on matching appropriate materials and tuning the 3D polymer printing technique in manufacturing flexible concrete shuttering. The second stage introduced a technique to bend bespoke steel rods robotically and assemble them into a structure employed as both formwork support and concrete reinforcement – thus alleviating the need for rigid external formwork components. Thirdly, a three-component full-scale prototype demonstrated how the augmented fabric formwork and the bespoke steel reinforcement techniques could merge into a coherent design- and manufacturing system.

3.1 Augmented Fabric Shuttering

The shuttering technique's rationale depends on the elastic properties and tensile strength of two components: The fabric and the thermoplastic print. While the fabric exhibits two-directional elastic properties with a finite maximum stretch, the thermoplastic pattern acts as a semi-rigid bracing. The printed bracing's shape and density thus affect the flexibility across the fabric. In areas with a high bracing density, the thermoplastic's elastic properties will dominate – and the shuttering will perform similarly to a rigid barrier. Conversely, in areas with a low density of printed material, the fabric's elastic properties will dominate, and the shuttering will perform as a soft membrane. The shuttering's deformation and structural performance thus depend on the toolpath's density and shape.

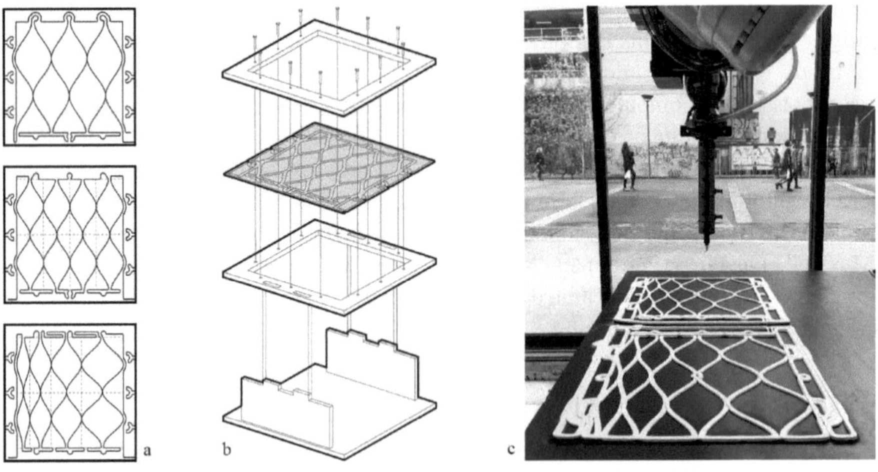

Fig. 2. Tiled pattern variations (a), horizontal rig assembly (b), thermoplastic extruder and printed shuttering (c).

The printing setup consisted of a large-scale thermoplastic extruder mounted on a six-axis industrial robot arm on a linear external axis. The extrusion die measured 6 mm in diameter and was calibrated to extrude a 10–12 mm wide layer of molten plastic at a layer height of 2 mm (Fig. 2). Initial experiments with acrylonitrile butadiene styrene (ABS) and various fabric types suggested a suitable match with coated "liquid foil" nylon and lycra/spandex stretch fabrics. When heated, the molten thermoplastic polymer print adhered very well to the foil coating (Fig. 2).

With a material match between thermoplastics and fabric established, the focus shifted towards the structural challenge confronting the hybrid shuttering during the concrete pour. This challenge involved a topological consideration of toolpath generation and a technical refinement of printing details to enhance the structural performance. We tested the features of both aspects on small-scale 40x40 cm horizontal concrete casts, which iteratively revealed the shuttering's weakest details. Two rigid form-ply frames clamped the hybrid shuttering between them, anchoring the polymer mesh as a rigid tensile structure against the concrete's weight (Fig. 3). This process prompted several features to emerge, such as double-layer bracing, carefully calibrated self-intersections, and a separate topology of attachment details inside the clamping rig.

Fig. 3. Example of a cast in tiled print. Print-reinforced shuttering clamped into a casting frame and resulting cast concrete form.

Toolpath pattern generation was based on Truchet-tiling with consistent point spacing on opposing edges, ensuring pattern connectivity across a tiled surface (Carlson 2018; Krawczyk 2011). In a grid of tiles, the curves can thus join into visually contiguous (tool-) paths. Since each tile is a topology of edge connections, this approach applies as a toolpath generator on faces across any tessellated surface. The local behavior and strength of the fabric membrane thus change through scaling and distorting tiles while potentially employing different internal tile designs (Fig. 2). Figure 3 shows an example of a tiling and print featuring a mirrored sine-wave design and the character of the cast concrete that it yields.

We converted the small-scale horizontal rig into a version that fit two membranes in a double-sided vertical cast. Connecting both sides of the fabric shuttering was achieved via a modification to the standard print intersection. By overshooting the intersection and bridging itself twice, the print formed a hole sized to fit around a bolt. This detail allowed the local thickness of the cast to be controlled via the length of the bolt. As seen in Sect. 4, the bolt spacer later became critical in attaching the fabric shuttering to the steel reinforcement structure. Additionally, closed-loop prints could clamp both sides together, creating block-outs to locate holes in zones between reinforcement bars (Fig. 4). These additions to the design space completed the fabrication method as a material and geometric system able to achieve double-sided casts of varying thickness with vast formal and structural-optimization potential.

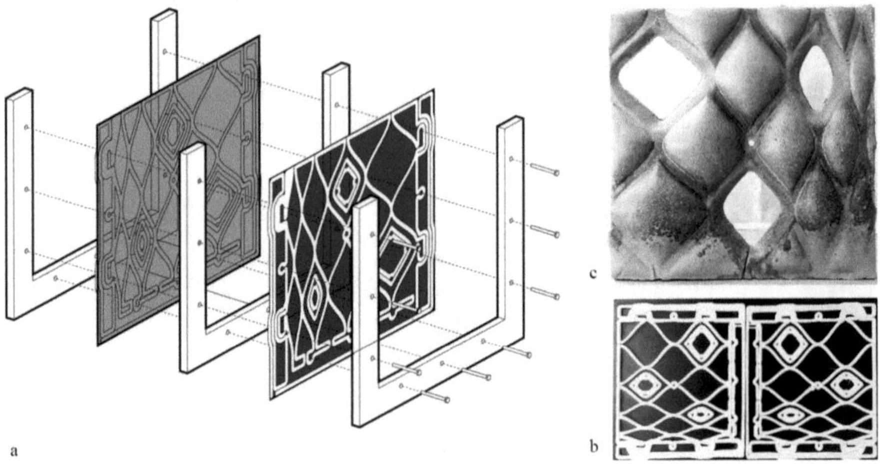

Fig. 4. Vertical casting rig assembly (a), printed vertical formwork with block-outs and central bolt fitting (b), resulting concrete cast (c).

The small-scale developments outlined in this section are described thoroughly in a previously published paper covering alternative toolpath generation strategies, printing detail development, and a multitude of casting experiments in variations of toolpath design. The earlier work did not include the combination of fabric formwork and rebar described here (Engholt and Pigram 2019).

3.2 Bespoke Steel Reinforcement

Parallel to the studies in printing media and vertical shuttering, a robotic steel rebar cold-forming (bending) method for reinforcement was developed. This development involved a robotic steel bending procedure and corresponding reinforcement and shuttering topologies. The robotic rod-bending end-effector was similar to that described in Smigielska 2018, building upon earlier work by Pigram and McGee 2011 and Maxwell et al. 2013. The steel rod fed through a floor-mounted rig consisting of a pneumatic collet gripper, a hydraulic rebar cutter, and a steel guide (Fig. 5). An ABB robotic arm performs a bend by rotating two grooved steel cylinders mounted along and parallel to the robot's sixth rotational axis. Once a bend is completed, the collet chuck opens. The tension between the bent steel rod and the end effector cylinders allows the robotic arm to pull the steel rod along its feed direction to control the distance between bends. A plane rotation between bends can be achieved by reorienting the rod in preparation for the next bend. The use of an industrial robot with end-of-arm tooling affords the ability to complete the bending process at a flexible distance from the collet gripper and a flexible orientation to its central axis. This flexibility provides additional collision (rod-arm, rod-floor, rod-work cell) avoidance possibilities, increasing the phase space of bendable rods.

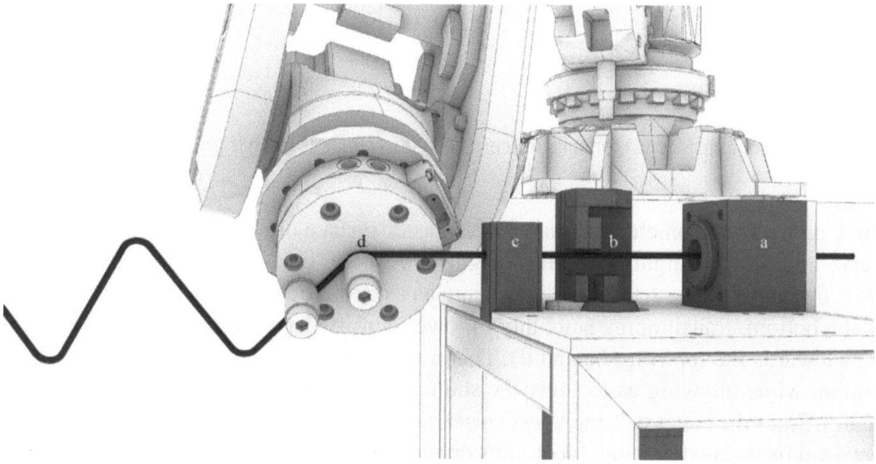

Fig. 5. Illustration of bending rig with collet gripper (a), rebar cutter (b), rod guide (c) and end effector (d).

4 Prototype Design and Production

The experiments described in Sect. 3.1 created a performance baseline between thermoplastic extruder, fabric formwork, and concrete. However, while the small-scale double-sided vertical cast had demonstrated tremendous formal potential, fully realizing this potential would depend on the scalability of material performance. A full-scale prototype combining steel reinforcement and polymer-supported fabric shuttering should thus demonstrate the system's feasibility and strength in controlling vastly increased hydrostatic pressure. At the same time, the prototype needed to be movable with small equipment. We estimated that an element (and, thus, fluid concrete column) height of 60 cm would provide a reasonable balance between these concerns.

4.1 Design

The mirrored sine wave tile employed in the small-scale vertical test translated into a similar rebar topology for a full-scale prototype, designed as a self-supporting wall in

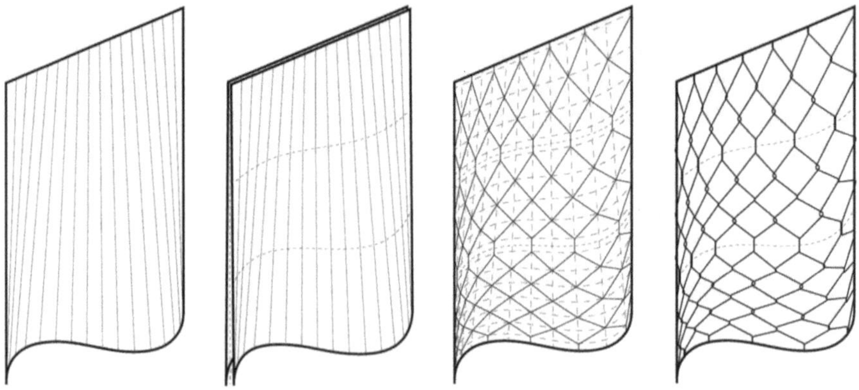

Fig. 6. Master surface (a), offset surfaces and subdivisions (b), rebar grid (c), rebar rods (d).

three parts. The geometry originated in a master surface drawn between a bottom sine curve and a top straight curve at a height of 180 cm, creating a self-supporting wall (Fig. 6a and b). This surface subdivides into an asymmetric grid with the highest density at the bottom, transitioning smoothly to lower density, larger spaces, and more block-outs towards the top (Fig. 6c and d). This transition concentrated the mass toward the bottom while allowing us to study the shuttering's strength and stretch across various scales. Since the concrete's thickness correlates directly with the unsupported fabric area between the printed bracing, block-outs dominate the design as these areas grow in size towards the top.

The three elements illustrate different balances between (and approaches to) bracing density and block-outs: The solid bottom element controls the thickness through a high-density shuttering and rebar grid. In contrast, the porous top element controls the thickness through a large grid with block-outs. In between, the middle element compares both techniques across similarly sized fabric areas. The rebar frame was modeled on the master surface and grid, and the shuttering's printing toolpaths originated in grids projected onto surfaces offset to both sides (Fig. 7).

By overlapping the rebar frame in eyelets corresponding to the clamping detail developed previously in the small-scale tests, both sides of the shuttering could attach to the reinforcement, locally keeping it in place (Fig. 8). The prototype was designed to demonstrate transitions and possibilities in the developed shuttering features while maintaining a tectonic expression as a self-supporting structure with the concrete mass concentrated towards the bottom. Figure 8 shows the distribution of block-outs on the unrolled grid.

4.2 Robotic Fabrication

The full-scale robotic fabrication process progressed along the two fabrication techniques in thermoplastic printing and steel rod bending. As seen in Fig. 9, the extruded thermoplastic adhered well to the non-coated side of the liquid-foil fabric, allowing the

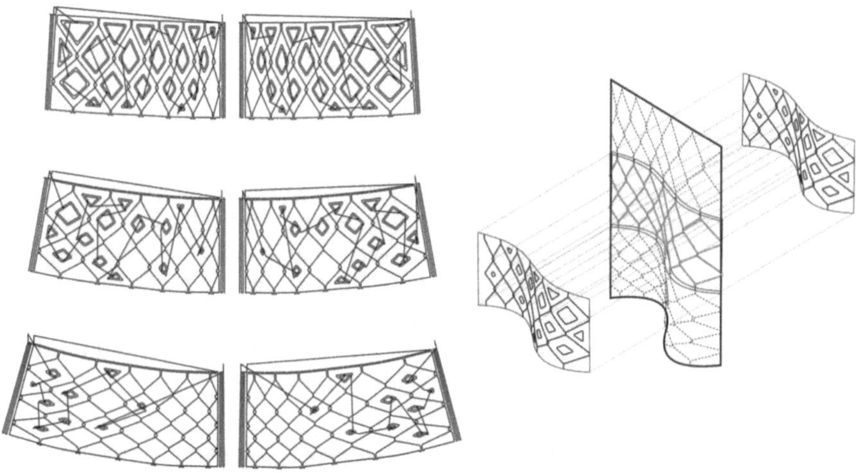

Fig. 7. Printing toolpaths for the three components and exploded view of the structure and the shuttering.

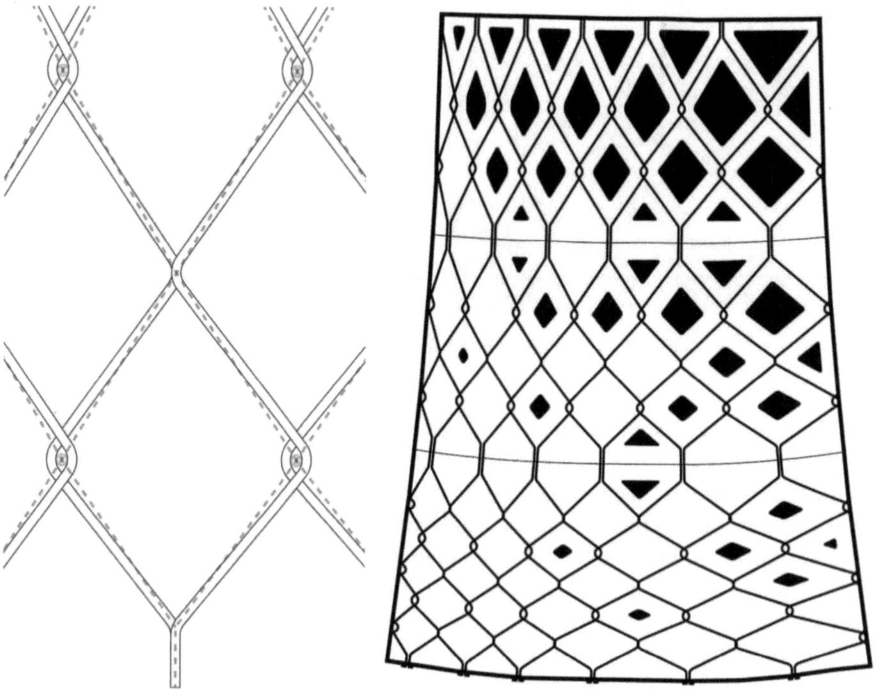

Fig. 8. Bent rebar structure overlaid on the structural grid and the block-out distribution on the unrolled grid.

coated side to face the concrete and achieve a smooth surface finish (Fig. 13). The vacuum table had ample space for each of the six large prints, measuring approximately 65 by 140 cm.

Each of the prototype's three components required a rigid frame to precisely secure the steel structure and control the concrete's outer dimensions. Figure 10 illustrates these rigid frames around the steel structure and tailored shuttering, defining the components' top and bottom profiles by securing the shuttering inside form-ply guides cut on a three-axis CNC router. The curvature along the prototype's bottom surface required the printed shuttering to bend considerably while fitting tightly between the form-ply guides. As outlined in Sect. 4.3, the developed thermoplastic blend turned out to be durable and flexible, allowing the printed shuttering to fit inside the curved rigid formwork and counter the concrete's hydrostatic pressure (Figs.10 and 14).

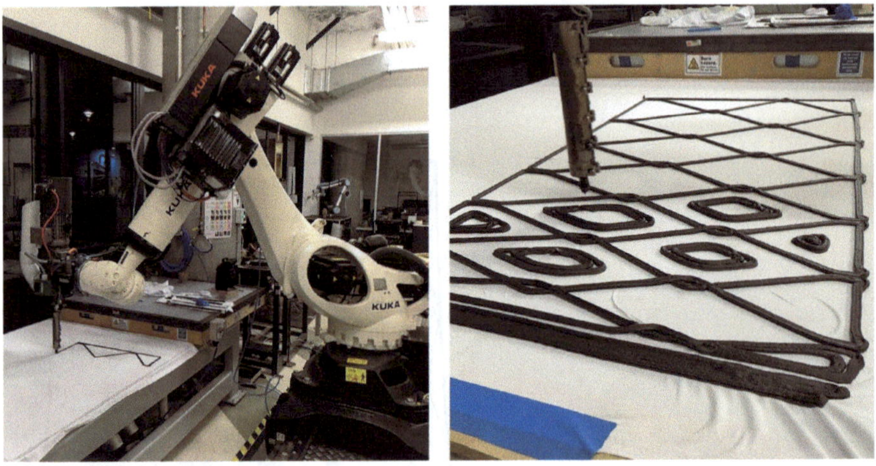

Fig. 9. Toolpath printing for the full-scale prototype

The prototype design required meticulous control in the steel bending procedure. The steel members inside the top component required only four bends across a fairly flat surface. However, each of the bottom component's steel rods vertically connected two sinusoidal curves through seven sharp bends (Fig. 11). In the most extreme case, one rod underwent bends totalling more than 850°, with bending planes rotating only a few degrees around the vertical. Figure 10 shows a part of the middle component's steel structure inside the rigid frame, where the rods' slight transition towards a more flat sinusoidal top curve can be seen. The photo also shows the fabric's glossy finish that would transfer to the concrete surface (Fig. 15) and the printed block-outs ready to connect to their corresponding counterparts on the second shuttering surface.

4.3 Formwork Assembly and Casting

While the thermoplastic polymer's tensile strength had little effect on small-scale testing, it became a significant issue as the fabrication method was scaled up. Initial tests with

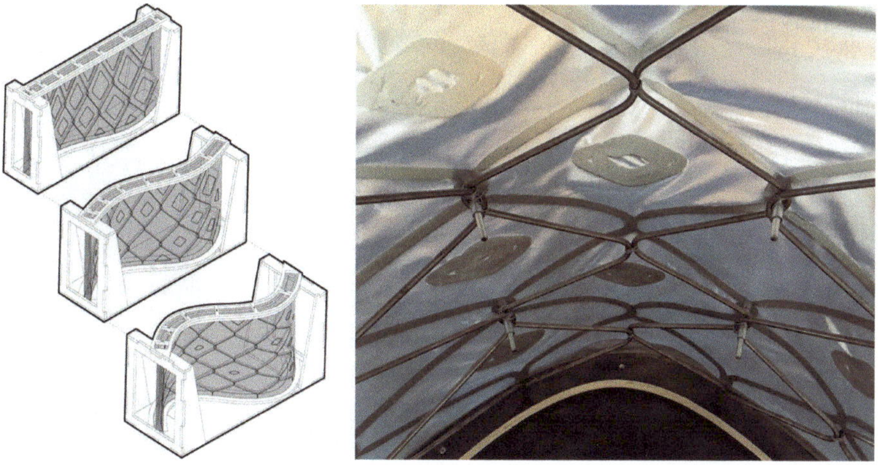

Fig. 10. The three formwork designs and a view inside the shuttering

Fig. 11. Steel bending rig and the prototype's steel reinforcement components.

Acrylonitrile Butadiene Styrene (ABS) thermoplastic exhibited high tensile strength but suffered from shrinking during cooling. The shrinking behaved unpredictably since the polymer distribution was uneven and asymmetric in each printed pattern – causing the formwork to twist and warp. The shuttering shown in Sect. 3.1 consists of white ABS; however, it did not deform noticeably due to its small scale (approximately 50x50 cm).

Turning to a Polylactic Acid (PLA) thermopolymer solved this issue; however, the increased hydrostatic pressure combined with vigorous vibration during pouring caused the PLA shuttering to shatter. While PLA performed well in tension, it was too stiff and brittle against the intense vibration and the concrete's weight (Fig. 12).

A third polymer, Thermoplastic Polyurethane (TPU), performed very well regarding adhesion and integrity but stretched excessively. A fourth iteration combined PLA

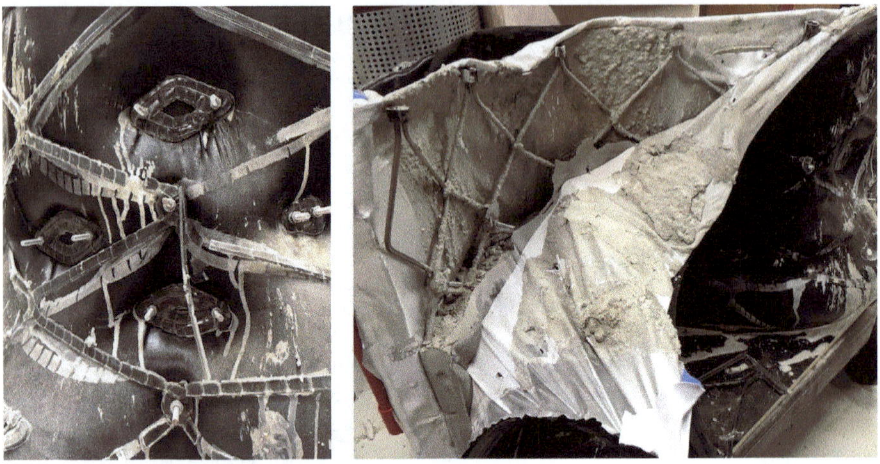

Fig. 12. Collapsed PLA print after the pour.

and TPU thermoplastic polymers to produce a hybrid polymer that would bend but not stretch. TPU and PLA pellets share similar extrusion temperatures and can easily combine inside the extruder, yielding the selected extrusion material. This thermoplastic blend performed well and proved exceedingly capable of withstanding the substantial hydrostatic force present when casting a 60-cm tall element (Fig. 13). With an 8:1 PLA to TPU ratio, the hybrid polymer can be seen as printed black shuttering in Figs. 9 and 13.

Fig. 13. Bottom element inside the formwork and the flattened shuttering after demoulding

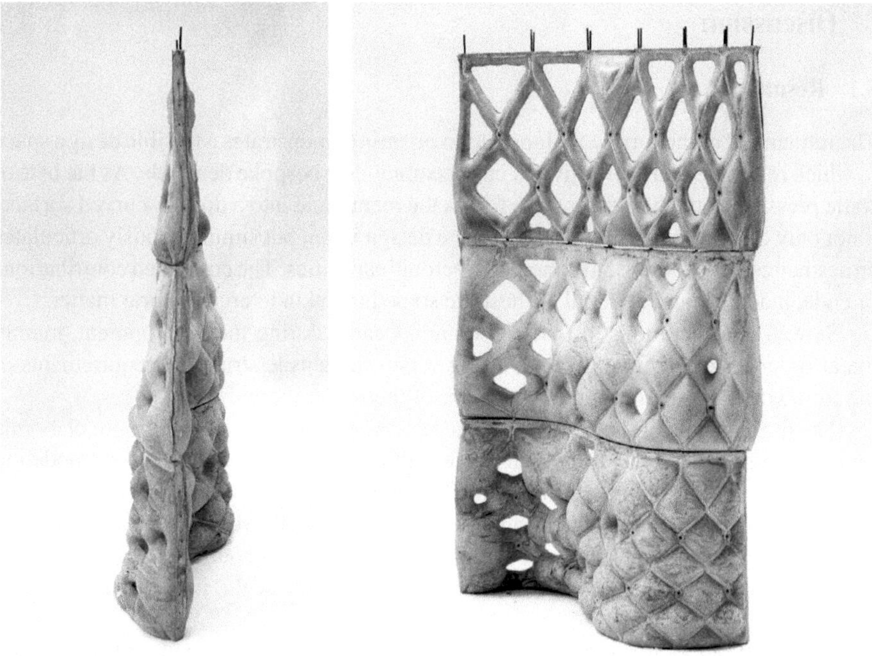

Fig. 14. The three-component full-scale prototype.

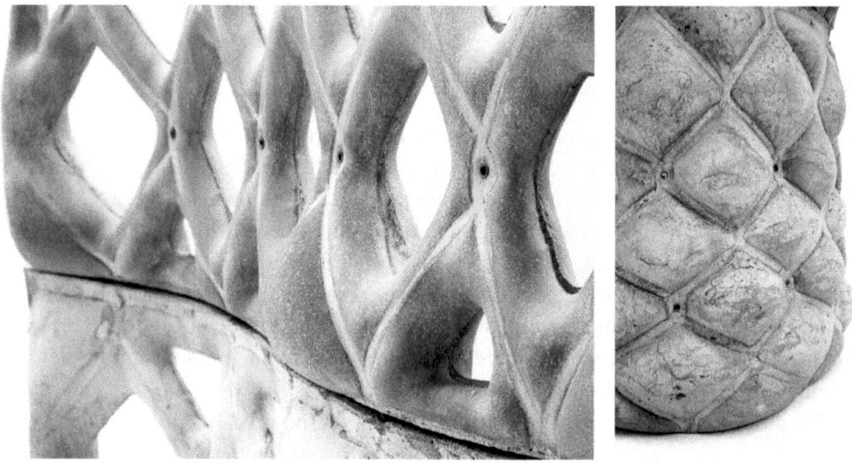

Fig. 15. Details from the finished three-component prototype

5 Discussion

5.1 Results

The refinement of the formwork fabrication process demonstrates a flexible design space in which reinforced fabric formwork can accommodate bespoke demands. As the hydrostatic pressure of the liquid concrete forces the membrane into a doubly-curved surface, it not only distributes mass according to the design intent but simultaneously articulates an aesthetics closely linked to concrete's tectonic capacities. The combined contributions of code, machine, and material orchestrate something akin to architectural matter.

Several considerations of material logic appeared during the development, mainly the affordances of additive manufacturing versus the tensile structural requirements of the formwork and the mechanical behavior of thermoplastic polymers.

The present state of the research suggests tests at a larger scale with more elaborate reinforcement structures. Further, the project aims to refine the feedback from modeling and simulation to produce a reliable prediction tool. This prediction should enable a process in which an overall shape can inform and generate instructions and drawings for machining, assembly, and casting. Presently, the most reliable source of prediction is the accumulated (human) knowledge and experience that follow the intricate involvement with physical experiments.

While the physical results hint towards technical refinement, the complex interactions of material, technique, and code might suggest further discussions of higher-order relations: How does a hybrid understanding of fabrication logic change and develop alongside the negotiated concrete formation as hierarchies and migrations emerge in the material systems of robotically augmented flexible concrete formwork?

5.2 Conclusion

Tuning the formal and structural capacities of a fabric formwork system via its attachment to robotically fabricated reinforcement steel and the addition of FDM 3D printed bracing is a method with seductive architectural potential. The evaluation of research output does not tie solely to structural performance, material economics, or formal complexity. Any realization of physical form - as in traditional and current building practice - is tied to a complex set of circumstances beyond merely material and structural bifurcations. A digital design and robotic fabrication strategy should exhibit levels of complexity similar to traditional modes of architectural realization, replacing appropriate fields of knowledge with others native to digital making.

The described method has been shown to enable the production of architectural-scale doubly-curved variable-thickness precast reinforced concrete panels with minimal material investment in shuttering and little additional falsework required. Combined, these outcomes represent significant potential embodied energy and cost-savings compared to conventional approaches, though further work remains to quantify the scale of these potential gains accurately. In addition, the formal characteristics imbued on the elements realized via this method are novel, compelling, diverse, and tunable, representing an additional contribution to architecture beyond optimization.

References

Ahlquist, S.A., Menges, A.M.: Frameworks for computational design of textile micro-architectures and material behavior in forming complex force-active structures. In: Proceedings of the 33rd Annual Conference of the Association for Computer-Aided Design in Architecture (ACADIA), pp. 281–292 (2013)

Block, P., et al.: NEST HiLo: Investigating lightweight construction and adaptive energy systems. J. Build. Eng. **12**, 332–341 (2017)

Carlson, C.: Multi-scale Truchet patterns. In: Proceedings of Bridges 2018: Mathematics, Art, Music, Architecture, Education, Culture. Tesselation Publishing, Phoenix (2018)

Engholt, J.K., Pigram, D.: Tailored Flexibility: Reinforcing concrete fabric formwork with 3D printed plastics. In: Proceedings of the 24th Annual Conference of the Association for Computer-Aided Architectural Design Research in Asia (CAADRIA): Intelligent & Informed, pp. 53–62 (2019)

https://pdfs.semanticscholar.org/7d7e/22cf69343f490cc1bb88ed5a8cd1488712b4.pdf

Hack, N., et al.: Structural stay-in-place formwork for robotic in Situ fabrication of non-standard concrete structures: A real scale architectural demonstrator. Autom. Constr. **115**(July), 103197 (2020). https://doi.org/10.1016/j.autcon.2020.103197

Hawkins, W.J., et al.: Flexible formwork technologies - A state of the art review. Struct. Concr. **17**(6), 911–935 (2016). https://doi.org/10.1002/suco.201600117

Krawczyk, R.J.: Truchet tilings revisited. In: Proceedings of ISAMA 2011, pp. 69–77 (2011)

Kudless, A.K.: Bodies in Formation: The material evolution of flexible formworks. In: Proceedings of the 31st Annual Conference of the Association for Computer Aided Design in Architecture (ACADIA), pp. 98–105 (2011)

Larsen, N.M., Pedersen, O.E., Pigram, D.: Realisation of complex precast concrete structures through the integration of algorithmic design and novel fabrication techniques. Adv. Archit. Geom. **2012**, 161–174 (2012)

Maxwell, I., Pigram, D., Mcgee, W.: The novel stones of Venice: The marching cube algorithm as a strategy for managing mass-customisation. In: Proceedings of ACADIA 2013: Adaptive Architecture (2013). https://doi.org/10.52842/conf.acadia.2013.311

Mendez Echenagucia, T.M.E., Pigram, D.P., Liew, A.L., Van Mele, T.V.M., Block, P.B.: A cable-net and fabric formwork system for the construction of concrete shells: Design, fabrication and construction of a full-scale prototype. Structures. **16**, 190–198 (2018)

Pigram, D., Larsen, N.M., Pedersen, O.E.: A method for the realization of complex concrete Gridshell structures in precast concrete, ACADIA 2012: Synthetic digital ecologies. In: Proceedings of the 32nd Annual Conference of the Association for Computer Aided Design in Architecture, pp. 209–216 (2012)

Pigram, D., Mcgee, W.: Formation embedded design: A methodology for the integration of fabrication constraints into architectural design. In: Proceedings of ACADIA 2011: Integration through Computation (2011). https://doi.org/10.52842/conf.acadia.2011.122

Smigielska, M.: Application of machine learning within the integrative design and fabrication of robotic rod bending processes. In: Humanizing Digital Reality, pp. 523–536. Springer (2018)

U.S. Geological Survey: Mineral Commodity Summaries: 2018. Available from http://minerals.usgs.gov/minerals/pubs/commodity/cement/ (accessed 8th July 2019).

Veenendaal, D.V., Coenders, J.C., Vambersky, J.V., West, M.W.: Design and optimization of fabric-formed beams and trusses: Evolutionary algorithms and form-finding. Struct. Concr. **12**, 241–254 (2011a)
Veenendaal, D.V., West, M.W., Block, P.B.: History and overview of fabric formwork: Using fabrics for concrete casting. Struct. Concr. **12**(3), 164–177 (2011b)
West, M.W.: Flexible fabric molds for precast trusses. BFT International Betonwerk + Fertigteil-Technik. **72**(10), 46–52 (2006)

3D Printing Pattern Effects on SCF-PLA Extrusion on Pre-stretched Textile Formwork

Ehsan Baharlou(✉)

University of Virginia, Charlottesville, VA 22903, USA
baharlou@virginia.edu

Abstract. This paper presents a novel approach for nonplanar robotic 3D printing of short carbon fiber-reinforced polylactic acid (SCF-PLA) pellets to reinforce a pre-stretched textile formwork. The study aims to explore the effects of 3D printing patterns to improve the mechanical properties of SCF-PLA to reinforce a double-curved shell structure. Cross-ply and quasi-isotropic patterns integrated the mechanical properties of SCF-PLA with robotic toolpath generation. In this paper, each stage of the development of the methods has been analyzed and evaluated methodologically to identify the fabrication capacities and constraints. The collected data was used to develop a computational framework to integrate structural performance, material properties, and fabrication constraints in design processes. This computational framework was developed to employ the 3D printing patterning on a nonplanar surface to generate robotic toolpaths. The implementation of these studies, from the design to the fabrication of a shell structure, is presented as a case study.

Keywords: Robotic 3D Printing · Short Carbon Fiber-reinforced Polylactic Acid (SCF-PLA) · Flexible Textile Formwork · Cross-ply Pattern · Quasi-isotropic Pattern

1 Introduction

The recent advancement in additive manufacturing (AM) is revolutionizing building construction. AM techniques, or 3D printing, enables designers and architects to rapidly realize their design while reducing construction time and improving resource efficiency. The implementation of sustainable additive manufacturing in the building industry is a promising approach to curb the amount of carbon emitted by this industry [1, 2].

In architectural design, fused deposition modeling (FDM) 3D printing has been employed to develop various building elements such as façade panels [3–5] and functionally graded façade systems [6], cast structural elements such as columns or slabs [7–10], construct stand-alone building structures [11], and support a free-form shell composite [12]. However, thermoplastic's poor mechanical performance is one of the obstacles to applying this material as the sole structural element. The combination of carbon fiber mixed with a Polylactic Acid (PLA) matrix results in a reinforced composite that benefits from the characteristics of both [13–15].

M. Yablonina et al. (Eds.): Robotic Fabrication in Architecture, Art and Design 2024, pp. 211–230, 2026.
https://doi.org/10.1007/978-3-032-22607-5_14

The rationalization of architectural geometries through robotic additive manufacturing can have different approaches, such as: flat layer-based printing [16]; non-planar layer-based printing [17–21]; curved-layer fused filament fabrication (CLFFF) [12, 22–27]; and spatial 3D printing [28–31]. Each one of these has a set of limitations, such as the need for supporting structures, the limitation of supporting surface curvature, or the speed of material solidification.

The aim of this project is to advance the current CLFFF techniques by investigating the effects of 3D printing patterns on flexible textile fabrics to reduce material waste and improve printing time. We studied the effects of cross-ply [0°/90°] and quasi-isotropic [0°/60°/90°] 3D printing patterns using SCF-PLA onto a textile membrane to enhance the mechanical properties of the printed materials. By infusing these two materials, we seek to develop a hybrid shell composite that minimizes the material waste in falsework and formwork applications while strategically reinforcing the structure. As a case study, we examine the effects of these printing patterns on a double-curved textile formwork (see Fig. 1).

Fig. 1. Final demonstration.

2 Background

In architectural design, one of the first examples of using fiber-reinforced plastic (FRP) is the Monsanto House of the Future (1957) [32]. With the advent of digital fabrication, carbon fiber-reinforced plastic (CFRP) and FRP have been considered in realizing complex geometries in architectural design. Advances in robotic fabrication enable designers to place carbon fibers on a compression shell [33], or to lay fiber tapes on a formwork for a kinetic system [34]. This fabrication process allows for the optimal placement of carbon fibers on formwork, based on the required structural performance.

The development of carbon fibers with a matrix of thermoplastic allows 3D printing CFRP directly on a neat polymer to manufacture a sandwich-panel element [13]. In concurrent research, continuous carbon fibers were infused with epoxy and resin into thermoplastic molds to serve as structural elements [14]. Comparing mechanical properties of carbon fibers in a polymer matrix such as SCF-PLA and neat polymers such as PLA shows a significant improvement in the tensile strength of PLA [15, 35]. PLA shows anisotropic properties; however, its strength depends on the printing direction [36, 37]. Additionally, the strength of SCF-PLA is influenced by various parameters, including the height and width of the printing [38], as well as the percentage of carbon fibers in the composite [39]. Similar to CFRP laminates, the development of cross-ply and quasi-isotropic patterns shows improvement of mechanical properties, particularly the flexural strengths [40]. This approach can improve the structural properties of SCF-PLA for reinforcing complex geometries.

2.1 Pre-Stretched Fabric Membrane

Flexible fabric formworks have been applied to create complex geometries for concrete reinforcement [41–45]. The development of a double-curved shell with flexible fabric requires falsework to temporarily support the pre-stretched fabric. Flexible formworks usually are used without internal falsework to reduce material waste but increase their complexity. Reinforcing a double-curved formwork with robotic 3D printing requires a method to generate nonplanar toolpaths for the robot. In FDM, curved-layered and nonplanar-layered techniques have been developed to address the nonplanarity of the print surface. These modifications include changing the z-axis and orienting the extruder direction to improve the quality of printing and reduce the printing time [18, 19, 22, 24, 25]. These techniques consider the geometric complexities in order to generate toolpaths regardless of the limitations of 3D printers. Generating these toolpaths pertains to the first layer of printing onto textile fabrics.

2.2 3D Printing on Textile Fabrics

3D printing onto textile fabrics has been widely employed in the fashion industry and architectural design, specifically utilizing the self-shaping properties of the 3D print on stretched fabric [46–49]. The hybridization of thermoplastic and textile fabrics in small-scale artifacts shows the potential of these techniques for use in large-scale architectural elements. The most important factors in 3D printing on textile fabrics relate to the physical and chemical bond formed between the 3D printed materials and the textile fabrics. Aside from the properties of printing materials [50] and textile substrate [51–53], several parameters determine the adhesion between the printing materials and the fabric substrate. These parameters include the fabric's woven texture or pattern [52, 54–57], the shorter z-distance between the printer nozzle and the fabric surface [58–63], higher printing temperatures [57, 59, 61, 63, 64], lower printing speeds [60, 61, 64], bigger printer nozzle size [59], the printing patterns of the first layer's design [65–67], and the alignment of the printer nozzle to the fabric surface [68]. These seven parameters play a crucial role in developing the design-to-fabrication process, especially for 3D printing patterns onto pre-stretched fabric membranes.

3 Materials and Methods

The methods employed in this research include: 1) the development of a material system that uses SCF-PLA to reinforce pre-stretched textile formwork; 2) the development of a robotic 3D printing process with a mobile printing platform; and 3) the development of a computational framework to generate 3D printing patterns that integrates both material properties and fabrication constraints, particularly for the design and reinforcement of a double-curved shell structure (see Fig. 2).

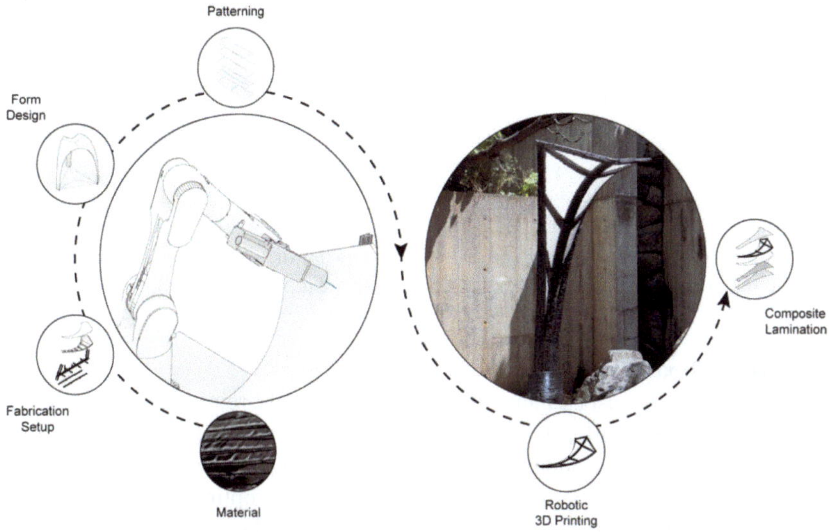

Fig. 2. Design-to-fabrication diagram.

3.1 Material System Development

Short Carbon Fiber-reinforced Polylactic Acid (SCF-PLA).
In order to reinforce the textile formwork under constant tension, we examined commercially available SCF-PLA (CF15 PLA Pellets-3DXTech). This material consists of 15% short carbon fibers and 85% PLA matrix in the form of pellets. At the microscopic level, the extruded SCF-PLA on a flat printing surface shows an anisotropic distribution of short carbon fibers that are slightly aligned with the direction of the printed filament (see Fig. 3). Similar to CFRP laminates, cross-ply [0°/90°] and quasi-isotropic [0°/60°/90°] patterns improve the mechanical properties, such as flexural strength, of the material [69]. This structural improvement of the materials can reduce the number of reinforcement layers on the textile formwork.

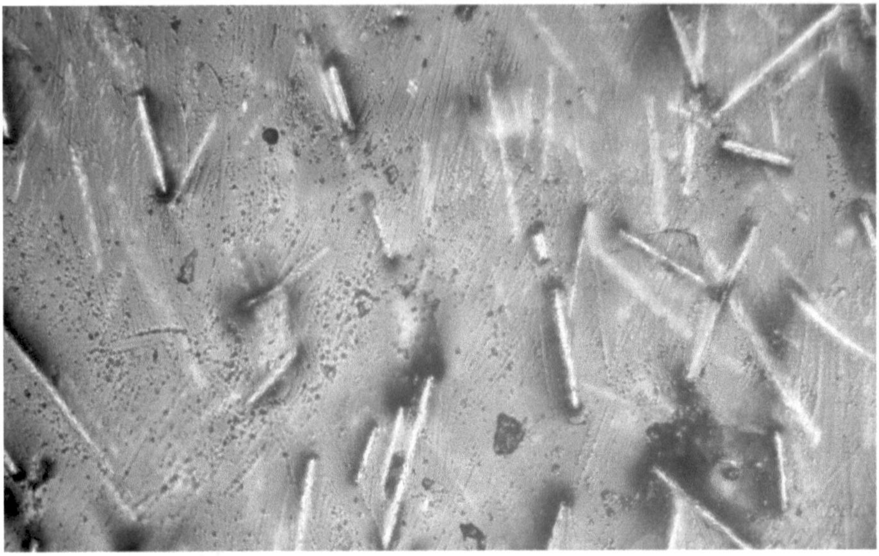

Fig. 3. Microscopic image of a printed SCF-PLA showing the distribution of carbon fibers in relation to the direction of printing (from the bottom to the top of the image).

Pre-Stretched Textile Formwork.
The textile formwork must maintain its shape throughout the reinforcement process with 3D printing. The formwork acts as a print surface without any internal support. Several flexible textiles—4-way spandex fabrics with nylon and polyester composition—were tested by 3D printing on fabric samples to determine the optimal fabric porosity and permeability. The textile formwork selected is a flexible fabric with 90% polyester and 10% spandex. One challenge of 3D printing on fabric is its deformation under a high temperature. Polyester has a melting point over 200 °C, which helps the fabric tolerate the temperature of molten SCF-PLA (190 °C) without changing its integrity. The spandex percentage adds enough flexibility to the fabric to carry nozzle pressure and the weight of the printed material and to achieve the intended form curvature.

3D printing SCF-PLA on textile formworks involves the lamination and delamination of two distinct material systems. One particular challenge associated with 3D printing on pre-stretched textile formwork is maintaining the bond between the printed layers and the textile substrate. Inadequate adhesion between the printed material and the formwork causes delamination, which changes the structural integrity of printed layers on the formwork and changes the coherence of the intended form.

To tackle this challenge, we conducted a series of tests to enhance the adhesion between the printed materials and the textile substrate. Based on the findings outlined in Sect. 2.2, we reduced the height of the first print layers (displacement in the z-axis of the print surface) to print materials with pressure on the textile, which penetrates partially into the pores of the fabric. We slowed down the printing speed for the first layer to deposit more material onto the textile. These modifications significantly improve the level of entanglement between the printing material and the textile substrate.

Furthermore, we utilized ironing as a post-printing technique to improve the bond between the 3D printed layers and the textile substrate. For this purpose, we reduced the z-distance of the extruder nozzle to iron the layers to enhance adhesion.

Composite System Development.
In order to increase the life span of the reinforced membrane, a composite system was developed. In this process, the textile formwork becomes part of the composite system. The microporosity of the textile with a limited void space was infused with epoxy resin to avoid absorbing water and prevent de-sizing effects. The infused epoxy resin rigidified the flexible textile and increased adhesion between the printed materials and the membrane. The composite includes the SCF-PLA-reinforced textile formwork, a layer of fiber glass (fabric style 120 E-Glass-FibreGlast), and a matrix of epoxy resin. The epoxy resin with the hardener (System 2000 Epoxy Resin-FibreGlast) was applied on the reinforced fabric formwork and the e-glass fabric, which were vacuum bagged for 14 h to cure (see Fig. 4). The e-glass fabric and the epoxy resin added an additional layer of strength to the reinforced fabric and improved its durability.

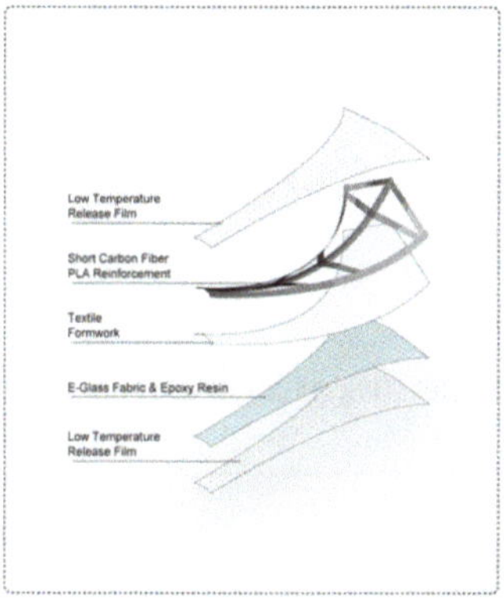

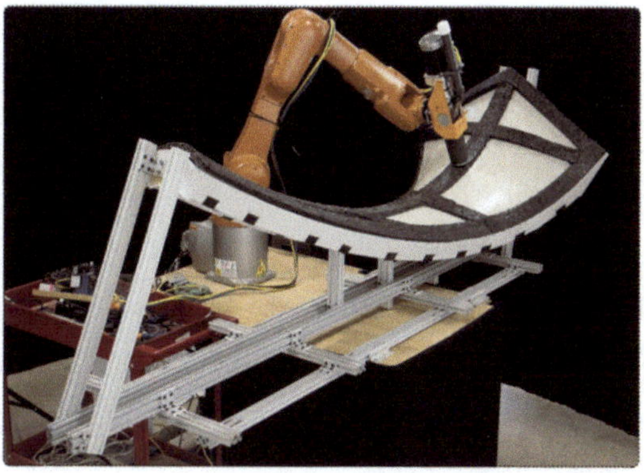

Fig. 4. Left image: the diagram of composite lamination layers for infusing epoxy resin; Right image: the final prototype located on the mobile platform.

3.2 Fabrication Process

Nonplanar Robotic 3D Printing Setup. The robotic 3D printing system includes industrial robotic arms KUKA KR10 R1100. This robot has a maximum reach of 1100 mm and a motion range of Axis 1 between ±170°. The workspace of the robot between 400 mm and 960 mm along the z-axis of the robot's world coordinate system (WCS) is between 810 mm and 1100 mm. The 3D printing end effector is a MDPH2 pellet head extruder (Massive Dimension) with a 3 mm nozzle diameter. The temperature of the nozzle was kept constant at 190 °C. This pellet extruder has a feeding mechanism that works with gravity, optimized for 3D printing aligned with the gravitational force. Accordingly, the main parameter of printing with this granular extruder is to preserve the alignment of the extrusion axis with the gravitational force. Any deviation of the angle between the extruder's x-axis (extruder direction) and the axis of gravity directly affects the print feed rate.

The range of the angle of deviation between ±30° was calculated through a series of tests to define a conical extrudable space for the extruder. Together these two systems also limit their operation (see Fig. 5). The extruder can hit the in-line wrist of the robot (Axis 4) at certain angles, which reduces the printable space. By increasing the height of the print surface and positioning it above 400 mm along the z-axis of the robot's WCS, the collision between the extruder and the robot's arm can be reduced. This range of robot's reach yields a printable space.

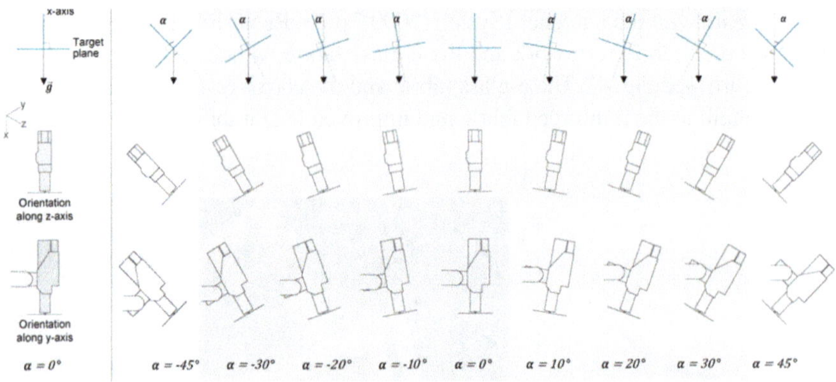

Fig. 5. This diagram illustrates the limitation of extrusion angles. The colored area shows the extrudable space.

Mobile Platform Development.

For the textile formwork with minimum falsework at the edges, a mobile platform was designed to address the limitations of the robotic 3D printing setup. A temporary frame system was developed to hold the prestressed fabric. This frame was assembled on a mobile platform. The mobile platform increases the height of the printing surface and expands the workspace of the robot in the y-axis of the robot's WCS. The height of 600 mm and the length of 1900 mm of the frame were derived from the robotic 3D printer's limitation. The platform was placed on two linear rails attached to the robot's table (see Fig. 6).

The maximum displacement was calculated based on the maximum reach of the robot and the extruder. Accordingly, the platform can be moved between –450 mm and + 450 mm (see Fig. 7). A fabrication process was defined to calibrate the sequence of the 3D printing regarding the mobile platform's displacement. Without the mobile platform, the robot has a spatial reach of 1600 mm in the y-axis of the robot's WCS. With this mobile platform, the robot can print filaments on a 2000 mm-long print surface. This printing space helps define the total height of 1800 mm for the double-curved shell.

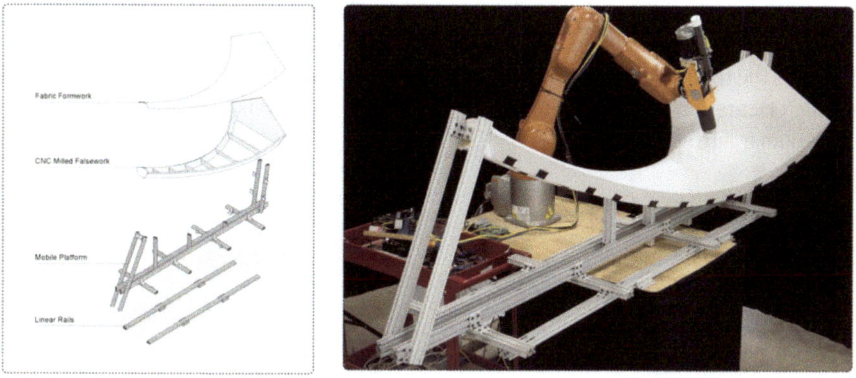

Fig. 6. Left image: the diagram of the mobile platform setup; Right image: the mobile platform setup with pre-stretched textile formwork and the robot.

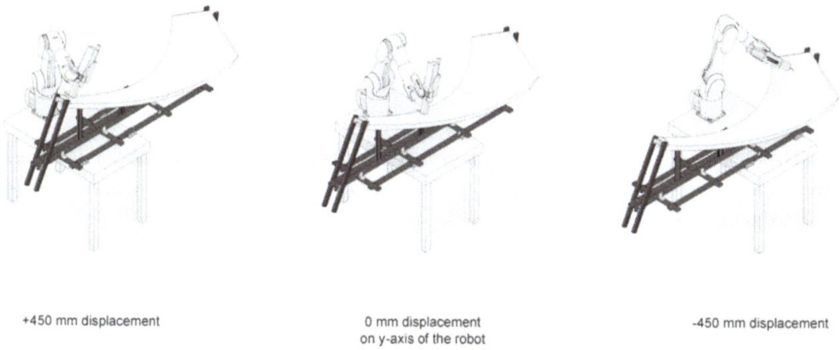

Fig. 7. Mobile platform displacements on *y*-axis of robot's WCS.

One main parameter for reinforcing the fabric formwork is controlling the adhesion level between the adhesive material and adherent platform. The frame system held the flexible textile while it was under tension to reach the intended shape. The simulation of the shell structure informed the physical shell. A tolerance between the simulated model and the physical model was expected. In order to reduce the tolerance, a feedback mechanism was applied to find the tolerance between the nozzle and the prestressed fabric formwork. By mapping the tolerance on the design process, the intended form was adapted to the physical model.

3.3 Computational Framework

A computational framework was developed to incorporate the collected data from the material investigation with the limitations of fabrication processes. This framework defines the location of the reinforcement by partitioning the formwork at the global level and generates the toolpaths for the robotic 3D printing process at the local level. This approach determines a printable space to control the computational framework as it generates the toolpath. This control mechanism modulates the location of the tool (XYZ) and the orientation of the tool (ABC) to tackle the limitations of robotic fabrication on a non-planar surface (see Fig. 8).

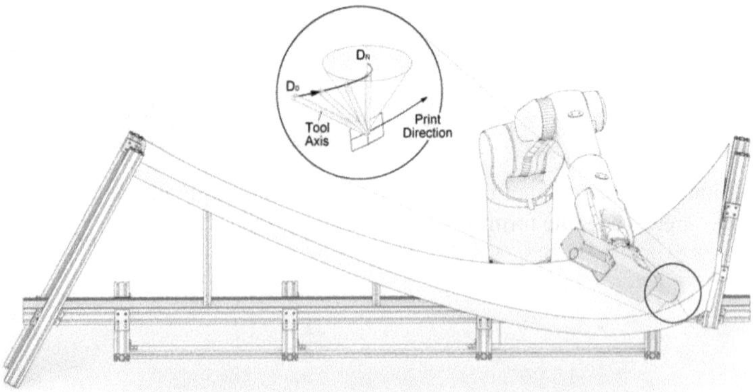

Fig. 8. The schematic diagram illustrates toolpath modulation. The control mechanism interpolates the rotational transformations $\{D_0, D_1, D_2, \ldots, D_n\}$ to ensure the tool remains within the conical extrudable space.

Design Implementation.

In this project, a part of a funnel surface as an axisymmetric column was developed for designing a shell structure. The design parameters of this shell were dependent on understanding the limitations of fabrication systems. Through an analytical approach, a printable space was developed to consider the limitation of the robotic 3D printing system. The collected data of this space enabled the development of an integrative computational design framework to adjust the geometric complexity of the initial design within the printable parameters. These printable parameters determined the height and the upper and lower opening radii of the funnel surface.

The height of the shell is associated with the robotic arm's maximum reach on two positions of the mobile platform. Although the mobile platform expands the robot's workspace in one direction (in *y*-axis of the robot's WCS), it also exceeds the acceptable range of the extruder's angle of deviation and the axis of gravity. This limits the upper and lower opening radii of the funnel surface.

In order to find the radii of the structure, a vault system was anticipated to connect the upper part of three and four axisymmetric columns. The intersection between these axisymmetric columns generated the upper edge of the vault system. The distance between the two corners of the upper edge must be less than 900 mm, which correlates to the maximum reach of the *x*-axis of the robot's WCS. The parameters of the robotic

3D printing setup, the reach of the robotic arm, and the extrudability of the extruder head indicated that the three columns are constructible. The design choice was based on the deviation angles of the tool and the axis of gravity on this axisymmetric column located within the extrudable angle range (see Fig. 9).

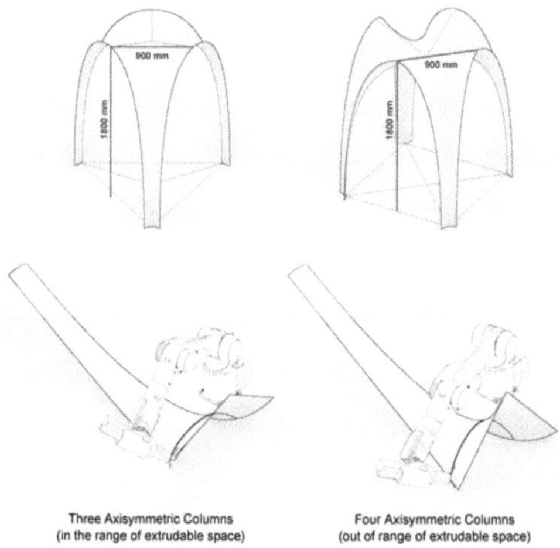

Fig. 9. Effects of the robot's extrudability on determining the number of axisymmetric columns.

The lower radius of this column is where it connects to the base. A finite element simulation was performed on a single column to test three radii of 375 mm (A), 245 mm (B), and 150 mm (C; see Fig. 10). In this simulation, the axisymmetric column was modeled as a shell system, featuring CFRP material properties with 20 mm thickness. The self-weight of the structure was considered as an applying load. The simulation suggested considering the radius of 245 mm for further development, due to the minimum and maximum displacements of the axisymmetric column. According to the scope of this research, this column was investigated further for fabrication.

Toolpath Generation.
To reinforce the fabric formwork, a topological optimization was performed to optimize the location of laying SCF-PLA on the structure without compromising the required performance. The topological optimization limited reinforcement areas to the edges and central spine, connected with ribs. Thus, the textile formwork reinforcement was split into primary edges, secondary central spine and tertiary ribs (see Fig. 11). The finite element simulation showed the principal stresses' directionality, suggesting the direction of extruding SCF-PLA on the formwork. In addition, the cantilever behavior of a single axisymmetric column indicated a need to develop cross-ply and quasi-isotropic patterns to improve the flexural properties of the SCF-PLA material.

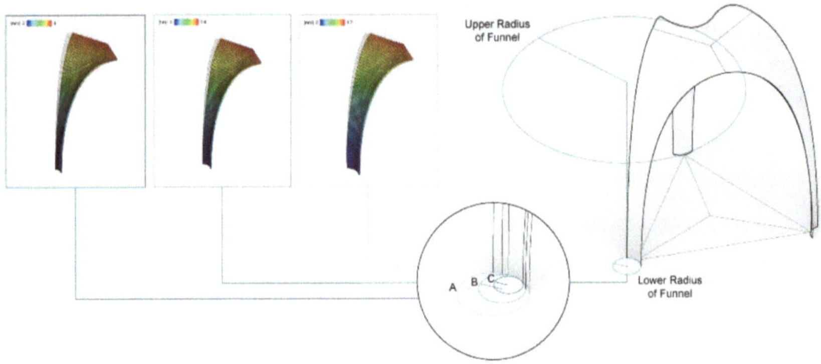

Fig. 10. Analyzing the lower radius of the shell structure. A = 375 mm (1.7 mm displacement), B = 245 mm (2.4 mm displacement), and C = 150 mm (4.0 mm).

Fig. 11. Intersection of two reinforcement parts (edges and rib).

The cross-ply and quasi-isotropic patterns defined the robotic toolpaths to lay SCF-PLA on the formwork. Different strategies were developed to address the structural properties of the textile formwork. Cross-ply [0°/90°] and quasi-isotropic [0°/60°/90°] patterns were developed in each reinforced section. For the edges and the central spine areas, the longer length was considered the main axis for developing the cross-ply [0°/90°] patterns. This means 0° is parallel and 90° is perpendicular to the main axis of the part. The orientation of robotic tool paths to strengthen the edges of the formwork follows the tangency of the edges. The parallel adjacent filaments were printed aligned with the edge, and then another layer of adjacent filaments was printed perpendicular to the edge.

For the rib areas, a quasi-isotropic [0°/60°/90°] was developed to connect the central spine to the edge of the formwork. The infill patterns of 0°/60°/90° improved the binding between adjacent and stacked printed filaments (see Fig. 12).

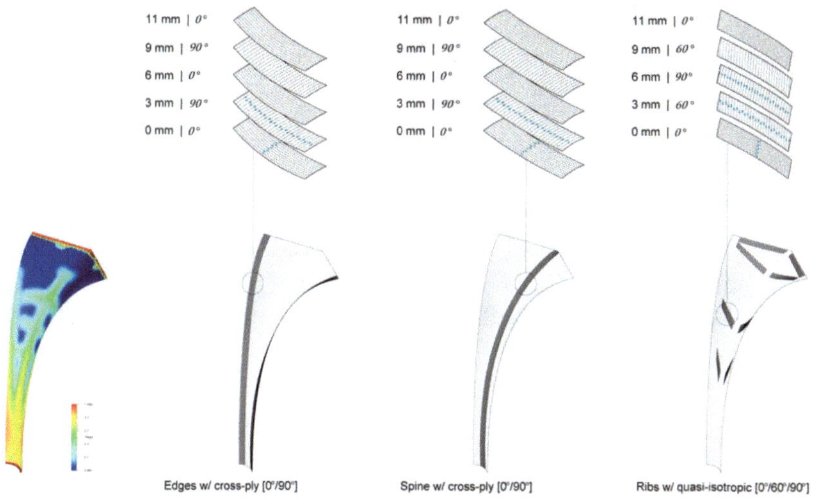

Fig. 12. Development of cross-ply [0°/90°] and quasi-isotropic [0°/60°/90°] toolpaths based on topology optimization in three parts (edges, central spine, and ribs).

A computational tool generated nonplanar toolpaths for extruding SCF-PLA on the formwork. The distance between each filament and the height of the extrusion was obtained by calibrating the 3D printer on a planar surface. The 3D printing calibration involved the size of the nozzle (3 mm), the velocity of the robot, and the extrusion speed, which set the height (2-3 mm) and the width (6 mm) of the filament (see Fig. 13). These parameters were used to define the distance between adjacent filaments and the height of the upper layer. The total reinforcement height for all three partitions is 11 mm. Some areas required additional reinforcement; in these locations the height of the extrusion reaches 18 mm.

Fig. 13. Cross-ply [0°/90°] patterns of reinforcing the edges of the fabric formwork.

Fig. 14. Flexible textile formwork as a double-curved shell structure reinforced with SCF-PLA.

4 Results

The developed nonplanar robotic 3D printing method was investigated by reinforcing a single axisymmetric column, which is a funnel structure section. The mechanical properties of SCF-PLA were enhanced by employing cross-ply [0°/90°] and quasi-isotropic [0°/60°/90] toolpaths. The association of these patterns with the structural performance of the formwork enhances the structural integrity of the shell structure. Preliminary observations indicate that this structure, after being released from the framework, has no displacement in any direction. Additionally, the thickness of 3D printed materials, ranging between 11 mm–18 mm, was significantly less than the 20 mm thickness assumed in the finite element simulation. It should be noted that the finite element simulation considered CFRP material properties, whereas this project applied PLA mixed with 15% short carbon fibers.

The energy stored in the pre-stretched textile is controlled by using the patterning central spine and rib systems as constraints. The patterns are generated by offsetting the surface and mapping the cross-ply and quasi-isotropic toolpaths on it. The height of the initial layer was adjusted to press printing materials into the textile substrates. Upon building the first layer, the other layers are printed with the same height. The width of the layers is controlled to achieve the interface bonds between the adjacent printed filaments.

The modifications of the initial layers create a strong bond between the melted SCF-PLA and the textile substrate. This bond is strong enough to hold the fabric in place. This multi-material infusion developed a hybrid composite that maintains the integrity of the form against both delamination and de-sizing. In addition, it helps rigidify the non-reinforced areas of the pre-stretched fabric (see Fig. 14).

This project, as a novel approach to large-scale 3D printing on a pre-stretched textile, challenges the anisotropic properties of the thermoplastics and presents the effects of patterning on CLFFF printing to develop a lightweight hybrid shell structure. The development of a patterning mechanism of cross-ply and quasi-isotropic toolpaths was effective for the reinforcement of the textile formwork.

Through an analytical approach, the robotic arm's reach and extrudability of the pellet extruder head were perpendicularly aligned to extrude materials with fixed feedrates onto the textile substrate. The reinforcement sequence was associated with the reach of the robot's arm and the coordination of the mobile platform. The height of the reinforcement layers ranged between 11 mm and 18 mm with a maximum width of 60 mm, resulting in a lightweight structural system. The reinforced formwork is a thin-shell structure with a complex geometry. The final demonstrator, including the shell structure and the base, has a height of 2300 mm with an across span of 900 mm. This demonstrator was reinforced with 10 kg of SCF-PLA pellets.

Fig. 15. Close shot of the final demonstrator.

5 Conclusion and Discussion

This research demonstrates the benefits of reinforcing textile formwork with SFC-PLA pellets. It involves the development of a composite system where textile formwork is part of the composite structure. To overcome the strength deficiency of SFC-PLA, a cross-ply and quasi-isotropic patterning is applied to strengthen the material in all directions, which also reduces the printing material. This approach offers an effective way of generating toolpaths for nonplanar robotic 3D printing. Specifically in the context of large-scale additive manufacturing, these patterning techniques can shorten the print time while also reducing printed materials in the structure (see Fig. 15).

This project entails a computational framework to address the limitation of robotic pellet 3D printing on a double-curved shell. This framework presents an alternative technique to simulate robotic fabrication systems. This approach interpolates the extrusion direction between the flexible formwork and the extrudability of the tool where the knowledge about the fabrication condition is limited.

This patterning system could be further developed by introducing a sensory system to control the deposition of material on the textile substrate in real-time. However, this requires uninterrupted communication between the sensory system, the decision-making system, and the operating system, which will be a challenge for rapid additive manufacturing.

Acknowledgements. The author would like to thank student research assistants Avery Edson, Juliana Jackson, Eli Sobel, and Tabi Summers for their invaluable assistance. Thanks also to Melissa Goldman, fabrication lab manager of the UVA School of Architecture; Dr. Trevor Kemp, fabrication facilities assistant manager of the UVA School of Architecture; and Andrew Spears, fabrication lab technician of the UVA School of Architecture for their profound support.

References

1. Peng, T., Kellens, K., Tang, R., Chen, C., Chen, G.: Sustainability of additive manufacturing: An overview on its energy demand and environmental impact. Addit. Manuf. **21**, 694–704 (2018). https://doi.org/10.1016/j.addma.2018.04.022
2. Dixit, M.K.: 3-D printing in building construction: A literature review of opportunities and challenges of reducing life cycle energy and carbon of buildings. IOP Conf. Ser.: Earth Environ. Sci. **290**, 012012 (2019). https://doi.org/10.1088/1755-1315/290/1/012012
3. Guerguis, M., et al.: High performance 3D printed façade with integrated energy: Built works and advancements in computational simulation. In: Advanced Buildign Skins (2017)
4. Sarakinioti, M.-V., et al.: Development and prototyping of an integrated 3D-printed façade for thermal regulation in complex geometries. J. Facade Des. Eng., Vol. 6 No. 2 029–040 (2018): Special Issue ICAE2018 (2018). doi:https://doi.org/10.7480/jfde.2018.2.2081
5. Leschok, M., et al.: 3D printing facades: Design, fabrication, and assessment methods. Autom. Constr. 152 104918 (2023). doi:https://doi.org/10.1016/j.autcon.2023.104918
6. Taseva, Y., Eftekhar, N.I., Kwon, H., Leschok, M., Dillenburger, B.: Large-scale 3D printing for functionally-graded facade. In: RE: Anthropocene, Design in the Age of Humans-Proceedings of the 25th CAADRIA Conference, pp. 183–192 (2020)
7. Jipa, A., Bernhard, M., Dillenburger, B., Ruffray, N., Wangler, T., Flatt, R.J.: skelETHon Formwork: 3D printed plastic formwork for load-bearing concrete structures. In: Proceedings of the 21st Congreso Internacional de la Sociedad Iberoamericana de Gráfica Digital, pp. 345–352 (2017)
8. Leschok, M., Dillenburger, B.: Dissolvable 3DP formwork. In: Bieg, K., Briscoe, D., Odom, C. (eds.) Ubiquity and autonomy. Paper proceedings of the 39th annual conference of the Association for Computer Aided Design in Architecture: ACADIA 2019, October 24–26, 2019, the University of Texas at Austin School of Architecture. Association for Computer Aided Design in Architecture (ACADIA), United States (2019)
9. Doyle, S.E., Hunt, E.L.: Dissolvable 3D printed formwork. In: Bieg, K., Briscoe, D., Odom, C. (eds.) Ubiquity and autonomy. Paper proceedings of the 39th annual conference of the Association for Computer Aided Design in Architecture: ACADIA 2019, October 24–26, 2019, the University of Texas at Austin School of Architecture. Association for Computer Aided Design in Architecture (ACADIA), United States (2019)

10. Burger, J., et al.: Eggshell: Ultra-thin three-dimensional printed formwork for concrete structures. 3D Printing and Additive Manufacturing. **7**, 48–59 (2020)
11. Bao, D.W., Yan, X., Snooks, R., Xie, Y.M.: Bioinspired generative architectural design form-finding and advanced robotic fabrication based on structural performance. In: Yuan, P.F., Xie, M., Leach, N., Yao, J., Wang, X. (eds.) Architectural Intelligence, pp. 147–170. Springer Nature Singapore, Singapore (2020)
12. Felbrich, B., et al.: A novel rapid additive manufacturing concept for architectural composite shell construction inspired by the shell formation in land snails. Bioinspir. Biomim. **13**, 026010 (2018). https://doi.org/10.1088/1748-3190/aaa50d
13. Kwon, H., Eichenhofer, M., Kyttas, T., Dillenburger, B.: Digital composites: Robotic 3D printing of continuous carbon fiber-reinforced plastics for functionally-graded building components. In: Willmann, J., Block, P., Hutter, M., Byrne, K., Schork, T. (eds.) Robotic Fabrication in Architecture, Art and Design 2018, pp. 363–376. Springer International Publishing, Cham (2019)
14. Mohamed, H., Bao, D.W., Snooks, R.: Super composite: Carbon fibre infused 3D printed tectonics. In: Yuan, P.F., Yao, J., Yan, C., Wang, X., Leach, N. (eds.) Proceedings of the 2020 DigitalFUTURES, pp. 297–308. Springer Singapore, Singapore (2021)
15. Maqsood, N., Rimašauskas, M.: Characterization of carbon fiber reinforced PLA composites manufactured by fused deposition modeling. Composites Part C: Open Access. **4**, 100112 (2021). https://doi.org/10.1016/j.jcomc.2021.100112
16. Song, H.-C., Ray, N., Sokolov, D., Lefebvre, S.: Anti-aliasing for fused filament deposition. Comput. Aided Des. **89**, 25–34 (2017). https://doi.org/10.1016/j.cad.2017.04.001
17. Cendrero, A.M., Fortunato, G.M., Munoz-Guijosa, J.M., de, M.,.C., Díaz Lantada, A.: Benefits of non-planar printing strategies towards eco-efficient 3D printing. Sustainability. **13**(4), 1599 (2021). https://doi.org/10.3390/su13041599
18. Mitropoulou, I., Bernhard, M., Dillenburger, B.: Print paths key-framing. In: Whiting, E., Hart, J., Sung, C., Peek, N., Akbarzadeh, M., Aukes, D., Schulz, A., Taylor, H., Kim, J. (eds.) Symposium on Computational Fabrication. SCF '20: Symposium on Computational Fabrication, Virtual Event USA, 05 11 2020 06 11 2020, pp. 1–10. ACM, New York, NY, USA (2020). doi:https://doi.org/10.1145/3424630.3425408
19. Mitropoulou, I., Bernhard, M., Dillenburger, B.: Nonplanar 3D Printing of Bifurcating Forms. 3D Printing and Additive Manufacturing. **9**(3), 189–202 (2022). https://doi.org/10.1089/3dp.2021.0023
20. Shi, X., et al.: A novel path generation approach for robotic spatial printing of branching geometry. Buildings. **12**(12), 2247 (2022). https://doi.org/10.3390/buildings12122247
21. Zhao, G., Ma, G., Feng, J., Xiao, W.: Nonplanar slicing and path generation methods for robotic additive manufacturing. Int. J. Adv. Manuf. Technol. **96**, 3149–3159 (2018). https://doi.org/10.1007/s00170-018-1772-9
22. Chakraborty, D., Aneesh Reddy, B., Roy Choudhury, A.: Extruder path generation for Curved Layer Fused Deposition Modeling. Comput. Aided Des. **40**(2), 235–243 (2008). https://doi.org/10.1016/j.cad.2007.10.014
23. Singamneni, S., Roychoudhury, A., Diegel, O., Huang, B.: Modeling and evaluation of curved layer fused deposition. J. Mater. Process. Technol. **212**(1), 27–35 (2012). https://doi.org/10.1016/j.jmatprotec.2011.08.001
24. Allen, R.J., Trask, R.S.: An experimental demonstration of effective Curved Layer Fused Filament Fabrication utilising a parallel deposition robot. Addit. Manuf. **8**, 78–87 (2015). https://doi.org/10.1016/j.addma.2015.09.001
25. Lim, S., Buswell, R.A., Valentine, P.J., Piker, D., Austin, S.A., de, K.,.X.: Modelling curved-layered printing paths for fabricating large-scale construction components. Addit. Manuf. **12**(B), 216–230 (2016). https://doi.org/10.1016/j.addma.2016.06.004

26. Chen, L., Chung, M.-F., Tian, Y., Joneja, A., Tang, K.: Variable-depth curved layer fused deposition modeling of thin-shells. Robot. Comput. Integr. Manuf. **57**, 422–434 (2019). https://doi.org/10.1016/j.rcim.2018.12.016
27. Xie, F., Chen, L., Li, Z., Tang, K.: Path smoothing and feed rate planning for robotic curved layer additive manufacturing. Robot. Comput. Integr. Manuf. **65**, 101967 (2020). https://doi.org/10.1016/j.rcim.2020.101967
28. Laarman, J., Jokic, S., Novikov, P., Fraguada, L.E., Markopoulou, A.: Anti-gravity additive manufacturing. In: Fabricate: Negotiating Design & Making, pp. 192–197 (2014)
29. Hack, N., Lauer, W.V.: Mesh-Mould: Robotically fabricated spatial meshes as reinforced concrete formwork. Archit Design. **84**, 44–53 (2014). https://doi.org/10.1002/ad.1753
30. Yuan, P.F., Meng, H., Yu, L., Zhang, L.: Robotic multi-dimensional printing based on structural performance. In: Reinhardt, D., Saunders, R., BURRY, J. (eds.) Robotic Fabrication in Architecture, Art and Design 2016, pp. 92–105. Springer International Publishing, Cham (2016)
31. Jimenez Garcia, M., Soler, V., Retsin, G.: Robotic spatial printing. In: Fioravanti, A., Cursi, S., Elahmar, S., Gargaro, S., Loffreda, G., Novembri, G. (eds.) Proceedings of the 35th International Conference on Education and Research in Computer Aided Architectural Design in Europe (eCAADe) [Volume 2]. eCAADe 2017: ShoCK! – Sharing of Computable Knowledge!, Rome, Italy, 9/19/2017–9/21/2017, pp. 143–150. eCAADe (2017). doi:https://doi.org/10.52842/conf.ecaade.2017.2.143
32. Faircloth, B.: Plastics now. In: On architecture's Relationship to a Continuously Emerging Material. Routledge Taylor & Francis Group, London, New York (2015)
33. Doerstelmann, M., et al.: ICD/ITKE Research Pavilion 2014–15: Fibre placement on a pneumatic body based on a water spider web. Archit. Design. **85**, 60–65 (2015). https://doi.org/10.1002/ad.1955
34. Körner, A., et al.: Integrative design and fabrication methodology for bio-inspired folding mechanisms for architectural applications. Comput. Aided Des. **133**, 102988 (2020). https://doi.org/10.1016/j.cad.2020.102988
35. Ferreira, R.T.L., Amatte, I.C., Dutra, T.A., Bürger, D.: Experimental characterization and micrography of 3D printed PLA and PLA reinforced with short carbon fibers. Compos. Part B. **124**, 88–100 (2017). https://doi.org/10.1016/j.compositesb.2017.05.013
36. Song, Y., Li, Y., Song, W., Yee, K., Lee, K.-Y., Tagarielli, V.L.: Measurements of the mechanical response of unidirectional 3D-printed PLA. Mater. Des. **123**, 154–164 (2017). https://doi.org/10.1016/j.matdes.2017.03.051
37. Grant, A., Regez, B., Kocak, S., Huber, J.D., Mooers, A.: Anisotropic properties of 3-D printed Poly Lactic Acid (PLA) and Acrylonitrile Butadiene Styrene (ABS) plastics. Results Mater. **12**, 100227 (2021). https://doi.org/10.1016/j.rinma.2021.100227
38. Dou, H., et al.: Effect of Process Parameters on Tensile Mechanical Properties of 3D Printing Continuous Carbon Fiber-Reinforced PLA Composites, Materials (Basel, Switzerland) (2020). https://doi.org/10.3390/ma13173850
39. Saeed, K., et al.: Characterization of continuous carbon fibre reinforced 3D printed polymer composites with varying fibre volume fractions. Compos. Struct. **282**, 115033 (2022). https://doi.org/10.1016/j.compstruct.2021.115033
40. Parmiggiani, A., Prato, M., Pizzorni, M.: Effect of the fiber orientation on thc tensile and flexural behavior of continuous carbon fiber composites made via fused filament fabrication. Int. J. Adv. Manuf. Technol. **114**, 2085–2101 (2021). https://doi.org/10.1007/s00170-021-06997-5
41. Cauberg, N., Tysmans, T., Adriaenssens, S., Wastiels, J., Mollaert, M., Belkassem, B.: Shell elements of textile reinforced concrete using fabric formwork: A case study. Adv. Struct. Eng. **9**(3), 189–202 (2012). https://doi.org/10.1260/1369-4332.15.4.677

42. Veenendaal, D., Block, P.: Design process for prototype concrete shells using a hybrid cable-net and fabric formwork. Eng. Struct. **75**, 39–50 (2014). https://doi.org/10.1016/j.engstruct.2014.05.036
43. Popescu, M., Rippmann, M., van Mele, T., Block, P.: Complex concrete casting: Knitting stay-in-place fabric formwork. In: Proceedings of the International Association for Shell and Spatial Structures (IASS) Symposium 2016, Tokyo, Japan (2016)
44. West, M.: The fabric formwork book. In: Methods for Building New Architectural and Structural Forms in Concrete. Routledge, London, New York (2017)
45. Popescu, M., Reiter, L., Liew, A., van Mele, T., Flatt, R.J., Block, P.: Building in concrete with an ultra-lightweight knitted stay-in-place formwork: Prototype of a concrete shell bridge. Structure. **14**, 322–332 (2018). https://doi.org/10.1016/j.istruc.2018.03.001
46. Neuß, J., Kreuziger, M., Grimmelsmann, N., Korger, M., Ehrmann, A.: s. In: Proceedings of the Aachen-Dresden-Denkendorf International Textile Conference, pp. 24–25, Dresden, Germany (2016)
47. Agkathidis, A., Berdos, Y., Brown, A.: Active membranes: 3D printing of elastic fibre patterns on pre-stretched textiles. Int. J. Archit. Comput. **17**(1), 74–87 (2019). https://doi.org/10.1177/1478077118800890
48. Jourdan, D., Skouras, M., Vouga, E., Bousseau, A.: Printing-on-fabric meta-material for self-shaping architectural models. In: Advances in Architectural Geometry (2020)
49. Jourdan, D., Skouras, M., Vouga, E., Bousseau, A.: Computational design of self-actuated surfaces by printing plastic ribbons on stretched fabric. Computer Graphics Forum. **41**, 493–506 (2022). https://doi.org/10.1111/cgf.14489
50. Sabantina, L., Kinzel, F., Ehrmann, A., Finsterbusch, K.: Combining 3D printed forms with textile structures - Mechanical and geometrical properties of multi-material systems. IOP Conf. Ser.: Mater. Sci. Eng. **87**, 012005 (2015). https://doi.org/10.1088/1757-899X/87/1/012005
51. Pei, E., Shen, J., Watling, J.: Direct 3D printing of polymers onto textiles: Experimental studies and applications. Rapid Prototyp. J. **21**(5), 556–571 (2015). https://doi.org/10.1108/RPJ-09-2014-0126
52. Loh, G.H., Sotayo, A., Pei, E.: Development and testing of material extrusion additive manufactured polymer–textile composites. Fash. Text. **8**, 2 (2021). https://doi.org/10.1186/s40691-020-00232-7
53. Spahiu, T., Piperi, E., Ehrmann, A., Shehi, E., Rama, D.: 3D printed geometries on textile fabric for garment production. In: Almeida, H.A., Vasco, J.C. (eds.) Progress in Digital and Physical Manufacturing Lecture Notes in Mechanical Engineering, pp. 271–276. Springer International Publishing, Cham (2020)
54. Korger, M., Bergschneider, J., Lutz, M., Mahltig, B., Finsterbusch, K., Rabe, M.: Possible applications of 3D printing technology on textile substrates. IOP Conf. Ser.: Mater. Sci. Eng. **141**, 012011 (2016). https://doi.org/10.1088/1757-899X/141/1/012011
55. Mpofu, N.S., Mwasiagi, J.I., Nkiwane, L.C., Njuguna, D.: Use of regression to study the effect of fabric parameters on the adhesion of 3D printed PLA polymer onto woven fabrics. Fash. Text. **6**, 24 (2019). https://doi.org/10.1186/s40691-019-0180-6
56. Demir, M., Seki, Y.: Interfacial adhesion strength between FDM-printed PLA parts and surface-treated cellulosic-woven fabrics. Rapid Prototyp. J. **29**(6), 1166–1174 (2023). https://doi.org/10.1108/RPJ-10-2022-0369
57. Kumar, R. et al.: Form exploration on different fabrics using material extrusion based additive manufacturing and end users experience. J. Manuf. Process. **101**, 959–973 (2023). https://doi.org/10.1016/j.jmapro.2023.06.053
58. Döpke, C., Grimmelsmann, N., Ehrmann, A.: 3D printing on knitted fabrics. Melliand. Int. **23**, 97–98 (2017)

59. Grimmelsmann, N., Kreuziger, M., Korger, M., Meissner, H., Ehrmann, A.: Adhesion of 3D printed material on textile substrates. Rapid. Prototyping. J. (2018). https://doi.org/10.1108/RPJ-05-2016-0086
60. Kozior, T., Blachowicz, T., Ehrmann, A.: Adhesion of three-dimensional print-ing on textile fabrics: Inspiration from and for other research areas. J. Eng. Fibers. Fabrics. (2020). https://doi.org/10.1177/1558925020910875
61. Gorlachova, M., Mahltig, B.: 3D-printing on textiles – an investigation on adhe-sion properties of the produced composite materials. J. Polym. Res. (2021). https://doi.org/10.1007/s10965-021-02567-1
62. Spahiu, T., Al-Arabiyat, M., Martens, Y., Ehrmann, A., Piperi, E., Shehi, E.: Adhesion of 3D printing polymers on textile fabrics for garment production. IOP Conf. Ser.: Mater. Sci. Eng. (2018). https://doi.org/10.1088/1757-899X/459/1/012065
63. Spahiu, T., Grimmelsmann, N., Ehrmann, A., Piperi, E., Shehi, E.: Effect of 3D printing on textile fabric. st International Conference "Engineering and Entre-preneurship" Proceedings **1**, 1–7 (2017)
64. Hashemi Sanatgar, R., Campagne, C., Nierstrasz, V.: Investigation of the adhe-sion properties of direct 3D printing of polymers and nanocomposites on tex-tiles: effect of FDM printing process parameters. Appl. Surf. Sci. (2017). https://doi.org/10.1016/j.apsusc.2017.01.112
65. Malengier, B., Hertleer, C., van Langenhove, L., Cardon, L.: 3D printing on textiles : testing of adhesion. In: International Conference on Intelligent Textiles and Mass Customisation, p. 6 (2017)
66. Kozior, T., Döpke, C., Grimmelsmann, N., Juhász Junger, I., Ehrmann, A.: In-fluence of fabric pretreatment on adhesion of three-dimensional printed material on textile substrates. Adv. Mech. Eng. (2018). https://doi.org/10.1177/1687814018792316
67. Singh, A., Yadav, P.K., Singh, K., Bhaskar, J., Kumar, A.: Design of 3D Printed Fabric for Fashion and Functional Applications. In: Singari, R.M., Mathiya-zhagan, K., Kumar, H. (eds.) Advances in Manufacturing and Industrial Engi-neering. Lecture Notes in Mechanical Engineering, pp. 729–735. Springer Nature Singapore, Singapore (2021)
68. Muenks, D., Eckelmann, L., Kyosev, Y.: Non-planar 3D printed elements on textile substrate using a fused filament fabrication 3D printer. Autex. Res. J. (2022). https://doi.org/10.2478/aut-2022-0026
69. Pham, L., Lu, G., Tran, P.: Influences of printing pattern on mechanical performance of three-dimensional-printed fiber-reinforced concrete. 3D Print. Addit. Manufact. (2022). https://doi.org/10.1089/3dp.2020.0172

ABB
AUTODESK
Technology Centers

Design for Additive Construction: Uncovering Challenges and Opportunities through the Robotic 3D Printing of a Pedestrian Bridge

Peter Storey[1](✉), Nathan King[1,2,3,4], James Donnelly[1], and Shadi Al-Ariss[5]

[1] Autodesk, San Francisco, California, USA
peter.storey@autodesk.com
[2] Virginia Tech, Center for Design Research, Blacksburg, Virginia, USA
[3] Harvard Graduate School of Design, Cambridge, Massachusetts, USA
[4] University of Pennsylvania, Philadelphia, Pennsylvania, USA
[5] Dar Al-Handasah, Beirut, Lebanon

Abstract. Based primarily on Additive Manufacturing Technologies (AMTs), Additive Construction (AC) is emerging as a viable construction methodology in a variety of contexts. While the full potential and consequences of these technologies are still far from understood, off and on-site solutions are being commercialized and adopted. Material extrusion is one AMT that has seen widespread scaling and has been used to create additively constructed buildings and building components from polymers and cementitious materials. Focusing on construction-scale thermoplastic polymer deposition, this paper presents a novel design experiment—the production of a five-meter pedestrian bridge—to explore opportunities and identify limitations of generative design techniques for AC, seven degree of freedom (7-DOF) robotic material extrusion, and hybrid processes for embedded components and feature addition. A custom robotic end effector was developed that incorporated a pellet-based polymer extrusion system, subtractive milling spindle, and infra-red temperature sensors. During this research, computational generative design is utilized to produce design iterations of the bridge based on load case, material properties, manufacturing parameters of the 7-DOF robotic work cell, and added evaluation criteria developed to enhance geometric complexity requiring discontinuous toolpaths. Each stage of production is expounded to uncover opportunities and limitations of the process. Based on lessons learned during production an evaluation of large format material extrusion for use in AC is described. Opportunities for further development are identified relating to limitations in generative design tools, thermal considerations for consistent layer bonding, strategies for build plate incorporation, labor consequences of limited automation of installation of embedded components, and mechanisms for further advancement of polymer-based AC outside of controlled manufacturing environments.

Keywords: Additive Construction · Construction Automation · Robotic Fabrication

M. Yablonina et al. (Eds.): Robotic Fabrication in Architecture, Art and Design 2024, pp. 232–249, 2026.
https://doi.org/10.1007/978-3-032-22607-5_15

1 Introduction

1.1 Additive Construction

While almost all construction methods could arguably be described as additive, a new distinction known as Additive Construction (AC), drawing heavily from Additive Manufacturing technologies (AMTs), is emerging. AC has the potential to leverage advanced computational design and robotic fabrication technologies to streamline design, manufacturing and assembly workflows, while reducing the dependence on manual labor. (ASTM International, 2023).

AC is experiencing rapid advancements, with a list of successful projects and companies growing year-on-year (Paolini et al. 2019, Wohlers Associates, 2022, Ma et al., 2022, Puzatova et al., 2022), and many notable bridge projects such as the MX3D bridge (MX3D, 2021), the Striatus bridge (Bhooshan et al., 2022), bicycle bridges in the Netherlands (Salet et al., 2018), and others. These processes are primarily based on material extrusion principles, but processes such as Particle-bed Binding (PBB) are also experiencing growth in research interest (Lowke et al., 2018). However, challenges still exist that limit widespread technology adoption. Principal among them is the need for standards that govern the implementation of additive manufacturing technologies in construction applications, defining best practices in design workflows and material quality control. (ASTM International, 2023).

1.2 Opportunities for Additive Construction

Few examples of construction-scale material extrusion approach the geometric complexity that is typical of the most basic desktop-scale material extrusion processes such as in Fused Filament Fabrication. In AC applications, a need for continuous extrusion typically limits design freedom, meaning that examples are often prismatic extrusions of 2-dimensional curves—similar to constructions using standard masonry practices. Often, continuous extrusion is necessitated by material rheology or by mechanical limitations of the equipment used (Paolini et al., 2019; Panda et al., 2023). A promise of AMTs is to reduce material usage while increasing part performance and enabling new design possibility through increased formal and functional geometric complexity. For AC to approach similar enhancements in Architecture, Engineering and Construction (AEC), new methods of understanding geometric complexity in construction must be developed in the context of additive processes.

This paper describes the production of a five-meter pedestrian bridge that was constructed using AC by way of a hybrid seven degree of freedom (7-DOF) robotic thermoplastic material extrusion system. At each stage of development, key digital and physical obstacles are uncovered and evaluated with the goal of capturing knowledge that advances next stages in the utilization of AC for the construction of complex geometries.

2 Design Experiment – Five-Meter Pedestrian Bridge

In this case, the pedestrian bridge is chosen due to its performance demands including span, slope, human interaction, as well as programmatic considerations including the need for a demonstration piece as an industry collaboration and showcase of a variety of technologies. As a demonstrator of advanced computational design and analysis tools that explores both potential and limitation of the robotic AC process, formal and functional geometric complexity is programmed into the conceptual bridge design. Digital part geometry is further developed based on feedback from the physical production of the bridge itself. The bridge is constructed from robotically deposited recycled fiber-reinforced polymers using a Fused Granular Fabrication (FGF) process (Woern et al., 2018; Roschli et al., 2019; Kalle et al., 2023), using a novel slicing method that leveraged the flexibility of the 7-DOF work cell. In addition to the FGF process, a secondary milling operation created a hybrid production environment and allowed the addition of features and embeds during the print.

Understanding the limitations of the production process including material properties and geometric limitations of FGF 3D printing is critical to Design for Additive Construction (DfAC). The conceptual design phase included establishment of design criteria to guide the generative design algorithm. Criteria were informed by the capabilities of the ABB IRB-4600 6-axis industrial robotic arm, and 10 m linear track providing a seventh degree of freedom. The constraints of the robotic work cell were factored into the design— a width of 1.5 m limited by the reach of the 6-axis robot, and a length of 5 m within the limits of the linear track. A streamlined digital workflow from initial concept to machine-ready code was developed (Fig. 1).

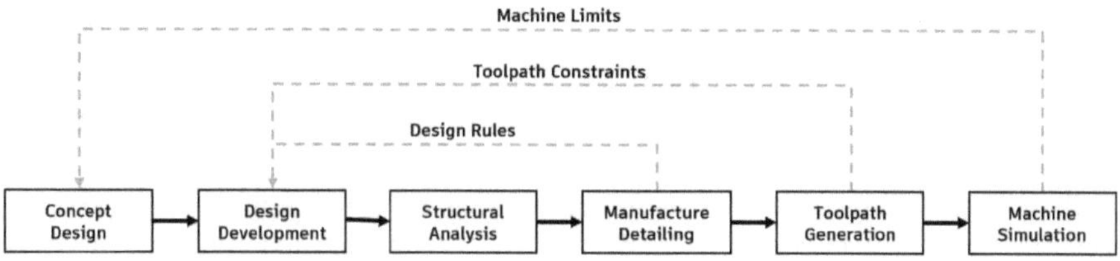

Fig. 1. Diagram describing the design-to-manufacture workflow, including iterative feedback loops employed for this project.

2.1 Process-Informed Generative Design

Post-Industrial PETG (PIPG) with 30% glass fiber content by weight, supplied by Mitsubishi Chemical Performance Polymers (MCPP) was the chosen material. PIPG balanced desirable mechanical properties (Technical Data Sheet), excellent printability (verified through in-house trials), and a roadmap towards sustainable sourcing. Testing revealed that PIPG was 46% weaker out-of-plane than in-plane (Storey et al., Forthcoming), which confirmed material anisotropy—a key structural consideration that should inform the computational generative design study. However, available topology optimization and structural simulation software tools could not account for anisotropy. As such, the design used the tensile strength value in the weaker direction, resulting in a product that is heavier than necessary because the weakest material parameters were assumed in all directions. Future studies would benefit from topology optimization and simulation tools that account for anisotropy of material extrusion AMTs.

With FGF, an overhang angle of 45° is achievable, meaning that the design problem can be presented as shown in the first diagram in Fig. 2.

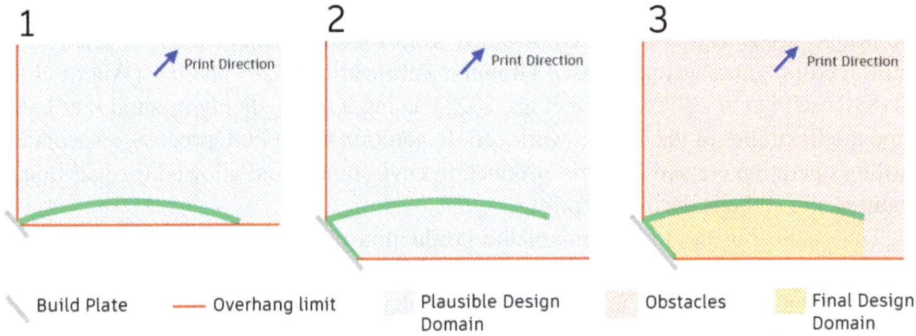

Fig. 2. Design domain for the 5 m bridge. The green regions represent the Preserve Geometry, which should not be altered by the generative design algorithm.

The 45° overhang angle limits the amount of geometry generated underneath the bridge deck, which offers negligible surface area for adhesion to the build plate, so a surface was added to expand the geometric domain of the model (Fig. 4, middle), thereby providing a larger surface for the cantilevering part to adhere to the build plate during printing. Finally, geometry exclusion regions are defined including the pedestrian path and the space occupied by the end conditions of the bridge (Fig. 4, right). The geometric limitations informed requirements for generative design studies, which were conducted using Autodesk Fusion.

Two load conditions consider the forces experienced both during printing and during end-use. The first load case, Fig. 3-left, considers maximum load due to self-weight during production with the component acting as a pure cantilever. The second load case (Fig. 3-right) considers the maximum load experienced after installation, where the bridge remains rigidly fixed at the "build plate" end, and simply supported at the other.

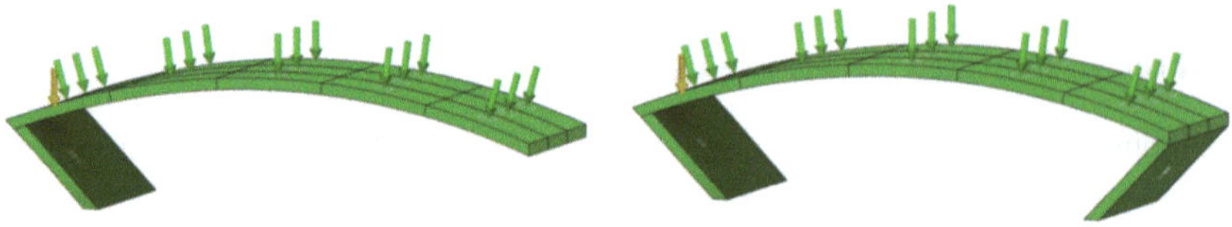

Fig. 3. Two load cases used during the design process. Left: Pure cantilever condition with I maximum self-weight during fabrication. Right: Simply supported condition with additional load during use.

2.2 Validation of Generative Design Results

The study aimed to reconcile oft-conflicting goals of producing an artifact that is rapidly manufacturable, minimizes mass, is aesthetically engaging, and offers sufficient geometric complexity to exemplify the strengths of the FGF manufacturing method. Monolithic designs are typically faster to produce because they require fewer pauses in material deposition. However, a highly branched design typically achieves a greater mass reduction, while offering the geometric intricacy that this project desired to demonstrate, as explained in Sect. 1.2.

Autodesk Fusion's generative design tool integrates manufacturing constraints, which guide the optimization towards feasible designs for the desired manufacturing method. For example, Additive Manufacturing constraints allow the user to specify a minimum thickness and maximum overhang angle. By considering multiple manufacturing constraints, and considering earlier iterations in a study, rather than necessarily selecting the final maximally optimized geometry, it is possible to achieve a suitable balance to the conflicting goals introduced earlier in this section.

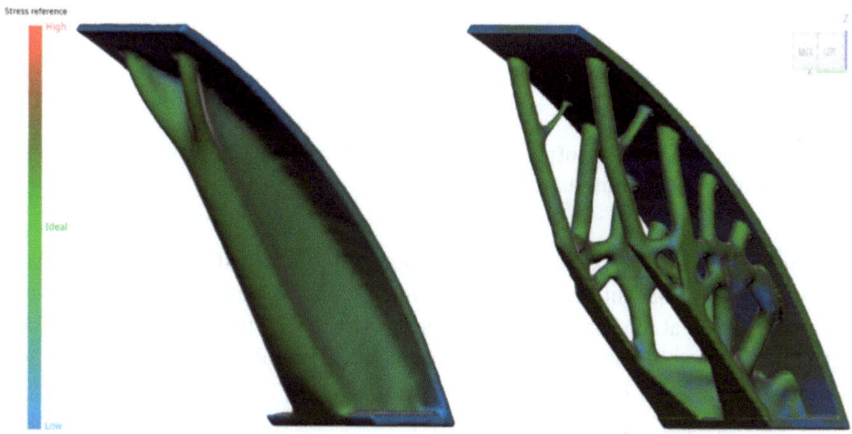

Fig. 4. Two contrasting outcomes of generative design studies from Autodesk Fusion, for the purpose of illustration. Left: No manufacturing constraints, iteration 20 of 42, volume 2.2m^3. Right: Additive Constraints, 45° maximum overhang angle, iteration 16 of 16, volume 1.6m^3.

2.3 Manufacture Detailing

Manual design adjustments were made to facilitate the installation of mounting hardware and reduce curvature of the pedestrian path to meet safety guidelines. As such, generative design acted as an aid for design exploration, with manual changes to meet more nuanced requirements as the design phase progressed (Fig. 5).

Additional manufacturing parameters are considered, including the width and height of the deposited polymer bead. Previous research suggests that the bead width for the FGF process should be 2 to 4 times larger than the bead height (Chesser et al. 2019), and guidelines that dictate appropriate flow rates have been published (Duty et al., 2018). This existing knowledge served as a baseline for trials with the PIPG material in this project, which concluded that a 3 mm bead height and 8 mm bead width yielded the

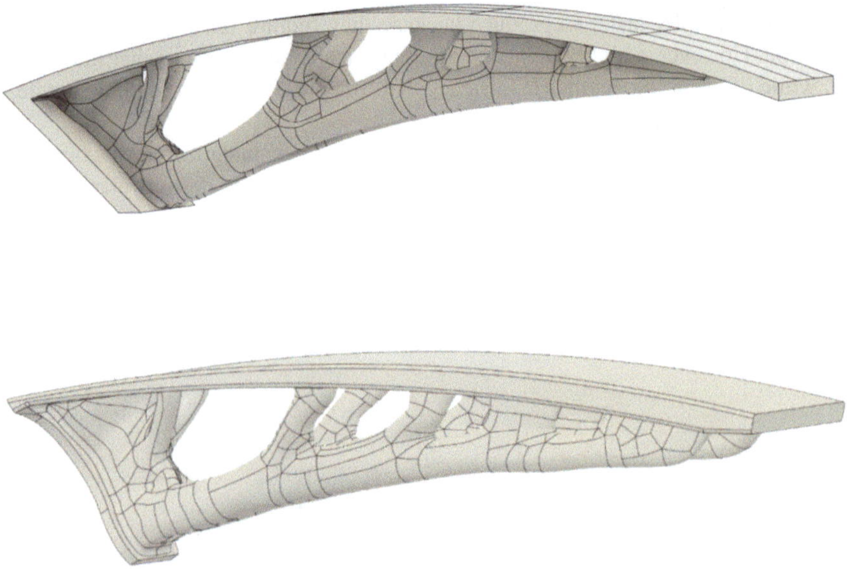

Fig. 5. Top: The exported result of the chosen generative design study. Bottom: The final geometry used for printing, following manual design edits.

best print quality for the fastest material throughput. Further, the outer surface would be printed as two adjacent beads, or two perimeter passes, which would ensure a solid outer surface even at internal overhang angles as shallow as 10°, compared to 20° for a single 3x8mm bead, while providing an additional structural safety factor if there were any defect in interlayer bonding.

Fig. 6. A comparison of gyroid infills for simulation and fabrication. Left: Gyroid lattice used for simulation, generated by the Autodesk Fusion Volumetric Lattice tool. Right: Gyroid Infill used in the printing toolpath, generated by the Autodesk Fusion FFF slicer.

Infill geometry can be distributed homogenously or heterogeneously based on performance needs. Due to their high strength to density ratio, gyroid infill patterns have seen increased adoption within Additive Manufacturing (Silva et al., 2021; Pais et al., 2022, Podroužek, 2019). Being continuously curved, gyroid infill patterns require significantly more geometric data than infills composed of straight lines, leading to larger toolpath file sizes. However, with increased target density comes an increased ability to control the toolpath and material distribution. Typically, gyroid infill patterns are generated by dedicated 3D printing slicers, which produce a complex data set that is computationally intensive to convert back into a solid model and to perform additional structural simulation. To avoid this translation, the gyroid infill created by the 3D printing slicer was first approximated using the Volumetric Lattice tool within Autodesk Fusion. Conversely, it is not practicable to directly create an FGF toolpath based on the latticed design because the toolpath would not be as smooth as a gyroid infill created with the dedicated slicer. This means that the model used for simulation was not precisely the same as the geometry that would be printed, which introduces an area for uncertainty between the real and simulated models. Future projects would benefit from either printing the Volumetric Lattice geometry directly via an alternative AC technology like particle bed binding (Panda et al., 2023, pp. 79–108) or developing a method to convert the stratified slice data into solid geometry for simulation.

2.4 Structural Verification

Although generative design algorithms perform structural analysis as part of their iterative process, manual design modifications and additional decisions relating to bead dimensions and infill patterns required optimization and validation through further structural analysis. An optimal combination of shell thickness and infill density is needed to maximize the bridge's stiffness while minimizing weight. Varying shell thicknesses and gyroid volumetric lattice densities were assigned and exported together as a solid model for independent simulation as a finite element model using Ansys 2022 R2. Figure 6 illustrates results using the minimum considered shell thickness and infill densities.

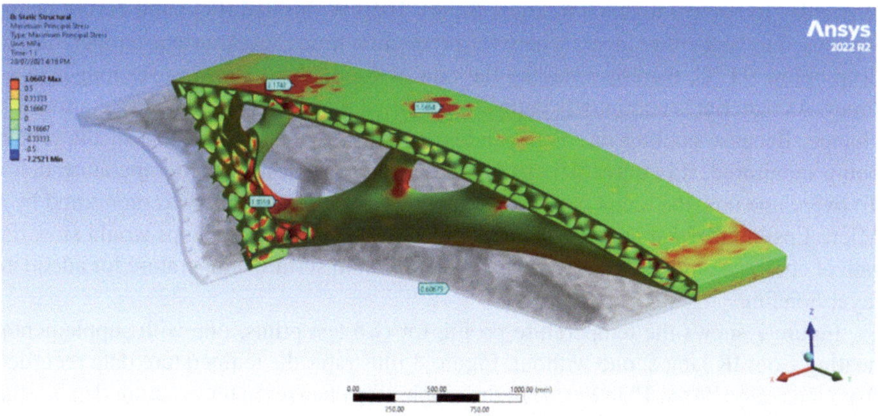

Fig. 7. Visualization of maximum principal stress data with a representative gyroid infill pattern.

The studies used a static load of 5kN per m^2 on the bridge deck. The results in Fig. 6 suggest a maximum compressive stress of 7.25 MPa and a maximum tensile stress of 3.06 MPa. These values are comfortably below the compressive and tensile yield strengths reported in Storey et al. (Forthcoming) of 52.9 MPa and 26 MPa respectively. A maximum shear stress of 7.96 MPa was reported, with a maximum deflection of 2.54 mm.

These results indicated that a single bead width of 8 mm could be used, with the lowest printable infill density that meets the printing constraints. However, as previously discussed, the outer shell was printed with two perimeter beads, for a total shell thickness of 16 mm, as was dictated by process considerations, rather than necessitated by these structural analysis studies. If a smaller bead width of fewer passes had been used, the bridge might have been lighter without compromising the required strength but would not have the added process-safety factor informed by test prints.

3 Methods for Thermoplastic Additive Construction

3.1 Temperature Control for large-scale AC

In AC, it is difficult to maintain a consistent temperature without an enclosed printing environment—a limitation that is exacerbated when considering in-situ printing under variable environmental conditions. With thermoplastics, this is problematic because a narrow temperature window must be maintained to ensure adequate bond between layers. Studies published in Storey et al. (Forthcoming) propose that an optimal surface temperature should be maintained between 95–150 °C to ensure adequate interlayer bonding and part strength when using post-industrial PETG with 30% glass fiber (PIPG).

Large and complex geometries, like the bridge, also result in long and inconsistent layer production times, which makes it difficult to maintain consistent surface temperature. Methods for local reheating have been explored by Kishore et al. (2017) and Wang et al. (2023), but these approaches are either ineffective or not yet commercially available. To overcome this limitation, Solary 1050 W infrared (IR) paint curing lamps were used to heat entire layers whenever production time causes surface temperature to drop below 95 °C, thereby reducing the rate of cooling and allowing for longer layer times. As material is continually deposited, thermal mass and part geometry continually change. Because cooling rate is dependent on surface area and volume of the region being monitored, it is unrealistic to predict accurate values for the temperature delta. To overcome this, the temperature over a set of typical print layers was monitored by a Micro Epsilon thermal imaging camera to determine whether IR lamps would slow the rate of cooling sufficiently to stay above the minimum surface temperature for adequate layer bonding.

Figure 7 shows the temperature profile for two test prints: one with supplemental heating from IR lamps, one without. Figure 8 illustrates the temperature data recorded from the region "Area 1" in Fig. 7. In the graph, step changes in temperature demark the deposition of a new layer of material within the measurement area. Differences in layer time are due to a slightly different toolpath ordering between the two test prints. The graph illustrates that the temperature drops below the minimum temperature during the layers tested without IR lamps, whereas the temperature remains above the minimum temperature for all but one layer with the IR lamps in use (Fig. 9).

Nevertheless, it was deemed critical to monitor real-time temperature of the part during final production. This was facilitated by two MLX90614 non-contact infrared thermometers mounted to the robot end effector. Data from these thermometers were processed by custom scripts on a Linux computer, which communicated with the ABB

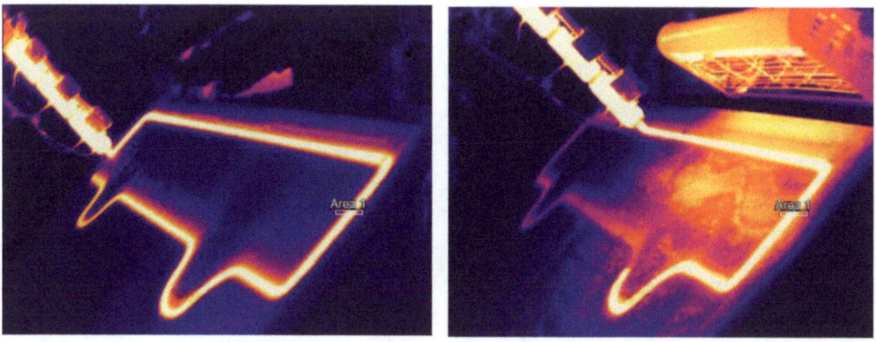

Fig. 8. Images from an infrared camera during test prints. Left: Without IR lamps. Right: With supplemental heat from IR lamps.

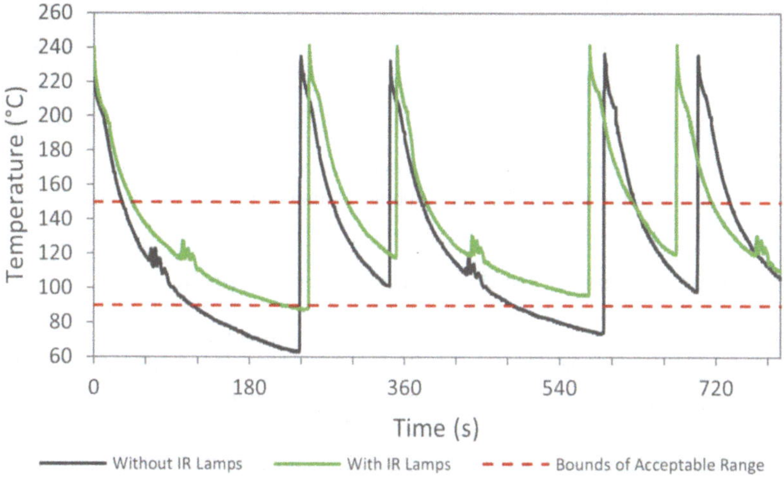

Fig. 9. Average temperature from "Area 1" (Fig. 8.) of two test prints, with and without IR lamps.

robot controller via Ethernet. This setup enabled the system to pause and resume printing based on temperature thresholds, providing a method for rudimentary closed-loop control. Specifically, temperature measurements were taken at the beginning of each layer and printing was paused if the surface temperature exceeded 150 °C. Printing was then automatically resumed once the surface temperature dropped below 150 °C. A fail-safe was implemented so the printing process paused if the surface temperature dropped below 95°, and IR lamps were manually positioned close to the part for an extended period to re-elevate the surface temperature. Once the surface temperature was high enough that a full layer could be completed without the temperature dropping below 95°, printing would be resumed. While these methods were successful for the given task, additional process methods for temperature control must be developed for thermoplastic AC to be viable, such as in Wang et al. (2023).

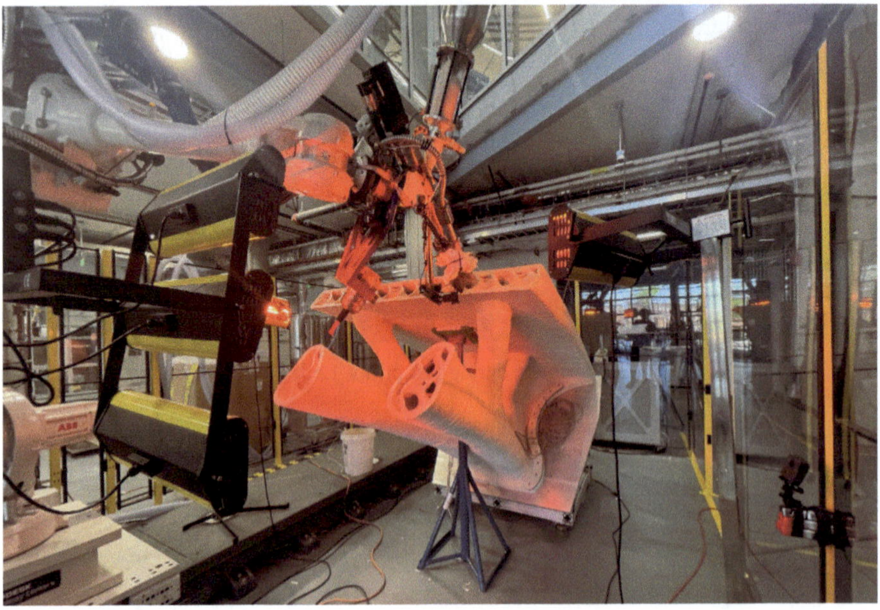

Fig. 10. Realization of the five-meter pedestrian bridge, with IR lamps providing supplemental thermal energy, and non-contact infrared thermometers mounted to the robot end effector.

3.2 Extrusion Parameters for Discontinuous Layer Deposition

When continuous toolpaths are not possible, extrusion should be paused to avoid the deposition of excess material between print segments, but even with an ability to start and stop extrusion via the NC program, poor quality is likely to occur due to the high material throughput and the inertia of material flow (Chesser et al. 2019). To overcome this limitation during the production of the bridge, techniques of creating lead in and lead out motions were developed to allow for discontinuous toolpaths with high print quality. When executing a 'lead in' motion, the length of the motion is optimized to ensure the material forms a full bead when the extruder reaches the first coordinate of the print path segment. Conversely, during 'lead out' motions, extrusion is stopped before the last target in the printing segment, and robot motion continues to wipe excess material on the previously printed bead from the beginning of the toolpath segment. This method serves structural and aesthetic purposes of ensuring proper bonding at the end of the path, while blending excess material from the extruder.

The strategy for infill segments of the toolpath requires extra consideration because infill segments are not closed loops. Straight lead in/lead out moves result in poor quality, as shown on the left of Fig. 10, because excess material is visible on the outside of the part. Reversing path direction during lead out can cause the recently deposited material to stick to the extruder, causing it to curl back up resulting in poor print quality. Therefore, the best solution is for the infill lead moves to follow the nearest perimeter, which provides cooled and consolidated material for the extruder to 'wipe' on without leaving the part boundary, thus blending any excess material on the interior of the print. The implementation of this strategy in software is illustrated in Fig. 11.

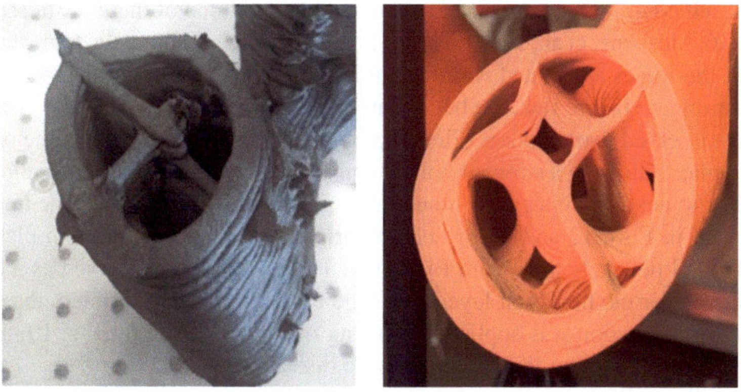

Fig. 11. Examples of lead moves for additive toolpaths. Left: With lead moves, but with no infill optimization. Right: With lead moves, and 'follow perimeter' infill optimization.

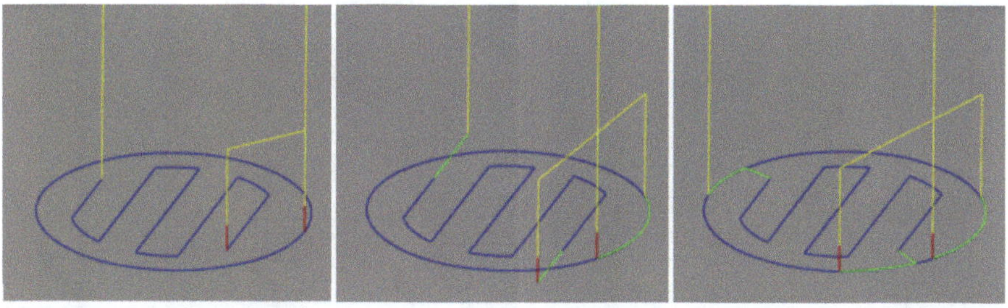

Fig. 12. Illustration of modifications to discontinuous additive toolpaths for improved quality. Left: Without lead moves. Middle: With lead moves. Right: With lead moves and 'follow perimeter' infill optimization.

To the knowledge of the authors, this is a novel approach to achieve high surface quality with discontinuous FGF print paths.

3.3 The Build Plate as a Foundation for AC

In most additive manufacturing processes, parts are printed onto a build plate and removed upon completion. For large format prints using non-planar deposition, the build plate can present both additional challenge and opportunity. If the build plate has a thermal coefficient significantly different to that of the printed material, part failure can occur. If the bond to the build plate is stronger than the bond between layers of the part, it is likely that the part will crack rather than warp. Given that cracks are more problematic than localized warping, it is critical that the difference in thermal coefficients is minimized, and that the weakest point is between the build plate and first layer of the part.

Previous work has demonstrated different methods of preventing delamination while still enabling post-print removal. These methodologies include the use of specific substrate materials and adhesives, as well as heating the build plate to limit thermal differences and control the cooling rate, reducing the likelihood of thermal stresses (Choi et al., 2016; Messimer et al., 2018). These strategies reduce the potential for print failure, however, the forces associated with the angled build plate used for this print were too high. To mitigate failure risk, the build plate became an integral part of the bridge, and in doing so, served two purposes. First, it eliminated the need to demount the part from the build plate, allowing a stronger bolted connection to be used. Second, it allowed the build plate to act as an integrated mounting plate for final installation. The build plate can be seen in Figs. 12 and 13, which comprised of a 25 mm aluminum plate and a layer of PIPG pellets bonded with PVA glue. These PIPG pellets served as a compatible interface for the first print layer. After the initial layers were printed, the stronger mechanical connection was applied, consisting of a series of M10 bolts, 10 mm-thick rubber gaskets, and 6 mm-thick steel plates to distribute the load (Fig. 14).

Fig. 13. Left: Installation of the build plate, with pellets adhered to the print area. Right: The full build plate assembly after printing up to layer 53, performing milling operations, and installing the rubber gaskets, steel load-spreading plates, and M10 bolts.

3.4 Hybrid Manufacture for Embedded Sensor Installation

Milling operations were used to create conduits and appropriate surfaces for the sensors and their accompanying cables. This was achieved with a milling router mounted to the robot end effector via a custom bracket, as shown in Fig. 12. The bracket was designed

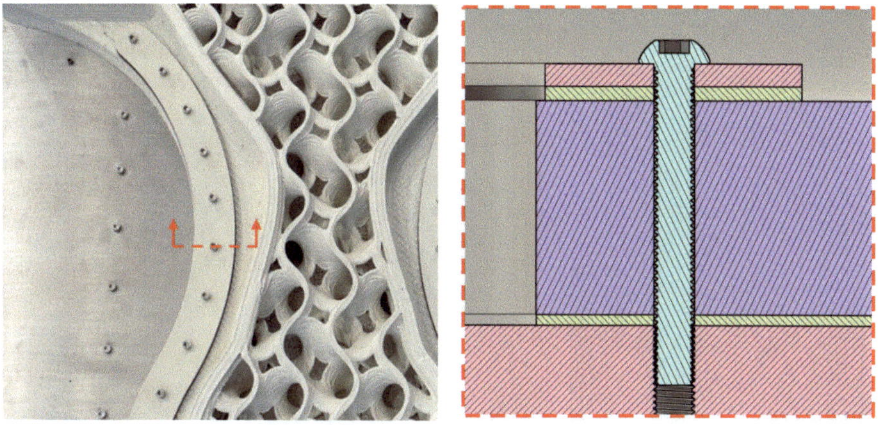

Fig. 14. Cross section of the build plate assembly. Right, from top to bottom: M10 bolt, steel load-spreading plate, rubber gasket, printed part with clearance hole milled for M10 bolt, layer of PIPG pellets, aluminum build plate with tapped hole for M10 bolt.

to allow milling and printing operations to be interleaved without one tool impeding the other, thereby removing the need for any manual adjustments to the end effector during fabrication, and allowing immediate switching between the two processes via robot code.

With finer resolution than possible with the printed bead, hybrid milling processes were successful at creating adequate cable chase and surface mills for proper adhesion and precise sensor placement. However, the introduction of cabling for these sensors within the printed part proved challenging and labor-intensive and was eventually abandoned for simpler installation. For cabling to be included within the part during or after production, special considerations during part infill design are needed and additional chase geometry included within the print. Embedded components do pose potential added value to the AC process, but further development is needed to reduce the labor associate with incorporating these parts during the printing process (Fig. 15).

Two types of strain gauges were used to monitor the behavior of the bridge during use. In the branches of the structure, HBM half-bridge strain gauge rosettes were adhered to milled pockets. In the deck of the bridge, Fisens fibre-bragg grating (FBG) sensors within fiber optic strands were embedded to monitor pressure and strain through movement within the part.

4 Results

The study culminated in the creation of a functional prototype pedestrian bridge, completed after 200 h of printing using 960 kg of material deposited over a 24.16 km toolpath. Post-fabrication, the bridge was mounted on a metal frame for transport, a step unnecessary for a project conducted entirely in-situ. While laboratory testing for structural integrity has yet to be performed, field tests during transport from Boston to New Orleans and then to London UK have shown no fatigue or delamination after a year of use (Figs. 16 and 17).

Fig. 15. Installation of cables for strain gauges in the branches. Left: Cables entering and exiting the part through milled conduits. Right: Final installation of all cables and strain gauges.

Fig. 16. The completed bridge after installation on a custom shipping frame.

Fig. 17. The completed bridge after installation at Autodesk University 2023

5 Conclusion

This paper discusses the computational design and fabrication of a pedestrian bridge through large-scale robotic additive manufacturing using fiber-reinforced polymers. The design experiment is an interdisciplinary synthesis of generative design, robotics, and additive manufacturing that serves as a framework to understand computational and mechanical challenges in polymer-based AC. The following bullets summarize the findings expounded within the paper.

- Industrial robotic manipulators present unique opportunities for non-planar printing, but robot reach limits the possible print volume. Utilizing an external axis that moves the manipulator can significantly increase print volume in one direction. To maximize geometric potential, novel slicers are needed.

•The integration of generative design algorithms informed design using constraints of material properties, fabrication limitations, and structural goals. However, the study highlighted the need for improved tools that consider material anisotropy and toolpath geometry for a holistic approach to generative design that considers the final printed artifact.

•Layer bonding was improved using localized infrared radiation. However, large-format factory-based or on-site production of thermoplastic AC components require carefully considered environmental control based on specific material characterization to ensure adequate layer bonding and appearance. Further development of local environmental control, specifically addressing stabilization of layer temperature is needed to advance this type of AC.

•Through this project novel strategies were developed to lead in and out of print segments to ensure acceptable appearance outcomes in large-format prints requiring discontinuous toolpaths. Further development and generalization of these tooling features is needed.

•Strategic incorporation of the build plate as a part of the final printed component is a valuable strategy for resolving complexity of build plate adhesion and removal. Further study of the opportunity and consequence of this strategy across multiple applications is needed.

•The utilization of hybrid robotic manufacturing processes presents opportunities to precisely incorporate added features and parts during the printing process. However, without further development of automated assembly processes this potential is limited due to increased labor.

The integration of computational design and innovative fabrication techniques demonstrates the capabilities of additive manufacturing in producing complex structures while adhering to structural requirements. Future research could focus on refining temperature control models, developing advanced methods for extrusion control, and exploring automated feedback loops between computational design and fabrication processes. This will further enhance the potential of AC to meet the needs of the construction industry.

Acknowledgements. The authors would like to express their gratitude to our colleagues and friends at Dar Al-Handasah, particularly Ghassan Zein, George Marj, and Charles Malek, for their commitment and advocacy for this project. We also extend our thanks to our colleagues at Autodesk Research and the wider company, especially Brandon Cramer, Stefanie Pender, Fope Bademosi, Taylor Tobin, Greg MacGlashing, Nic Carey, and Samer Bakdach. Also, to the teams at the Autodesk Technology Center in Boston, Autodesk Technology Centre in Birmingham, and the Autodesk Materials Lab in Kilsyth. The skills and expertise of the teams at these facilities were paramount in the success of this project.

References

ASTM International: Advanced Technologies for Digitalization of Construction Industry Roadmap. https://amcoe.org/research-development/publications/ (2023).

Paolini, A., Kollmannsberger, S., Rank, E.: Additive manufacturing in construction: A review on processes, applications, and digital planning methods. Addit. Manuf. **30**, 100894 (2019). https://doi.org/10.1016/j.addma.2019.100894

Wohlers Associates: Building an additive construction future. In: Wohlers Specialty Report on Construction. https://wohlersassociates.com (2022). ISBN 978-1-6220-4964-6.

Ma, G., Buswell, R., Leal da Silva, W.R., Wang, L., Xu, J., Jones, S.Z.: Technology readiness: A global snapshot of 3D concrete printing and the frontiers for development. Cem. Concr. Res. **156**(1), 106774 (2022). https://doi.org/10.1016/j.cemconres.2022.106774

Puzatova, A.V., Shakor, P., Laghi, V., Dmitrieva, M.A.: Large-Scale 3D Printing for Construction Application by Means of Robotic Arm and Gantry 3D Printer: A Review. Buildings. **12**(11), 2023 (2022). https://doi.org/10.3390/buildings12112023

MX3D: MX3D Bridge. https://mx3d.com/industries/infrastructure/mx3d-bridge (2021).

Bhooshan, S., et al.: The Striatus bridge. Archit. Struct. Constr. **2**(4), 521–543 (2022). https://doi.org/10.1007/s44150-022-00051-y

Salet, T.A., Ahmed, Z., Bos, F., Laagland, H.: Design of a 3D printed concrete bridge by testing. Virtual Phys. Prototyp. **13**(3), 222–236 (2018). https://doi.org/10.1080/17452759.2018.1476064

Lowke, D., Dini, E., Perrot, A., Weger, D., Gehlen, C., Dillenburger, B.: Particle-bed 3D printing in concrete construction – Possibilities and challenges. Cem. Concr. Res. **112**, 50–65 (2018). https://doi.org/10.1016/j.cemconres.2018.05.018

Panda, B., Shakor, P., Laghi, V.: Additive Manufacturing for Construction. Emerald Publishing Limited (2023) ISBN 978-0-7277-6641-0.

Woern, A.L., Byard, D.J., Oakley, R.B., Fiedler, M.J., Snabes, S.L., Pearce, J.M.: Fused particle fabrication 3-D Printing: Recycled materials' optimization and mechanical properties. Materials. **11**, 1413 (2018). https://doi.org/10.3390/ma11081413

Roschli, A., et al.: Designing for Big Area Additive Manufacturing. Addit. Manuf. **25**, 275–285 (2019). https://doi.org/10.1016/j.addma.2018.11.006

Kalle, J., Joni, K., Alexander, S.: Potential and challenges of fused granular fabrication in patternmaking. Inter Metalcast. **17**, 2469–2476 (2023). https://doi.org/10.1007/s40962-023-00989-9

Redwood, B., Schöffer, F., Garret, B.: The 3D Printing Handbook: Technologies, design and applications. In: 3D Hubs, 1st edn, (2017)

Somireddy, M., Czekanski, A.: Anisotropic material behavior of 3D printed composite structures – Material extrusion additive manufacturing. Mater. Des. **195**(1), 108953 (2020). https://doi.org/10.1016/j.matdes.2020.108953

Storey, P., Groom, A., Perumal, V., Al-Ariss, S.: Interlayer temperature optimisation: An approach to maximising strength in the large-scale additive manufacture of fibre-reinforced polymers. Forthcoming.

Kishore, V., et al.: Infrared preheating to improve interlayer strength of big area additive manufacturing (BAAM) components. Addit. Manuf. **14**(1), 7–12 (2017). https://doi.org/10.1016/j.addma.2016.11.008

Wang, T., Consul, P., Bublitz, D., Drechsler, K.: Continuous directed laser preheating of big area additive manufacturing. In: Proceedings of the Twenty-Third International Conference on Composite Materials (2023) https://www.iccm-central.org/Proceedings/ICCM23proceedings/papers/ICCM23_Full_Paper_9.pdf

Chesser, P., et al.: Extrusion control for high-quality printing on Big Area Additive Manufacturing (BAAM) systems. Addit. Manuf. **28**(1), 445–455 (2019). https://doi.org/10.1016/j.addma.2019.05.020

Silva, C., Pais, A.I., Caldas, G., et al.: Study on 3D printing of gyroid-based structures for superior structural behaviour. Prog. Addit. Manuf. **6**(1), 689–703 (2021). https://doi.org/10.1007/s40964-021-00191-5

Pais, A., Alves, J.L., Belinha, J.: Functional gradients of the gyroid infill for structural optimization. Mater. Proc. **8**(1), 72 (2022). https://doi.org/10.3390/materproc2022008072

Podroužek, J., Marcon, M., Ninčević, K., Wan-Wendner, R.: Bio-Inspired 3D infill patterns for additive manufacturing and structural applications. Materials (Basel). **12**(3), 499 (2019). https://doi.org/10.3390/ma12030499

Caiazzo, F., Guillen, D.G., Alfieri, V.: Simulation of the mechanical behaviour of metal gyroids for bone tissue application. Materials (Basel). **14**(17), 4808 (2021). https://doi.org/10.3390/ma14174808

Duty, C., et al.: What makes a material printable? A viscoelastic model for extrusion-based 3D printing of polymers. J. Manuf. Process. **35**(1), 526–537 (2018). https://doi.org/10.1016/j.jmapro.2018.08.008

Choi, Y., Kim, C., Jeong, H., Youn, J.: Influence of bed temperature on heat shrinkage shape error in FDM additive manufacturing of the ABS-engineering plastic. World J. Eng. Technol. **4**(3), 186–192 (2016). https://doi.org/10.4236/wjet.2016.43D022

Messimer, S.L., Patterson, A.E., Muna, N., Deshpande, A.P., Rocha Pereira, T.: Characterization and processing behavior of heated aluminum-polycarbonate composite build plates for the FDM additive manufacturing process. J. Manuf. Mater. Proc. **2**(1), 12 (2018). https://doi.org/10.3390/jmmp2010012

As-Built/As-Found

RUBBLE #10

Rising From Rubble Leveraging Existing Construction Tools for Upcycling Concrete Waste into Slender Walls

Maxence Grangeot[1,2(✉)], Qianqing Wang[3], Katrin Beyer[3], Corentin Fivet[1], and Stefana Parascho[2]

[1] EPFL SXL, 1700 Fribourg, Switzerland
maxence.grangeot@epfl.ch
[2] EPFL CRCL, 1015 Lausanne, Switzerland
[3] EPFL EESD, 1015 Lausanne, Switzerland

Abstract. In this paper, we present a new method for upcycling concrete rubble waste into slender walls through the lightweight digital augmentation of mainstream construction machines. Our method presents the potential to alleviate the environmental impact of concrete construction and demolition through leveraging existing accessible tools. In our proposed method, pieces of rubble are scanned from photos, robotically drilled to insert a lifting anchor, and assembled along their larger surface plane with a digitally augmented overhead crane. The highlight of this method is the orientation of irregular concrete debris through gravity, thus only requiring a lifting and a drilling tool for 3D orientation. Our method benefits a 2D stacking algorithm which accounts for the logistical constraints of construction sites. We document the construction process with augmented tools and implement them for the design and construction of two full-scale wall demonstrators. We also detail prospects of scalability for prefabrication, for onsite applications and potential use cases.

Keywords: concrete rubble · upcycling · wall

1 Introduction and Research Goal

Although concrete is known to be one of the most environmentally damaging construction materials [1], it remains as popular as ever. All over the world, new concrete buildings continue to be built, and demolition debris from the old ones are the second largest source of waste globally after excavated soils [2]. Current strategies for reuse of concrete waste generally require cutting, grinding, or crushing concrete parts into smaller elements, often used as aggregates for casting of new concrete objects. These strategies are environmentally insufficient and logistically constraining [3]. Reusing concrete rubble generated by conventional demolition techniques without excessive material processing remains complex and underexplored [4], partly due to the high variability and irregularity of rubble geometry. Variability of geometry of concrete post-demolition stock can be

M. Yablonina et al. (Eds.): Robotic Fabrication in Architecture, Art and Design 2024, pp. 254–265, 2026.
https://doi.org/10.1007/978-3-032-22607-5_16

harnessed through digital imaging and classification processes, enabling new strategies for upcycling construction waste for novel load-bearing applications [5]. Our research seeks to upcycle large flat concrete rubble units, as seen in Fig. 1, into slender walls while making the most out of market-ready tools and aiming for scalability.

Fig. 1. Pile of concrete rubble and segmented debris oriented along their largest surface plane

2 Background

Using concrete demolition debris for the design of new structures has been explored in academia for trusses [6], walls [7, 8] and retaining walls [9]. However, their implementation in industry remains difficult as they have no built demonstrators, or they rely on expensive machinery such as automated excavators or carving robots.

Precedent research projects use 3D scanning to capture the geometry of complex as-found materials. Examples include LiDAR [10], structured light [11] or photogrammetry [12] to build walls and columns out of irregular stones. However, the 3D manipulation of non-standard materials is computationally intensive and requires high expertise, thus preventing their scalable adoption in the construction industry. To bridge digital processing to construction by leveraging existing tools of construction sites, other studies show the potential of augmenting cranes [13] or manual tools [14], while historical practices provide hints to orient irregular stones simply through lifting [4].

3 Methods

Due to the flat property of most large concrete debris, we consider rubble pieces as 2D geometries and define their flat orientation as largest surface plane. To assist the design and construction planning of the walls, we implemented a 2D scanning and stacking method. Aiming for scalability and accessibility, this project uses tools commonly used on construction sites, such as cranes and hammer drills, augmenting them for accessible digital processing. Instead of orienting heavy concrete fragments of non-standard geometries in 3D with expensive and payload-constrained robots, we achieve their desired orientation using any hoisting machinery. By lifting rubble units from precisely drilled lifting points computed from their desired orientation and their centre of mass, they are oriented simply by gravity. The concrete debris are drilled with a robotically augmented hammer drill, lifted with a digitally-augmented overhead crane, stacked along their largest surface plane, and connected with mortar or reclaimed rebar, to construct single-leaf masonry walls. We construct two full-scale demonstrators as case studies to identify the construction time and robustness of our upcycling process.

3.1 Design Tools

To sustainably construct walls for buildings, their embodied energy and footprint size need to be minimized by optimizing their thickness and material mass. Unlike traditional masonry, we aim to stack large flat rubble pieces along their largest surface plane into single-leaf masonry walls. This strategy requires care for stability throughout the construction. To avoid large storage areas and cumbersome logistics, we only consider the concrete fragments of the topmost layer of a pile for scanning and stacking. Each rubble unit is placed flat on a pallet and photographed from an augmented overhead crane. The contour 2D outline geometry of each concrete piece is obtained through canny edge detection [15] from the top-view images (Fig. 2), while thickness is acquired through a distance sensor of the augmented overhead crane. To maximize the success rate of edge detection, we increase the image contrast using a black backdrop placed underneath the rubble unit and its pallet.

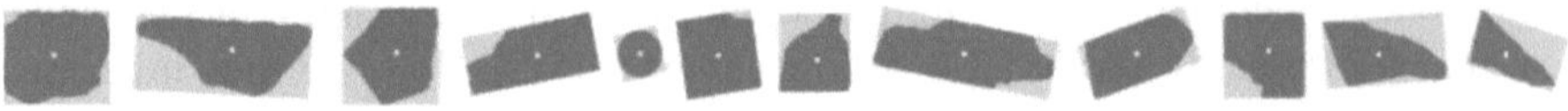

Fig. 2. Rubble pieces 2D contour geometry obtained from edge detection from top images.

To identify the optimal placement for each candidate rubble unit, we implemented a raster-based stacking algorithm [16] to obtain a tightly packed arrangement within given boundaries. Computing the object-oriented minimum bounding box of 2D outlines provides the orientation of straight edges if any, informing stacking stability and aesthetic design intents. Oriented in 4 orthogonal directions, the best out of 5 random candidates is selected using a hierarchical method that prioritizes large lateral force capacity, uniform course height, interlocking, and small void area. The details are given in an external article [16]. The lateral force capacity is represented by the maximal tilting angle of the wall in a tilting table simulation using kinematics analysis [17].

Fig. 3. First 8 ranked stacking options using a stock of (a) 15 and (b) 27 rubble pieces.

Aiming for robustness in case of stock or geometry alteration, the sequential algorithm allows for on-the-fly changes. Ranking wall stacking solutions by void area and wall height highlights optimum solutions (Fig. 3). Using the centroid of the 2D outline as the estimated centre of gravity and knowing its intended orientation in the wall, the locations of lifting points are identified as above the centre of gravity of each rubble unit and transmuted to the ground position of the rubble unit to be drilled and lifted.

3.2 Hardware Tools

Augmenting Heavy-Lifting Hoisting Machinery

For acquiring the geometry of concrete debris and their location, we use an overhead crane with lifting capacity of 10 tons. By augmenting it with three optical distance sensors, an industrial RGB camera, a compact computer, and an acquisition module, we gain access to the x and y position of the trolley based on a virtual static point, the vertical distance to the nearest surface below the trolley, and top-view image streams. This sensing unit shown in Fig. 4 is clamped to the crane trolley and can be easily dismantled and installed on another overhead crane or a tower crane. The distance sensor pointed at the ground allows to detect the thickness of each rubble unit by subtracting the distance to the floor and the pallet height. We also use the visible red beam of the vertical distance sensor as a digital pointing device, to calibrate all tools and coordinate systems to one reference. A load cell between the swivel hook and lifting slings allows to record the weight of the rubble pieces for documentation but this data is not used in the fabrication planning. The data of the sensing unit is locally processed in LabView, a systems engineering software, and streamed to a ground computer.

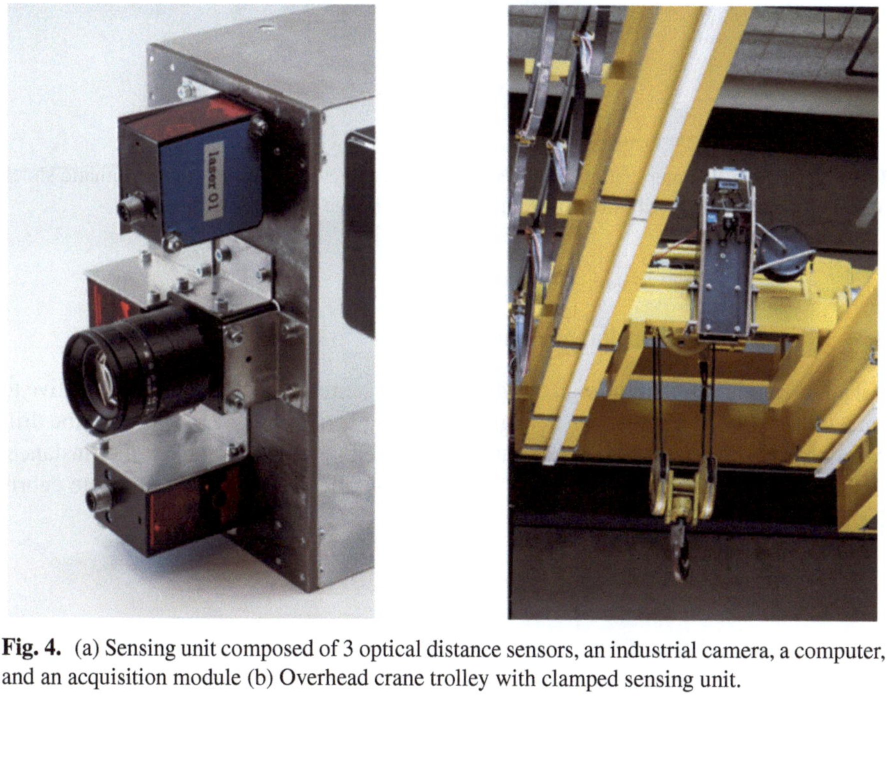

Fig. 4. (a) Sensing unit composed of 3 optical distance sensors, an industrial camera, a computer, and an acquisition module (b) Overhead crane trolley with clamped sensing unit.

The overhead crane is positioned by the human operator based on a cartesian target and live feedback while avoiding the swinging of loads and physical collisions. Manual operation of the overhead crane ensures safe and swift operation.

Robotically Augmenting Concrete Hammer Drills
We achieve the desired orientation of irregular concrete fragments through gravity when lifted from specific anchor points. The lifting anchors of each rubble unit are drilled using an augmented hammer drill. Leveraging the fast-prototyping capabilities and 3 m reach of an industrial robotic arm on a linear track, we convert a hammer drill to a robotic end-effector via a bespoke pneumatic shock absorber (Figs. 5, 6). The off-the-shelf concrete drill accepts a wide range of drilling bits for varying hole diameters and length. For deep holes that require long bits, a holder extends during predrill to minimize eccentricities. Extending the holder and turning on the drill is controlled automatically via the I/O signals of the robot controller. Robotic control is achieved using compas_RRC [18], while compas_FAB [19] is used for path planning and simulation.

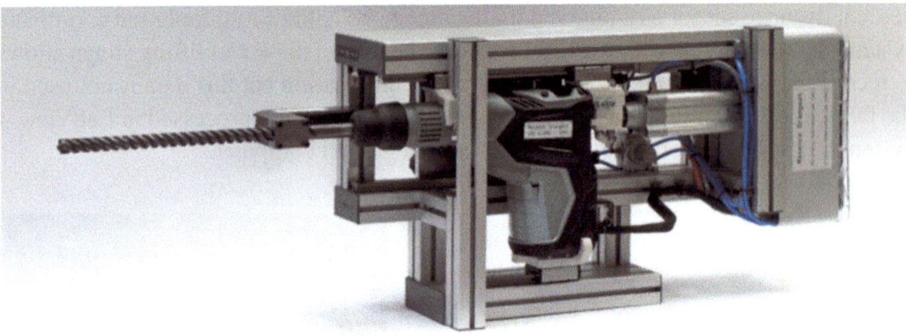

Fig. 5. Concrete hammer drill acting as a robotic end-effector, mounted on a pneumatic shock absorber for dampening vibrations and relieving stress on the robotic arm joints.

3.3 Human-Machine collaboration

All augmented tools and the wall building support are referenced in 3D space relative to each other when located within a single fabrication facility (Fig. 7). Operating the drill below the augmented crane allows data to be shared across several tools. For instance, the camera of the crane is also used for referencing the position of the concrete debris for robotic drilling.

Fig. 6. The positions of the lifting points are precisely drilled on the middle axis of the side faces of flat rubble pieces. For heavy concrete fragments, two anchors ensure redundancy.

At the interface of all tools, a human operator acts as a critical agent of the upcycling process (Fig. 8). Like a mason, the human operator executes design decisions, ensures safe operation, and adjusts concrete debris positioning for stability during assembly. This ability to react instantly and adapt efficiently is our rationale behind avoiding full automation. A metal detector operated manually allows for the detection of potential drilling collision with rubble rebar and the adaption of drilling points in consequence. During assembly, we manually verify the position of the rubble pieces using measuring tape. The hardware limitations to 2D silhouettes narrow connection strategies to those that provide tolerances and stability when interfacing the debris: dry stack, cementitious mortar, concrete rubble wedges, and reclaimed rebar as alignment and support rods. The choice between such connections is dependent on requirements for airtightness, load-bearing capacity, and rubble geometry as it influences stability during assembly.

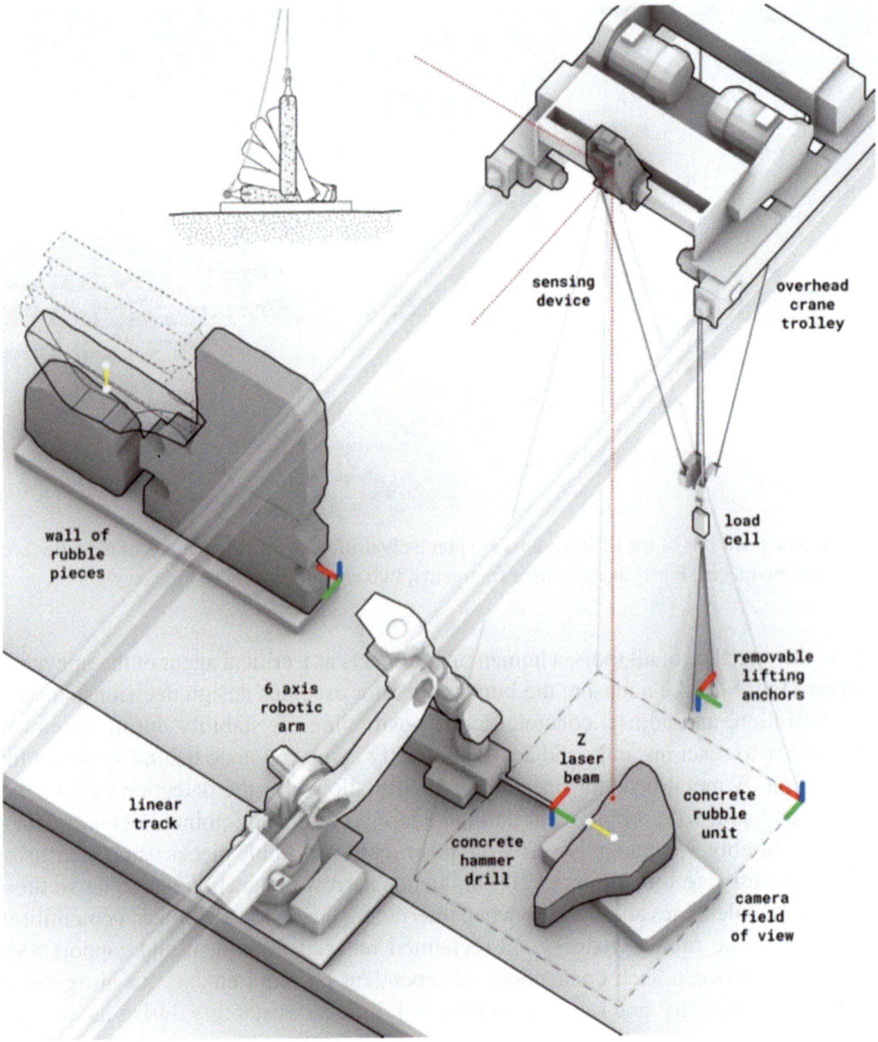

Fig. 7. Tool augmentations and relevant planes used for calibration and operation. The Z laser beam of the sensing unit of the overhead crane acts as a reference for calibrating the camera, the concrete hammer drill end-effector, the wall base, and the lifting hooks.

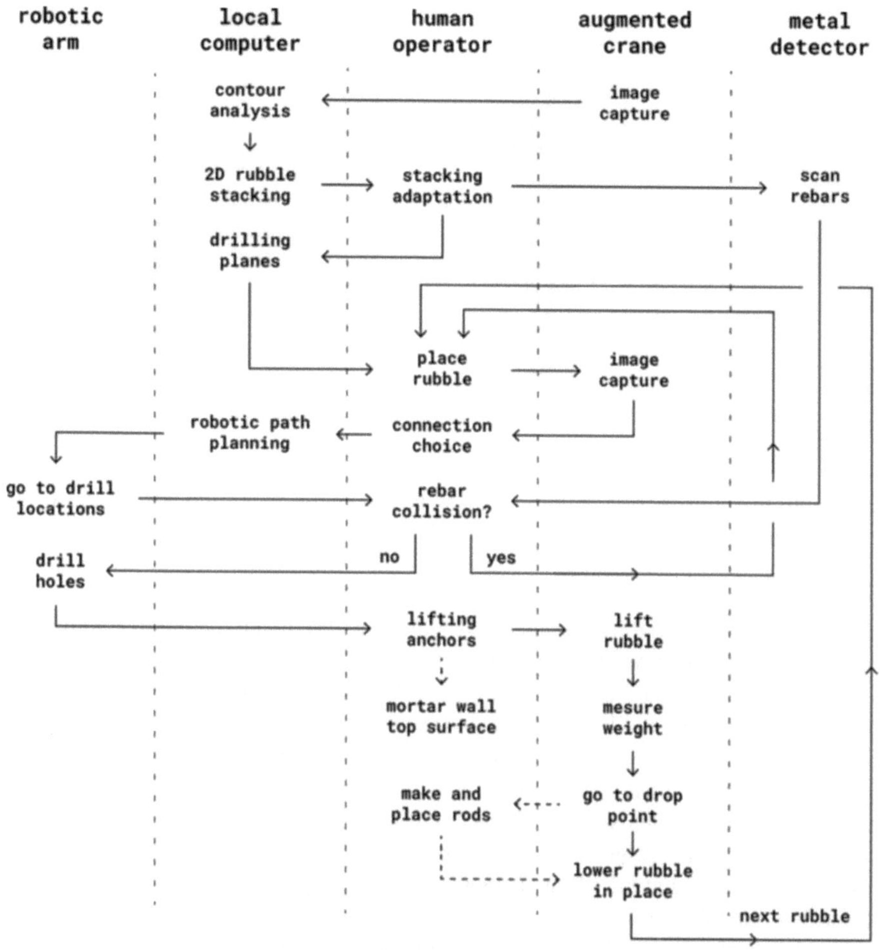

Fig. 8. Machines and human interactions within the upcycling construction process, starting from the image capture of debris. Dashed arrows are optional connection types.

4 Case Studies

To assess the performances of our upcycling method and improve it, we designed and built two full-scale slender single-leaf walls using equal boundary geometries (270 × 250 cm) but different stock sizes and different connection types (Fig. 9).

For the first prototype, we scanned 12 large rubble pieces as the first extracted from a bin. We chose the stacking solutions containing the smallest void area (15%) (Fig. 3) and manually adjusted it, due to software limitations of raster resolution. Adjustments prevented physical collision of fragments and improved stability and aesthetics. To identify the most rapid processing method, we tested all 4 connection types independently and in combination. Mainly due to the numerous dowel holes and their 8.5 min drilling

Fig. 9. Walls built out of large off-the-pile concrete rubble units ranging from 80 kg to 730 kg.

time each, the construction of the wall is executed in 14 h. We did not fill the voids, leading to a mortar use of 1% of the wall volume.

Learning from these results, we refined the upcycling process for the second prototype. We scanned more rubble units, 27 in total. The best stacking solution contained 21% of void, which we further reduced by adding one fragment to minimize infill material and related labour. We manually filled the remaining void with smaller debris and casted concrete using a localized formwork. This process provides airtightness but is more time consuming (40 h of construction) and increased mortar quantity to 15%.

Based on this refined prototype and processing method, its embodied energy is lower than stone masonry and recycled concrete. However, its structural capacity lies between these alternatives, while its 7:1 slenderness outperforms irregular stone masonry walls.

During lifting and rotation of rubble pieces from horizontal to vertical position, we observed no splitting at the lifting anchors. Concrete fragments thinner than 15 cm are not drilled but lifted from the same point using a lifting clamp, reducing processing time and risk of splitting. For some fragments with lifting anchors, we reoriented the robotically augmented drill to the normal of the drilling point during predrill to avoid the slippage of the drilling bit. We also solved this issue by jackhammering the surface at the lifting point to create a flat surface normal to the drilling direction.

5 Discussion

The tools of our upcycling process are augmented with mainstream components for accessibility and scalability, at the expense of auxiliary data. The absence of automatic detection of rebar layout and of side faces 3D geometry requires human intervention in non-design tasks which could otherwise be automated. Moreover, differences of rubble thickness are not considered in 2D stacking, neglecting pragmatic stability principles.

The tools augmented in this research project can be found in industrial prefabrication facilities, thus making our upcycling process suitable for prefabricating walls from concrete rubble, or in parts. Our stacking strategy, which also targets minimal storage area, increases the prospects of scalability for prefabrication and on-site construction.

Furthermore, our upcycling fabrication process can be easily adapted to construction sites using tower cranes and concrete drills. Most tower cranes are already equipped with position sensors, load cells, and occasionally with cameras. Concrete drills are frequently present during the construction of structures, yet they lack the positional capabilities of robotic arms. While other projects have shown the feasibility of operating robots on construction sites [20], the required drilling precision is achievable with guiding devices such as positional lasers. In future work, we will use the Z laser sensor of the sensing unit of the crane to guide human operators in manual drilling.

Compared to traditional masonry and concrete walls, our method, when applied to prefabrication, is expected to save embodied energy by minimizing mortar quantities, while maintaining the average amount of needed labour and construction time. Due to the experimental and academic nature of this study, and the variance of labour costs throughout the world, we do not provide a price comparison. However, concrete rubble from demolition is a cheap commodity in large quantity often given to recycling centres alongside compensation to avoid the high costs of landfilling.

6 Outlook

Concrete rubble from demolition remains a largely unexploited construction material due to its irregular geometry and is systematically downcycled as a result. In our upcycling method, large concrete rubble pieces are stacked along their largest surface plane into slender single-leaf walls of controlled dimensions using a 2D stacking algorithm considering logistic constraints, quantity of labour and material, and stability. We digitally augment tools of construction sites such as cranes and hammer drills to achieve orientation of irregular rubble pieces through gravity during lifting. The construction process, assessed with two full-scale demonstrators, involves human intervention to safely operate tools and make adjustments to ensure stability. Application-wise, our upcycling method is suitable for prefabrication and can be easily adapted for onsite construction. Compared to stone masonry walls, the resulting walls are faster to build, more space-efficient, have comparable structural performances and use less mortar. Consequently, we offer an approach that not only has a practical and economic path to meaningful adoption in industry, but also diverts the environmentally burdensome concrete waste stream away from consumptive practices of recycling and landfilling.

Acknowledgements. We warmly acknowledge the contribution of Eric Duong for proof-reading, Sika AG and Tinguely Recyclage SA for supplying materials, Lullin Engineering for robotic fabrication hardware and GIS EPFL for technical assistance.

References

1. Monteiro, P.J.M., et al.: Towards sustainable concrete. Nature Mater. **16**, 698–699 (2017)
2. Giorgi, S., et al.: Guidelines for effective and sustainable recycling of construction and demolition waste. In: Benetto, E., Gericke, K., Guiton, M. (eds.) Designing Sustainable Technologies, Products and Policies: From Science to Innovation, pp. 211–221. Springer International Publishing, Cham (2018)

3. Küpfer, C., Fivet, C.: Panorama of approaches to reuse concrete pieces: Identification and critical comparison. J. Phys. Conf. Ser. **2600**(19), 192006 (2023)
4. Clifford, B.: The Cannibal's Cookbook: Mining Myths of Cyclopean Constructions, 2nd edn. Oro Editions, Novato, California (2021)
5. Grangeot, M., Fivet, C., Parascho, S.: From concrete waste to walls: An investigation of reclamation and digital technologies for new load-bearing structures. J. Phys. Conf. Ser. **2600**, 192019 (2023)
6. Al Khayat, L.K.Y.: Fibers and Fragments-Weaving Local Resources into the Arabian Gulf's Modern Material Culture. Massachusetts Institute of Technology, Cambridge (2023)
7. Marshall, D., Mueller, C., Clifford, B., Kennedy, S.: Computational arrangement of demolition debris. Detritus. **11**, 3–18 (2020)
8. Clifford, B., McGee, W., Muhonen, M.: Recovering cannibalism in architecture with a return to cyclopean masonry. Nexus Netw. J. **20**(3), 583–604 (2018)
9. Johns, R.L., Wermelinger, M., et al.: A framework for robotic excavation and dry stone construction using on-site materials. Sci. Robot. **8**(84), eabp9758 (2023)
10. Settimi, A., Wang, Q., Andò, E., et al.: Projector-based augmented stacking framework for irregularly shaped objects. Constr. Robot. **7**, 159–175 (2023)
11. Furrer, F., et al.: Autonomous robotic stone stacking with online next best object target pose planning. In: International Conference on Robotics and Automation, pp. 2350–2356 (2017)
12. Wibranek, B., Tessmann, O.: Digital rubble compression-only structures with irregular rock and 3D printed connectors. In: Proceedings of IASS Annual Symposia, pp. 1–8 (2019)
13. Joachim, L. et al. Evaluation of the quality of real-time mapping with crane cameras and visual slam algorithms. In The International Archives of the Photogrammetry, Remote Sensing and Spatial Information Sciences, 43, B2, 545–552 (2022)
14. Settimi, A., Gamerro, J., Weinand, Y.: Augmented-reality-assisted timber drilling with smart retrofitted tools. Autom. Constr. **139**, 104272 (2022)
15. Bradski, G.: OpenCV Canny Edge Detection https://opencv.org/ (2024). Accessed 07 Feb 2024
16. Wang, Q., et al.: An image convolution-based method for the irregular stone packing problem in masonry wall construction. Eur. J. Oper. Res. **316**(2), 733–753 (2024)
17. Wang, Q., Dos Santos, K.R.M., Beyer, K.: Micro modeling of irregular stone masonry walls using mathematical programming. Str. Anal. Hist. Constr. **46**, 591–602 (2024)
18. Fleischmann, P., Casas, G.: COMPAS RRC: Online Control for ABB Robots Over a Simple-to-Use Python Interface, V 1.1.0. https://compas-rrc.github.io/compas_rrc (2021). Accessed 03 Oct 2023
19. Rust, R., Gonzalo C., et al.: COMPAS FAB: Robotic Fabrication Package for the COMPAS Framework, V 0.28.0. https://github.com/compas-dev/compas_fab/ (2018). Accessed 03 Oct 2023
20. Bock, T., Linner, T.: Site automation: automated/robotic on-site factories. Cambridge University Press, New York, NY (2016)

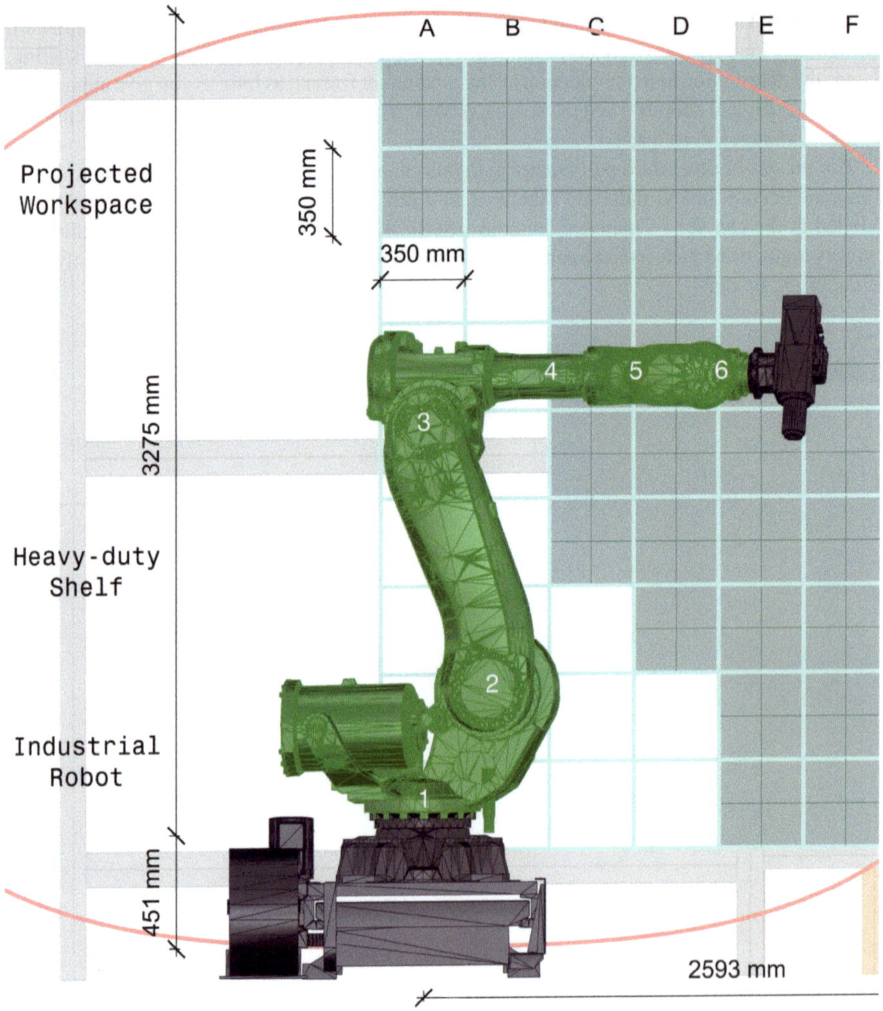

A
B
C
D
E
F
Projected Workspace
350 mm
350 mm
3275 mm
Heavy-duty Shelf
Industrial Robot
451 mm
2593 mm
1
2
3
4
5
6

Feasibility Check of a Workflow with Industrial Robots and Laser-Tracking for Timber Machining Across the Workspace

Marc Pantscharowitsch(✉) and Benjamin Kromoser

Department of Civil Engineering and Natural Hazards, Institute for Green Civil Engineering, University of Natural Resources and Life Sciences Vienna (BOKU), Vienna, Austria
{m.pantscharowitsch,benjamin.kromoser}@boku.ac.at

Abstract. In recent years, interest in industrial robots for machining timber construction elements has visibly increased. The main advantages, when compared to widely-used joinery machines, are their higher flexibility and larger workspace—particularly in the vertical direction. This study forms part of a larger investigation of machining accuracy, with a focus on the feasibility and potential of a workflow that combines industrial robots with laser-tracking technology to reference workpieces and enable machining at various positions. The workflow comprises (A) laser-tracking-based workpiece referencing, (B) coordinate transformation (3D-displacement and 3-axis-rotation), (C) approach of the start position and (D) machining. Steps (A) through (C) were conducted for fifty different positions within a two-dimensional work plane of the robot's workspace, while step (D) was investigated at three positions with three machined specimens each. The feasibility pre-tests are rated by function, applicability, process time, and machining accuracy. The results indicate that laser-tracking systems and industrial robots can be combined to machine workpieces in any position while offering reasonable measuring times. Further, they validate the possible application of the workflow, with only programming restricting the robot configuration. The machining accuracy is within legal and industry standards, paving the way for further, more detailed examinations. The applicability of laser-tracking for new or existing robots can counteract the shortage of skilled workers and bear the potential to increase the efficiency of timber construction.

Keywords: Automation · Construction · Glued Laminated Timber

1 Introduction

The state of the art of wood machining for construction purposes are joinery machines (JM), available in various designs specific to workpiece dimensions and aspect ratios. For beam elements, for example, the Hundegger k2i (Hans Hundegger Maschinenbau AG 2011) can be seen as representative. For large beams and slab elements, gantry type JMs, e.g. Technowood TW-Mill M (TechnoWood AG 2021), are typically used, while compact JMs, such as the Weinmann Beamteq B-660 (HOMAG/Weinmann AG 2019),

M. Yablonina et al. (Eds.): Robotic Fabrication in Architecture, Art and Design 2024, pp. 267–284, 2026.
https://doi.org/10.1007/978-3-032-22607-5_17

are available for compact workpieces or small companies. Recent trends tend towards industrial robots (IR) equipped with machining spindles, which offer novel machining possibilities. IRs are not only cheaper than all JM types, but are also universally applicable and simultaneously offer higher flexibility and larger workspaces. It is also shown that the obtained machining quality in terms of dimensional accuracy and surface quality for construction timber elements is sufficient and furthermore comparable to those of the JMs. (Pantscharowitsch and Kromoser 2022c, 2022a, 2022b)

In recent years, the growing interest in digital and robotic solutions to achieve design goals can be quantified by an increased number of projects, events and publishing platforms. Examples of growing platforms include the International Conference on Computational Design (Yuan et al. 2020), the Association for Robots in Architecture, and the ROB|ARCH conference itself.

The possibility of high-accuracy machining within the advantageously large vertical workspace of IRs creates valuable design opportunities for architects. To enable the full utilization of the workspace, information on feasibility, machining performance, and accuracy is required. The effects of working direction and robot position on machining results have been investigated by the authors, among others, in previous studies (Pantscharowitsch and Kromoser 2023; Furtado et al. 2017). It is further shown that the occurrence of vibrations decreasing machining quality is position-dependent (Bisu et al. 2011). To obtain high machining quality, the IR's stiffness is of major importance and depends on (1) the absolute stiffness of the IR, as defined by its design and parameters, and (2) the relative stiffness given by the position of the robot and the workpiece (Ji, Yin, and Wang 2019). Contrary to expectations, the highest stiffness is not observed at the smallest cantilever of the robot arm, but at mid-range positions expressed by the tool center point (TCP) relative to robot base (Bisu et al. 2011); the resulting machining accuracy varies as high as a factor of 14 relative to different positions (Caro et al. 2013). For Kuka KR-type IRs, the optimal workpiece placement and thus the robot's position related to the workpiece was determined when milling aluminum (Klimchik et al. 2016). To enable optimal workpiece placement and reachability of the workpiece, it needs to be referenced. For referencing workpieces and determining TCP positions within the workspace, there are already existing solutions applicable for different types of robot kinematics and stationary and mobile robot platforms. (Zi et al. 2008; Gambao, Balaguer, and Gebhart 2000; Möller et al. 2017)

In the aircraft engineering industry, there exists already research work and equipment with mobile robotic platforms for machining within an extremely large workspace referenced by a predecessor of the laser-tracker used in this study (Susemihl et al. 2016, 2017).

Given the upward trend of IRs for machining timber construction elements, it is a logical next step to transfer existing knowledge from other manufacturing fields. By combining this knowledge with research on timber machining accuracy, the goal to investigate the machining accuracy of IRs across the entire workspace can be reached and a more efficient utilization of IRs can ultimately be achieved. Prior to the investigation of machining accuracy, a workflow that enables the referencing of the workpiece within the workspace must be developed. The objective of this paper is to determine the feasibility of referencing laminated timber workpieces in different positions by laser-tracking within the 2D work plane of an ABB IRB 7600–325. To further validate this approach, pre-tests for machining are conducted in three selected positions with three specimens each. The workpieces have been positioned in grid along the workspace (see Fig. 1). The positions of the individual workpieces were determined by a Leica ATS600 laser-tracking system. Subsequently, the workpiece coordinate system was transformed into the robot workobject for easy adaption of the machining program (TCP trajectories) to the individual workpiece positions.

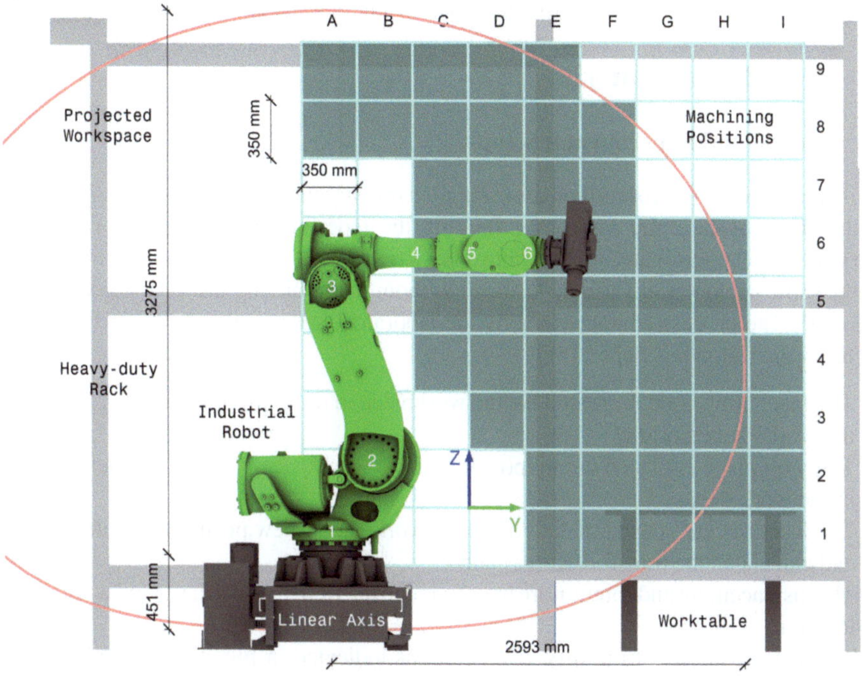

Fig. 1. Side view on the experimental setup with grid indicating the workpiece positions and the projection of the IR's nominal workspace sphere on a work plane set 1,600 mm away (without end effector). The numbers represent the robot axes. The grid size is 350 mm x 350 mm.

By exploring the feasibility of the presented workflow (laser-tracking – coordinate transformation – machining), the validation of the accurate use of the IRs workspace was investigated. As a result, this study is the basis for detailed investigations of machining accuracy within the workspace, as well as a detailed determination of how these work-piece positions can be implemented in an automated industrial workflow. Ultimately, better knowledge of machining accuracy across the workspace will enable all users of IRs to maximize their capabilities. When keeping in mind the lower acquisition costs of IRs compared to joinery machines, this may level the field for various parties involved in architecture and timber construction.

2 Research Questions

In the present study, the feasibility of laser-tracking based robotic machining is investigated through the following research questions:

1. Is the combination of an IR and a laser-tracking system a suitable and valuable tool for machining timber construction elements? Does this pave the way for more detailed investigations of machining accuracy within the entire workspace?
2. Are there limitations or opportunities for improvement of the hardware, software and the presented workflow?
3. Is the workflow presented sufficiently user-friendly to be applied in practice?

3 Methods and Material

3.1 Process and Experimental Setup

This study is a pre-test for further, more detailed investigations of the machining accuracy of the IR across the entire workspace. Initially, the underlying idea is a transformation of the workobject coordinate system using a laser-tracking system to enable the machining of the same geometry at various positions without manipulating the TCP trajectories in the code. The process characterized by the laboratory setup is illustrated in Fig. 2 and includes the following steps:

1. The machining code is created for the workpiece in programming position located at the workobject's origin.
2. The workpiece is placed at any position of the workspace within a pre-defined grid in the work plane.
3. The laser-tracker is used to determine the workpiece's new position (or referencing) and calculates the transformation matrix ${}^{\text{Workobject}}T_{\text{Workpiece}}$.
4. The displacement and rotation are transferred to the IR to modify the coded workobject location.
5. The start positions of the fifty grid positions indicated in Fig. 1 are approached by the IR. Initial preliminary tests on three specimens, for three positions (A9, E5 and I1 in Fig. 1), are machined.

3.2 Machines and Tools

Industrial Robot (IR), Tools and Workpiece Mount. The experiments were conducted in the robot laboratory at University of Natural Resources and Life Sciences Vienna (BOKU) as displayed in Fig. 3.

The laboratory comprises an ABB IRB 7600–325/3.1 mounted on an ABB IRBT 6004 external linear axis with a length of 8.7 m. It is equipped with an HSD ES 951 A 1112 S machining spindle and has a maximum payload of 325 kg, a reach of 3.1 m, and a maximum wrist torque of 2,680 Nm. According to manufacturer's data, the position and path repeatability are 0.19 mm and 0.40 mm, respectively (ABB 2019). It is important to note that while absolute accuracy focuses on the robot's ability to reach specific points

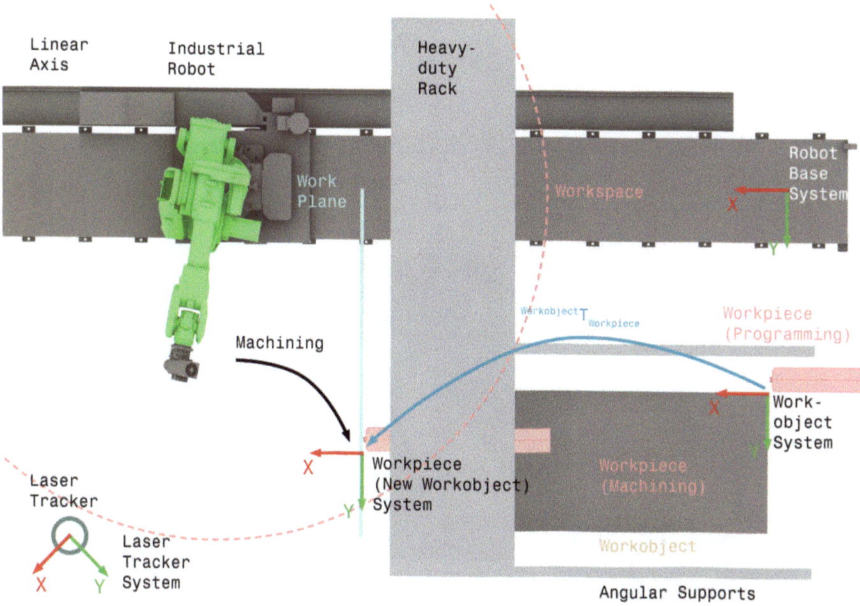

Fig. 2. Top view on the BOKU robot laboratory setup including an ABB IRB 7600–325/3.1 mounted on an ABB IRBT 6004 external linear axis, heavy-duty shelf, reference workobject (heavy steel welding table), laser-tracker and workpiece (programming and machining position). This figure also illustrates the coordinate transformation process indicated by the different coordinate systems and the main transformation matrix ${}^{\text{Workobject}}T_{\text{Workpiece}}$.

accurately, path repeat accuracy evaluates its ability to consistently reproduce predefined motions or paths. The spindle itself offers a maximum power of 13.2 kW in the range from 12,000 min^{-1} to 15,000 min^{-1} and a maximum torque of 10.5 Nm at 12,000 min^{-1}(HSD Mechatronics 2021). The robot laboratory contains a steel worktable, where one of the worktable's corners is defined as the workobject (or coordinate system). This is the reference of the computer aided manufacturing (CAM) system and serves as origin and for orientation when programming and machining further. The robot system is specifically calibrated by ABB's Absolute Accuracy calibration method and the TCP is calibrated at the linear axis position of 1,200 mm.

The tool used for the present study (Fig. 4) was an insert cutter with a diameter of 125 mm, an edge height of 80.9 mm and an HSK-F63 holder, with new cutting edges installed before the trials.

A heavy-duty shelf with the dimensions of 1,100 mm x 4,810 mm x 3,800 mm (length x width x height) and a field load of 6,600 kg was set up within the IR's workspace to allow for the stable mounting of specimens in any required position (Fig. 2 and Fig. 3). It was leveled using the laser-tracking system described in Sect. 3.2 and anchored to the ground at each column to maximize stability and vibration resistance. Additionally, it was laterally stabilized by angular supports anchored to the ground and by extra reinforcements mounted to the wall.

Fig. 3. BOKU robot laboratory comprising an ABB IRB 7600–325/3.1 on an ABB IRBT 6004 external linear and, as an end effector, an HSD ES 951 A 1112 S machining spindle. For this study, a heavy-duty shelf and Leica ATS600 laser-tracking system were added.

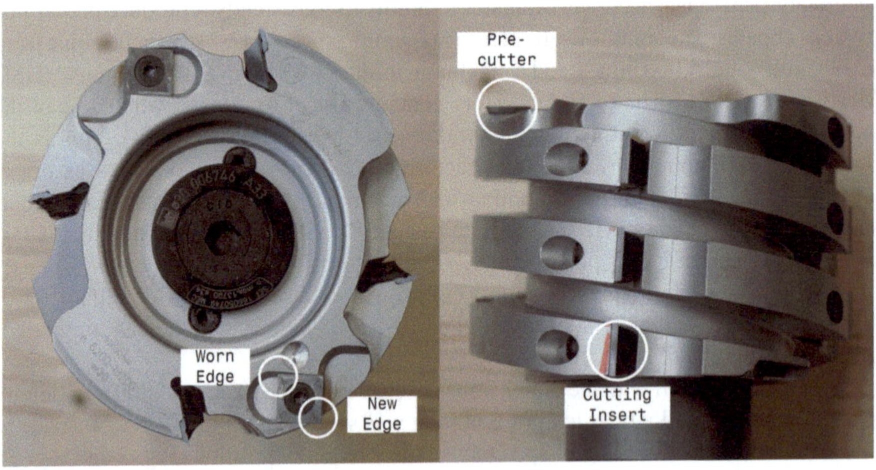

Fig. 4. Insert cutter used for this study: diameter = 125 mm, edge height = 80.9 mm and an HSK-F63 holder. In the left side a worn and new edge is presented.

Laser-Tracker. The engineering geodetic tasks were carried out using a Leica ATS600 laser-tracking system. The device was used to measure single points on the workobject and workpiece surface represented by a 1.5″ red ring reflector with the accuracy (at a 10 m distance) of ±38 µm at location test, ±53 µm at length test and ± 72 µm at ranging test (Hexagon AB 2021). Leica's Inspire software was used to operate the system and calculate the results. To maintain accuracy during the investigation, regular drift tests were conducted and the system's position was referenced to an industrial measurement network: four fixed reference points located on adjacent walls. The measuring device was referenced to the robot coordinate system by measuring the workobject (Fig. 5).

Fig. 5. Left: Leica ATS600 laser-tracker in the BOKU robot laboratory with ABB IRB 7600–325/3.1 and heavy-duty shelf including a workpiece. Right: Leica 1.5″ red ring reflector while referencing a workpiece.

In Leica's Inspire software, the laboratory environment—including the heavy-duty shelf and table—was imported as computer-aided design (CAD) data and registered using the ATS600 laser-tracker and 1.5 "red ring reflector. A measuring routine was also set up to allow for convenient and error-proof repetitive measurements and automated construction of the workobject coordinate system. Per specimen, a minimum of seven points were measured to define the new workpiece coordinate system: three points to construct the x-y plane, and two points for the y- and z- direction, respectively. The intersection of the plane with the lines resulted in the origin of the new workpiece coordinate system. The workpiece coordinate system, expressed in the previously-defined workobject coordinate system, is the ${}^{\mathrm{Workobject}}T_{\mathrm{Workpiece}}$ information required for the transformation of the code. This is further explained in Sect. 3.4.

3.3 Material and Specimens

The specimens (Fig. 6) were tenons machined on 200 mm x 200 mm beams made of glued laminated timber specified as GL 24c (Norway Spruce, Picea abies L. Karst.), as regulated in standard EN 1408 (CEN 2013), with lamellae orientation perpendicular to the tenon. The blank parts were 3,000 mm long and processed to a minimal length of 750 mm to guarantee measurement accuracy (especially rotation around the z-axis) and stable mounting to the heavy-duty shelf.

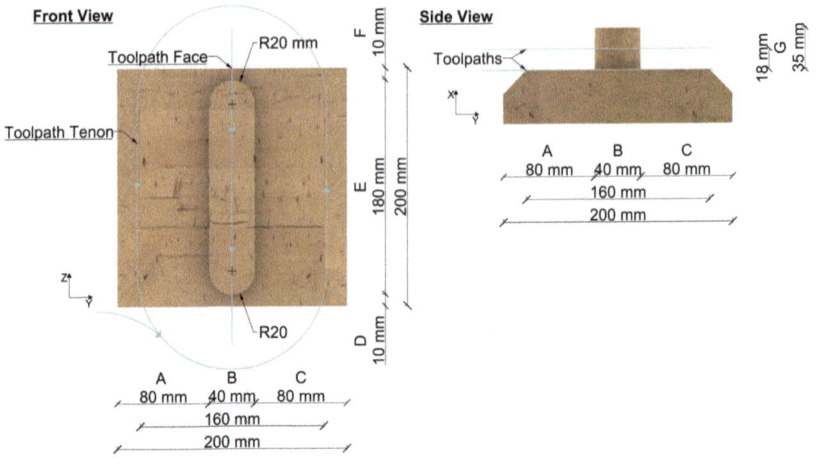

Fig. 6. Specimen dimensions and industrial robot (IR) tool path trajectories.

3.4 Code and Coordinate Transformation

The code was created using Autodesk PowerMill with a robot plugin and robot system-specific custom post-processor. Based on the workpiece geometry, a pre-defined machining routine was selected for both the face machining (using PowerMill's face-machining routine) and the tenon (using PowerMill's model area clearance routine). The toolpath is indicated in Fig. 6. The machining parameters were chosen based on results from previous publications (Pantscharowitsch and Kromoser 2022a, 2022b), however, the axial depth of cut was reduced by 50% to minimize tool wear and maintain more uniform results. The machining parameters were set to a spindle speed of 9,200 min^{-1}, feed rate of 4,000 mm/min and axial depth of cut of 18 mm. The radial depth of cut was set to tool radius and can subsequently be neglected due to the tool diameter exceeding the geometries margins.

The CAM software output files, as described in Fig. 7, comprise a module file and one process file for each of the two toolpath's face machining and tenon. The module file includes metadata such as the workobject shown in Fig. 2. In the presented robot system, the workobject's origin is defined as the corner of the solid steel welding table that is anchored to the ground and standard laboratory inventory. The x- and y-axis direction is defined by the table's edges and the z-axis results from them.

The workobject definition includes the base frame, that defines the workobject's position and orientation, and the object frame (Oframe). The Oframe uses the base frame as the origin and allows the programmer to offset and rotate the workobject. This

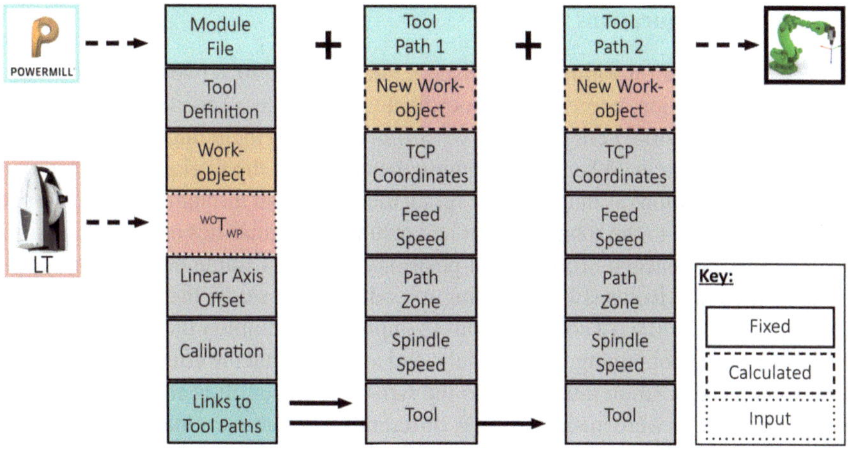

Fig. 7. Code hierarchy, content and process structure. LT = Laser-tracker; WO = Workobject; WP = Workpiece

characteristic of ABB's RAPID code is utilized to insert the measurement data from the laser-tracker to transform the workobject according to the new workpiece position. As indicated in Fig. 2, the key information is the transformation matrix ${}^{\text{Workobject}}T_{\text{Workpiece}}$. The main elements of the module file are highlighted in Fig. 8. Given the proposed workflow and code, there is no need to change the individual points of the toolpath, as these refer to the programming or machining workobject, respectively.

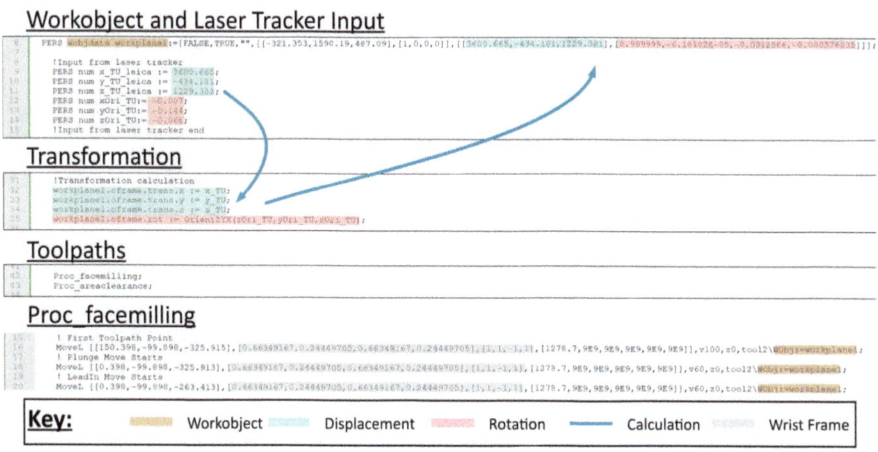

Fig. 8. The main elements of the module file and transformation calculation. Also there are exemplary lines from the face milling process file.

3.5 Robot Configurations

To reach every position within the workspace, the robot requires two pieces of information: the TCP position, expressed by coordinates relative to a specific workobject, and the robot configuration, defined by the wrist frame (Fig. 8) including the rotation of the tool coordinate system. When programming the code with the workpiece located in the programming position (as indicated in Fig. 2), then moving it throughout the workspace, its relative position to the robot changes and results in singularities or specific positions which cannot be reached. To reach all positions, the wrist frame needs to be adapted to orient the end effector (EE) according to code. In this study, the goal was to use a minimum of different wrist configurations to improve and simplify the process. A further aim was to prevent obtaining unsafe results and to ensure that the effects are attributable to the workpiece position rather than to the wrist orientation. It is crucial that no other changes in the code were made. In Fig. 9, different wrist configurations are displayed; in positions I-II, the IR machines from above in different positions with the wrist located above the EE. In positions III and IV the wrist is located below the EE (i.e. machining from below), while in positions V and VI the axes 3 and 5 are very close to their limits, making individual wrist orientations necessary. In positions VII and VIII the EE is positioned in one line with the wrist, machining leveled.

To reach the three grid fields, A9 was machined close to axis limits, since it nearly matched the height of axis 3. As a result, the wrist frame was programmed specifically for this grid field; E5 was machined from above and I1 was machined with a leveled wrist frame.

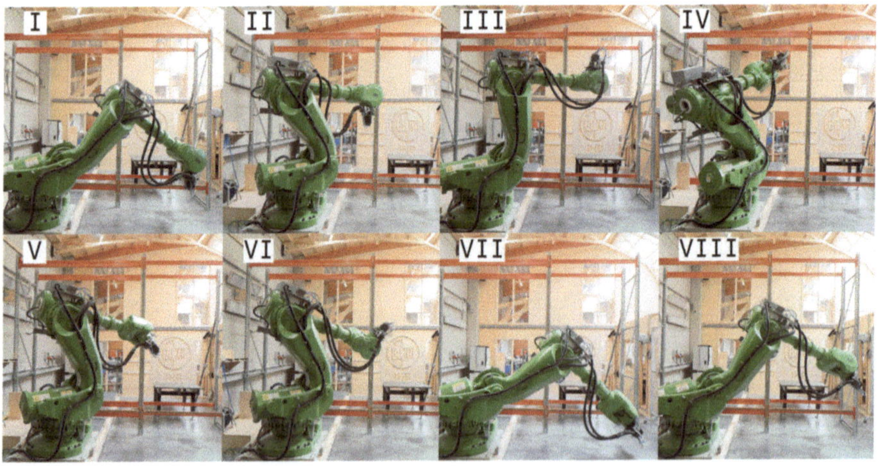

Fig. 9. Different wrist configurations used to reach all workpiece positions. I + II: machining from above; III + IV: machining from below; V + VI: machining close to axes limits due to machining at height of axis 3; VII + VIII: machining leveled.

4 Results

4.1 Laser-tracking and Coordinate Transformation

First, the laser-tracking and coordinate transformation was investigated by approaching every start point of the fifty workpiece grid positions with the robot. This process defines the position of the blank workpiece within the IR's workspace. Subsequently, the transformation matrix ${}^{\text{Workobject}}\text{T}_{\text{Workpiece}}$ enables the transformation of the workobject to react to the workpiece position and orientation.

The laser-tracker is not capable of recognizing which side of the red ring reflector sphere is in contact with the specimen surface. Therefore, the projection can be flipped either to one or the other side of the sphere. This must be validated and adjusted as necessary. Further, for some measuring points the line of sight between laser-tracker and red ring reflector was interrupted or difficult to maintain, resulting in limitations on the choice of measuring points. The obstructions mainly took place when measuring the top side of the high work piece positions in grid levels 6 through 9. A similar problem occurred when measuring the long workpiece side (x-direction) in column 9 of the grid.

While the construction of the coordinate system was calculated automatically, it was not possible to specify the orientation of the z-axis of the coordinate system as it changed its direction randomly. This specific problem had to be observed throughout the investigations and needed to be adjusted manually. The resulting output was the workpiece coordinate system which was then manually added to the module file of the code.

To maintain high quality laser-tracking results, a drift test, as described in Sect. 3.2, was performed regularly requiring an industrial measurement network. Based on the drift test's results, the tracker was recalibrated. Drift tests were of particular importance because machining, workpiece handling and other activities in the working environment caused vibrations and the laser-tracker was exposed to possible contacts. Furthermore, when working in dust-exposed environments such as a timber machining facility, both the laser-tracking device and the red ring reflector had to be cleaned on a regular basis. The mean observation during drift checks of four points on eight days resulted in a mean deviation of 0.757 mm, minimum deviation of 0.192 mm and maximum deviation of 1.537 mm. Every drift check was followed by a correction of the position.

As described above in Sect. 3.5, it was observed during the examinations that the IR configuration, predominantly expressed by the wrist frame, is crucial to define the tool orientation. However, the transformation of the workobject coordinate system into the workpiece coordinate system is not sufficient to reach every position. The initially programmed wrist frame in the code is only valid for similar spatial workpiece-robot relations (e.g. working from above in Fig. 9). Consequently, the wrist frame of the grid position indicated in Fig. 9I can be used for the grid position Fig. 9II and the transformation ${}^{\text{Workobject}}\text{T}_{\text{Workpiece}}$ is sufficient information to machine both positions. In contrast, it is not possible to just apply the transformation ${}^{\text{Workobject}}\text{T}_{\text{Workpiece}}$ when machining the grid position Fig. 9III with the wrist frame used for Fig. 9I and II because the robot cannot reach the position with the configuration "working from above". Such limitations caused by the wrist frame occurred more often at very short or long reach and when machining around the height of axis 3. By either reaching one of the axis limits, reaching a singularity, or when the components of axes 4 through 6 are aligned horizontally, the robot's flexibility will not be sufficient without adapting the wrist frame.

4.2 Overall Process

In addition to the tracking system, the proposed workflow was examined using preliminary tests where three workpieces in the each of the positions A9, E5 and I1 as indicated in Fig. 1 were machined.

Overall, the process including the steps, check/ transformation/ post-processing, machining cutting and repositioning of the blank beam lasted from 09:39 min to 16:12 min with a mean of 12:45 min (see Table 1) depending on the different positions and associated effort.

Table 1. Minimum, maximum and mean overall process times per workpiece for grid positions A9, E5 and I1.

Grid Position	Min. Duration [mm:ss]	Max. Duration [mm:ss]	Mean Duration [mm:ss]
A9	09:39	10:11	09:56
E5	15:52	16:12	16:02
I1	11:58	12:55	12:19
Total	**09:39**	**16:12**	**12:45**

The raw active process time for measuring, check/ transformation/ post-processing, machining, cutting and repositioning of the blank beam was 05:31 min (Table 2). The sum of raw laser-tracker measurements and post processing time accumulated to 02:21 min.

Table 2. Process steps and raw duration.

Process Steps	Duration [mm:ss]
Measuring points x-y-plane	00:38
Measuring points y-direction	00:23
Measuring points z-direction	00:25
Check/ Transformation/ Post-Processing	00:55
Machining	02:20
Repositioning	00:50
Total	**05:31**

4.3 Accuracy

In Fig. 10, the machined workpiece I1.3 is shown. The measurands A through G (indicated in Fig. 6) of specimens in the grid positions A9, E5 and I1 were measured by hand, resulting in the actual dimensions and nominal-actual deviations (absolute values) listed in Table 3. The largest mean nominal-actual deviation was observed for specimen A9, with a sum of 5.67 mm, while the smallest was registered for I9, with 3.33 mm. It is important to note that every individual measurand is within the legal margins of +4 mm and − 2 mm regulated in ÖNORM EN 14080 (CEN 2013) as well as any listed industry requirements as communicated by partner companies.

Fig. 10. Machined specimen in grid position I1, version 3.

Table 3. Nominal results, mean (*M*) values per specimen and nominal-actual deviations of measurands A through G (indicated in Fig. 6) of specimens in grid positions A9, E5 and I1 measured by hand.

Grid	Specimen	A	B	C	D	E	F	G
Position		Nominal [mm]						
		80.00	40.00	80.00	10.00	180.00	10.00	35.00
		Actual [mm]						
A9	Nom-*M*	1.67	0.33	0.67	1.67	0.67	0.67	0.00
	M	78.33	39.67	80.67	8.33	179.33	10.67	35.00
	1	78.00	40.00	81.00	8.00	179.00	11.00	35.00
	2	78.00	40.00	81.00	9.00	179.00	11.00	35.00
	3	79.00	39.00	80.00	8.00	180.00	10.00	35.00
E5	Nom-*M*	0.00	0.00	0.33	0.33	0.33	3.33	0.33
	M	80.00	40.00	79.67	9.67	179.67	6.67	34.67
	1	80.00	40.00	80.00	10.00	179.00	0.00	34.00

(continued)

Table 3. *(continued)*

Grid	Specimen	A	B	C	D	E	F	G
	2	80.00	40.00	80.00	9.00	180.00	10.00	35.00
	3	80.00	40.00	79.00	10.00	180.00	10.00	35.00
I1	Nom-*M*	1.00	0.00	1.33	0.67	0.33	0.00	0.00
	M	81.00	40.00	79.00	9.33	179.67	10.00	35.00
	1	82.00	40.00	78.00	9.00	179.00	10.00	35.00
	2	80.00	40.00	80.00	10.00	180.00	10.00	35.00
	3	81.00	40.00	79.00	9.00	180.00	10.00	35.00

The minimum, maximum and mean nominal-actual deviations along the three coordinate axes x, y, z, as indicated in Fig. 6, are presented in Table 4. The largest mean deviation of 0.89 mm occurs in z- and the smallest of 0.11 mm in x- direction.

Table 4. Minimum, maximum and mean nominal-actual deviations along the three coordinate axes x, y, z of specimens in grid positions A9, E5 and I1 measured by hand.

Direction	Nominal-Actual Deviation		
	Minimum	Maximum	Mean
	[mm]	[mm]	[mm]
x	0.00	0.33	0.11
y	0.00	1.67	0.56
z	0.00	3.33	0.89

5 Discussion

Is the combination of an IR and a laser-tracking system a suitable and valuable tool for machining timber construction elements? Does this pave the way for more detailed investigations of machining accuracy within the entire workspace?

The proposed workflow is a valuable tool and enables users to machine workpieces in any position within the robot's workspace without mounting specific calibrated reference stops or fasteners. This is of major importance, as this study is a pre-test for further detailed investigations of the machining accuracy of the IR at fifty different positions within the entire workspace. There are benefits from this approach in industry practice, also, like when machining heavy or large workpieces that cannot be moved to position easily and precisely. In particular, the proposed technique may be convenient for companies with a laser-tracking system, e.g. for quality control. However, it is important to note that there are other possibilities to reference the workpieces that may be more cost-efficient, such as the implementation of a 2D-line laser scanner mounted to the IR (Pantscharowitsch and Kromoser 2023) or robot vision technology (Dörfler et al. 2019). When mounting the 2D-scanner directly to the robot, any inaccuracies due to robot wear or damages that emerge over time can be compensated. This is because the 2D-scanner works in the same kinematic and coordinate system. Further, advantages result when producing small series or when frequent changes between different machining jobs occur. The presented workflow requires less adjustment to the robot environment (e.g. reference stops and fixtures) to meet the requirements of individual workpieces of different series, which can significantly speed up the process. The measured process times are reasonably short. especially when keeping in mind the application of the process to machine large or complex workpieces with relatively long machining durations. The approach is even more important when the workpiece is not mobile at all and the robot must approach the workpiece. In this case, the proposed workflow can be a foundation for mobile robot platforms or multi-robot systems.

The nominal-actual deviations along all three axes of the coordinate system indicate sufficient accuracy to meet both legal standards (CEN 2013) and industry requirements.

5.1 Are there Limitations or Opportunities for Improvement of the Hardware, Software and Presented Workflow?

The workflow is constrained by line-of-sight requirements between the tracking system and the red ring reflector. To address this limitation, the tracker could be mounted to the ceiling. This solution not only improves line-of-sight when measuring the top surface plane of the workpieces, but also keeps the delicate tracking system out of the way to prevent contact, damage and position errors. The repositioning of the tracker is a viable option in cases when an industrial measuring network is available.

The use of a 1.5“ red ring reflector can introduce challenges related to the orientation of the measured surface relative to the center of the reflector, impacting measurement accuracy. Replacing the 1.5” red ring reflector with a Leica T-Probe offers a solution to this limitation. The Leica T-Probe not only addresses the orientation issue but also enables remote control of the Inspire software, facilitating a faster transition to subsequent steps in the measuring protocol. Further, it enables users to measure around corners, which additionally mitigates the line-of-sight limitation.

Environmental factors, such as the presence of wood dust or chips, can potentially interfere with the accuracy and reliability of the tracking system and measurements. Implementing a cleaning and maintenance routine or automated air pressure system (especially when out of reach e.g. mounted to the ceiling) can help mitigate the impact of dust on the system's performance. Regular cleaning, drift tests and recalibration are essential to ensure precise measurements.

The required wrist frame adaption is potentially the limitation with the highest impact. It drastically limits the flexibility of the process and the IR and requires manual adaption of the code when changes of the workpiece location require changes of the robot configuration. This can be mitigated by programming an algorithm, like the post-processor of PowerMill, that runs in the module file of the code and adapts the wrist frame to the transformation input provided by the laser-tracking system. In addition to this, it is crucial to highlight the effect of different wrist frames to the machining accuracy. First, during the standard TCP calibration procedure, the same point is approached from different angles and with different wrist positions to achieve maximum accuracy. However, there is evidence in previous studies (Pantscharowitsch and Kromoser 2023; Furtado et al. 2017) showing the influence of the wrist position relative to the machining direction on the resulting machining quality. The IR used in this study did not compensate for machining forces nor any resulting deflections.

The transfer of the laser-tracker measurements to the module frame of the code was managed manually. Automating this process by fully implementing the laser-tracker within the IR ecosystem will increase the practicability as well as the efficiency.

5.2 Is the Workflow Presented Sufficiently User-friendly to be Applied in Practice?

The laser-tracker setup procedure in the robot environment is characterized by its simplicity and user-friendliness, facilitating swift deployment and configuration. The Leica Inspire software offers a clearly arranged interface and enables the user to set up automated measuring routines as well as efficient control and setup of the system.

Given the previously mentioned limitations, the system was operated by a team of two individuals. It can be assumed that in industrial implementations operational feasibility is sufficient with a sole machine operator.

The previously-explained wrist configuration limitation must be mitigated for successful applications in practice.

The process's benefits extend to producers of small-series, highly-complex workpieces, as well as those tasked with the processing of exceptionally large workpieces. Furthermore, developers and users of mobile robotic platforms and implementors of multi-robot systems can benefit from the utilization of this process.

6 Conclusions

- The proposed workflow comprising an industrial robot and laser-tracking system to machine timber construction elements is a user-friendly and valuable tool.
- Incorporating a laser-tracking system enables the convenient utilization of the industrial robot's entire workspace to gain advantage of the higher flexibility when compared to widely used joinery machines.
- The measuring process is not remarkably time consuming and provides the opportunity to streamline production processes especially when machining small series.
- The programming of a wrist frame adjustment algorithm would, with high probability, pave the way for a more practicable and industrial application. Machining accuracy within the presented pre-trials is sufficient but the results need to be verified by a more comprehensive study.

7 Outlook

The outcomes of this feasibility study not only provide a foundation for utilizing the entire workspace of an IR, when combined with a laser-tracking system for machining timber construction elements, but could also have broader implications for robotics in construction.

In the future, more detailed research can be conducted on the machining accuracy of the robot when tracing the tool center point with a laser-tracker. This can lay the foundation for wood-machining specific compensation methods. Ultimately, it can be anticipated that automation improvements in timber construction can increase the amount of timber used in buildings and contribute to more sustainability in construction.

Acknowledgements. We highly appreciate the initial support of the TU Wien - Research Unit Engineering Geodesy. Furthermore, we would like to acknowledge the generous funding provided by the Austrian Research Promotion Agency (FFG) for the project Sys.Wood (grant number FO999896268), which has been instrumental in the successful execution of our work.

References

ABB. 2019. "ABB IRB 7600 Data Sheet."

Bisu, C., Cherif, M., Gerard, A., Yves, J., K'nevez.: Dynamic behavior analysis for a Six axis industrial machining robot. Adv. Mater. Res. **423**. (December, 65–76 (2011). https://doi.org/10.4028/www.scientific.net/AMR.423.65

Caro, S., Dumas, C., Garnier, S., Furet, B.: Workpiece placement optimization for machining operations with a KUKA KR270-2 robot. In: 2013 IEEE International Conference on Robotics and Automation, pp. 2921–2926. IEEE, Karlsruhe, Germany (2013). https://doi.org/10.1109/ICRA.2013.6630982

CEN: ÖNORM EN 14080: 2013 08 01: Timber Structures - Glued Laminated Timber and Glued Solid Timber - Requirements. European Committee for Standardization, Brussels (2013)

Dörfler, K., et al.: Mobile robotic fabrication beyond factory conditions: Case study Mesh Mould wall of the DFAB HOUSE. Constr. Robot. **3**(1–4), 53–67 (2019). https://doi.org/10.1007/s41693-019-00020-w

Furtado, L.F.F., Villani, E., Trabasso, L.G., Sutério, R.: A method to improve the use of 6-DOF robots as machine tools. Int. J. Adv. Manuf. Technol. **92**(5–8), 2487–2502 (2017). https://doi.org/10.1007/s00170-017-0336-8

Gambao, E., Balaguer, C., Gebhart, F.: Robot assembly system for computer-integrated construction. Autom. Constr. **9**(5–6), 479–487 (2000). https://doi.org/10.1016/S0926-5805(00)00059-5

Hans Hundegger Maschinenbau AG. 2011. "Hundegger K2i Booklet."

Hexagon AB. 2021. "Leica Absolute Tracker ATS600 Datasheet."

HOMAG/Weinmann AG. 2019. "Homag/Weinmann BEAMTEQ Booklet."

HSD Mechatronics. 2021. "HSD Catalouge ES950 E-CORE ES951 e-CORE."

Ji, W., Yin, S., Wang, L.: A big data analytics based machining optimisation approach. J. Intell. Manuf. **30**(3), 1483–1495 (2019). https://doi.org/10.1007/s10845-018-1440-9

Klimchik, A., Ambiehl, A., Garnier, S., Furet, B., Pashkevich, A.: Experimental Study of Robotic-Based Machining. IFAC-Papers OnLine. **49**(12), 174–179 (2016). https://doi.org/10.1016/j.ifacol.2016.07.591

Möller, C., et al.: Machining of large scaled CFRP-parts with mobile CNC-based robotic system in aerospace industry. Procedia Manuf. **14**, 17–29 (2017). https://doi.org/10.1016/j.promfg.2017.11.003

Pantscharowitsch, M., Kromoser, B.: Milling tenons on GLT beams using an industrial robot and joinery machine – comparison based on surface quality and 3D scans. In: International Mass Timber Conference, Portland, OR (2022a)

Pantscharowitsch, M., Kromoser, B.: Influence of machining parameters on subtractive manufacturing of elementary geometries in glued-laminated timber using an industrial robot. Wood Mater. Sci. Eng. **4**(1–3) March, 472–490 (2022b). https://doi.org/10.1080/17480272.2022.2051734.

Pantscharowitsch, M., Kromoser, B.: Investigation of industrial robots vs joinery machines for milling pockets in glulam: Comparison based on surface quality and 3D scans. Civ. Eng. Des. **4**(1–3), 25–34 (2022c). https://doi.org/10.1002/cend.202100049

Pantscharowitsch, M., Kromoser, B.: On the utilisation of the workspace of an industrial robot for milling structural timber components – experimental trials with different arm positions and evaluation of machining quality—Abstract. In: World Conference on Timber Engineering - WCTE, Oslo (2023)

Susemihl, H., et al.: Referencing strategies for high accuracy machining of large aircraft components with mobile robotic systems. SAE Int. J. Aerosp. **9**(2), 231–238 (2017). https://doi.org/10.4271/2017-01-2166

Susemihl, H., et al.: High accuracy mobile robotic system for machining of large aircraft components. SAE Int. J. Aerosp. **9**(2), 231–238 (2016). https://doi.org/10.4271/2016-01-2139

TechnoWood AG. 2021. "Technowood TW Mill - M Booklet."

Yuan, P.F., Xie, Y.M., Leach, N., Yao, J., Wang, X. (eds.): Architectural Intelligence: Selected Papers from the 1st International Conference on Computational Design and Robotic Fabrication (CDRF 2019). Springer, Singapore (2020)

Zi, B., Duan, B.Y., Du, J.L., Bao, H.: Dynamic modeling and active control of a cable-suspended parallel robot. Mechatronics. **18**(1), 1–12 (2008). https://doi.org/10.1016/j.mechatronics.2007.09.004

S191256
LOOK UP AND LIVE
RETROPOLIS

Mixed Reality Carpentry

Gwyllim Jahn(✉), Cameron Newnham, and Nick van den Berg

Fologram, Melbourne, Australia
gwyll@fologram.com

Abstract. We demonstrate a fast and low-cost method for designing and fabricating doubly curved reciprocal frame structures from reclaimed timber offcuts. In contrast to fully automated robotic fabrication and assembly processes that require exact digital models of available material and as-built structures, our approach enables large-scale reciprocal frame structures to be assembled quickly and approximately, with minor deviation in part geometries caused by material defects or fabrication imprecision made up for by improvised adjustments in joint positions during assembly. We argue that mixed reality guides eliminate the need for drawn instructions and leverage the discretion of saw operators to work around material defects, reducing secondary waste and simplifying cataloguing of available materials prior to fabrication. We show that our method achieves an average error of 2.6 mm from digital models of parts with minimal accumulative error across the completed structure.

Keywords: Mixed Reality · Digital Fabrication · Circular Economy

1 Introduction

1.1 Context

Construction from timber can have a lower embod Design in the age of humans
ied carbon cost than alternatives of concrete and steel [1] and acts as a store of carbon absorbed during the lifetime of the tree [2]. However, 10–15% of all timber in new construction goes to the waste stream, with timber from demolition contributing to 10% of all landfill waste globally [3]. Timber in landfill decomposes to release the captured carbon, and so there are environmental incentives to extend the productive utility of these materials. Offcut timber can be reprocessed as standardized lengths using techniques like finger-jointing or cross-lamination, however this introduces the embodied costs of transporting materials for processing and necessitates use of polyurethane—all with high embodied carbon costs [4]. Working with reclaimed timber materials that are not processed in such a way is challenging because design and construction practices are optimized for standard dimensional materials [5].

There is an opportunity to rethink design practices to accommodate available materials salvaged from timber-framed buildings whereby the apparent limitation of short lengths of material is advantageous for describing curved geometry [6]. Computer-aided

M. Yablonina et al. (Eds.): Robotic Fabrication in Architecture, Art and Design 2024, pp. 286–298, 2026.
https://doi.org/10.1007/978-3-032-22607-5_18

Design and Manufacturing (CADCAM) processes can simplify the fabrication of large numbers of unique parts, as variation in parts can be generated from design models parametrically and then automatically translated to machine code [6]. However, working with irregular stock material such as offcut timber introduces various challenges to robotic fabrication due to the necessity of scanning, cataloguing and matching parts to available materials [7].

Mixed reality environments allow fabricators to apply tacit knowledge to the simple challenge of how to respond to the deviation of materials, parts and assemblies from digital models. Because these responses are improvised rather than anticipated and drawn or programmed, they offer a comparatively simple approach to design and fabrication from non-uniform materials compared to computer vision and robotic automation. Improvising techniques for aligning physical and virtual designs may enable working with wider tolerances than in fully automated systems or from 2D drawings, and thereby also lead to a new capacity for rapid but approximate construction of complex digital designs such as reciprocal frame structures.

1.2 Goals

This research postulates that the life of timber building materials can be extended by developing new design and fabrication systems specifically for working with short lengths of reclaimed timber. The aim of this research is to identify productive uses for offcut and reclaimed timber that would normally be discarded, without the need for measuring available materials and using methods that can easily be deployed on existing construction sites with minimal cost and complexity. By demonstrating that typically discarded materials can be used in such a way, we aim to extend the productive lifecycle of timber buildings and illustrate the potential to reduce the material and labor cost of constructing novel double-curved structures (Fig. 1).

Fig. 1. Assembling timber offcuts using the HoloLens 2

This research proposes a method for working with reclaimed timber offcuts by matching and cutting available material stock to required geometry using digital models displayed in mixed reality. Inferring fabrication information from mixed reality environments is typically less precise than explicit measurements or CNC fabrication, though is well suited to tasks where design tolerances are within the expected deviation of fabricated parts or when working with materials that are not readily described by digital models due to deformation, defects, or other inconsistencies. We aim to increase the expected precision of mixed reality environments to within typical construction tolerances by combining explicit and inferred dimensions and using completed design models to reduce accumulative error during assembly.

1.3 Mixed Reality Fabrication

Mixed Reality refers to a continuum of display technologies that blend digital information with a viewer's physical environment [8]. The Microsoft HoloLens 2 is a stereoscopic display capable of rendering digital models at a fixed location in physical space [9]. There are several advantages to displaying fabrication information in mixed reality. When the placement of virtual models in physical space is precise and reliable, part dimensions can be inferred visually from part geometry rather than communicated through explicit notation. This enables fabricators to select from offcut lengths of timber to best match the geometry of a part without the need for laborious manual measurements that might ordinarily lead to offcuts simply being discarded. Mixed reality environments may provide some efficiencies when working with non-standard structures and materials, thereby extending their productive lifecycle and contributing to a circular economy [10].

2 Background

There are many examples of researchers exploring how to design new architectural typologies to suit the constraints of reclaimed timber. Re-Emerge, by HASSEL and the AA EmTech program, is a small structure demonstrating that boards reclaimed from timber pallets can be forced into curvature using kerf cutting [11]. Unprocessed timber logs can also be assembled into spanning structures through 3D scanning and precise robotically cut joints [12]. Variation in the lengths of reclaimed timber from a demolished building form the screen in March Studio's Grand Stair in Canberra [13]. Decades of work by Gramazio Kohler Research at ETH in Zurich has explored the design and fabrication of structures with large numbers of parts and variation in joints using 6-axis robotic cutting [14]. The use of robotic arms to assist with fabricating reciprocal frame structures has also been demonstrated extensively by Gramazio and Kohler's lab [15, 16] and others [17, 18]. Many projects also utilize mixed reality environments to communicate to fabricators how to assemble or form unique parts. These include assembling robotically cut reciprocal frame structures [19], guiding during assembly of standard lengths of timber [20, 21] or marking up joints on large-scale timber beams [22, 23].

2.1 Mixed Reality in Architecture and Construction

Applications of mixed reality in architecture and construction are typically focused on design visualization or project administration in the form of clash checks and building inspections [24]. The benefits of mixed reality for architectural fabrication become more apparent when designs consist of large numbers of unique parts and joints [25], are made up from non-standard materials [26], or derived through active forming processes [27]. There are many examples in literature of small scale, sculptural projects completed by students in short spaces of time and to wide tolerances such as in workshops by Kristof Crolla and Garvin Goepel [28, 29]. These also include projects specifically working with reclaimed timber, such as Wibranek's reconfiguring of lumber blocks into architectural columns [30], or research at Cornell University exploring the use of mixed reality to assembly unprocessed logs [31] or to template timber boards reclaimed from a local barn [32]. However outside of masonry construction [33, 34], the literature provides no examples of building scale structures assembled to construction tolerances from standard construction materials using only mixed reality documentation.

3 Methods

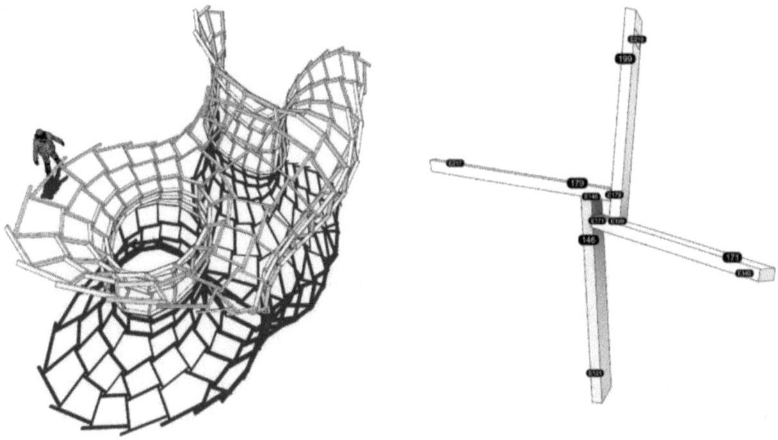

Fig. 2. An aerial view of the demonstrator project design and typical labelled joint

To evaluate the feasibility of our method, we designed and built a double-curved shell structure at the scale of a small building from reclaimed lengths of 90x35mm framing timber donated by a local carpentry school. The structure consists of 382 parts that are fabricated by making a cut at each end of the part using a miter saw. Joint locations describing the topology of the structure are marked directly on the face of timber parts following a digital template viewed in mixed reality. Connectivity information described on the physical part is then used in combination with the completed design model viewed in mixed reality to assemble the structure without accumulative error from small inaccuracies in fabricated parts. The structure was built over 4 days by a team of three people, demonstrating the speed and simplicity of our method.

3.1 Modelling Reciprocal Frames

The design of the structure is modelled as a network of lines in McNeel and Associates' Rhinoceros 3D. A graph representing the topology of this line network is used to generate parts and joints. Each end of any individual length of timber forms a "T" joint with a neighboring part. The line network representing the design model describes a graph of node-edge relationships, and the topology of this graph is modified to describe edge-end to edge relationships. Because T joints are always defined by the end of a part and the face of the next clockwise edge around a node, this topology can be constructed by finding the list of edge indexes for the node at an edge-end, getting the index of the edge-end in this list and incrementing this index by 1.

Joints between parts are distributed around each node by defining a circle perpendicular to each node with radius R, finding the closest point intersection between edges and this circle, moving these points in the normalized parameter space of the circle by T then re-drawing edge geometry by substituting the original end points with the points evaluated in the parameter space of the circle. The approximate spacing between joints is a product of the radius R and the offset T. A T value of 0.25 rotates each joint by approximately 90° to its neighbor and creates the appearance of a typical reciprocal frame (see Fig. 2).

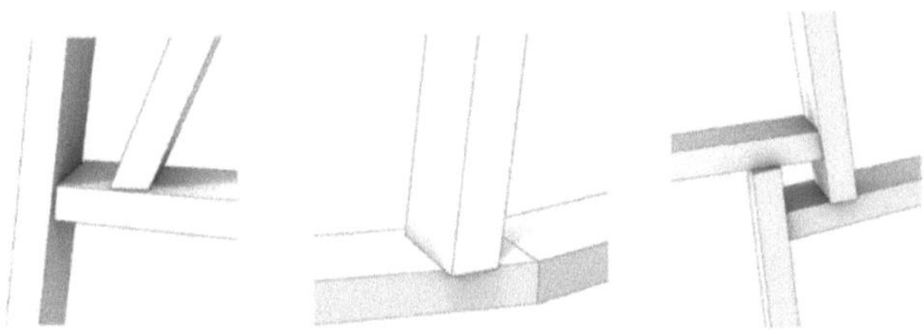

Fig. 3. From left to right: boolean heuristics for an unevenly distributed node, a miter joint and offset T at the base of the structure, and a typical four-way node

The topology of the graph is also used to determine optimal joint types for timber parts. When a node has only two edges, parts are not distributed using the reciprocal method but are instead mitered using the plane that bisects the two edges. When edges are unevenly distributed around a node such that they cannot form a reciprocal joint, a central part is identified, and neighboring parts are mitered using the intersecting faces of this part (see Fig. 3).

3.2 Fabrication

The geometry of parts in the design are cut to the correct angle and length using a single saw cut at each end of a piece of 90x35mm framing timber. To display the miter and bevel angles of these cuts to a fabricator, the part geometry is oriented in 3D space such that the long edge of the part is perpendicular to the X-axis of a cutting plane that represents the saw blade. The digital model of the part is then overlaid on the physical saw using two fiducial markers fixed at precise coordinates relative to the saw blade and pivot points. Miter and bevel angles are displayed as camera-aligned text dots above the saw pivots, with line and arc geometry indicating these angles visually for quality control checks (see Fig. 4).

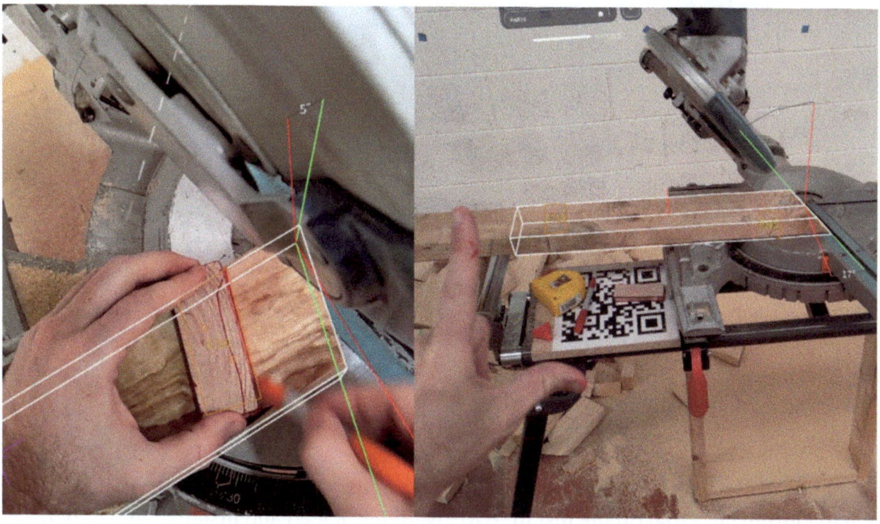

Fig. 4. Left: Templating joint locations on the face of a part using a mixed reality guide. Right: A part after one end has been cut with the saw aligned to miter and bevel angles displayed in mixed reality

Selecting a length of material and determining the cut sequence is left to the discretion of the saw operator. The virtual model of the part displayed in mixed reality provides a guide to the length of material required, and an appropriate offcut is found from available material. The part is then placed on the saw table and visually checked for lengths. The saw operator can then use a mixed reality user interface to toggle between cut geometry for each end and rotate the part around its central axis to switch miter and bevel angles if these exceed the limits of the saw.

The saw operator then pivots the saw blade using the mixed reality guide and cuts one end of the part. Any joints or labels on the exposed face of the part are immediately marked to ensure maximum precision relative to the cut face (see Fig. 4). The saw operator then uses the mixed reality user interface to re-orient and rotate the part to optimally align the opposing end with the cutting plane. The part is slid along the cutting guard such that the end of the virtual model aligns with the newly cut geometry of the length of timber and any joints and labels are marked. The pivot of the saw is then readjusted, and the second cut completes the part.

3.3 Assembly

Assembly was completed by locating a virtual model displaying all parts, labels and joint locations within the construction environment. The virtual model was aligned to

Fig. 5. Still from the HoloLens 2 RGB camera showing alignment between physical structure and virtual model (white lines)

the physical site using two fiducial QR codes fixed exactly 4000 mm apart on the ground plane. During an initial offsite test fit, parts were fabricated and installed in parallel. The assembly team would request a sequence of parts to be fabricated by reading labels of required parts in the virtual model. After these parts were cut, they were then installed in the structure by orienting the part to the virtual model and screw fixing in the location indicated by the pre-marked joints. Although parts were expected to self-register with templated joints, we anticipated some tolerance in both the geometry of parts and the location of manually marked joints. The virtual model therefore served as a visual indication of the degree of accumulative error and deviation from the design model during assembly (see Fig. 5), and the assembly team could make informed decisions about when to deviate from the virtual model or physical templates to minimize local or global error.

4 Results and Reflection

The completed free-standing structure measures approximately 9500 mm long by 5500 mm wide by 4300 mm high and consists of 382 hand-cut timber parts (see Fig. 6). The average time for a team of 3 people to select, label, cut and install a part was 3 min. The structure was then dismantled into 1 5-sided frame, 84 quads, 2 tris and 35 individual lengths and re-built for exhibition by a team of 6 people in 3 h.

Describing fabrication information in this manner reduced documentation complexity by replacing multiple 2D drawings with a single interactive 3D model. Removing the need to read and remember dimensions and schedules also simplified fabrication tasks for carpenters operating the miter saw. When it was difficult to infer dimensional information from geometry, such as estimating bevel angles from the short cut edge of stock material, dimensions were displayed as virtual models in mixed reality and aligned with saw protractors. This simplified the exercise of reading dimensions and ensured

Fig. 6. Completed structure in the workshop space

that information could always be checked and confirmed visually. We found this method to eliminate all human error in misreading dimensions over the course of cutting 382 parts, though some imprecision and deviation from digital models was still experienced.

4.1 Precision

Our mixed reality interface for describing part geometry, joints, labels and cut angles was sufficiently accurate for all parts to be installed in the structure within the tolerances of joint locations and material flex. An empirical measure of precision was determined by taking a sampling of 62 parts and comparing physical edge lengths to those in the digital model using a tape measure. This sampling found an average of 2.6 mm of error with a standard deviation of 1.7 mm. This error could be introduced by tolerances in miter and bevel protractors, tolerance in display resolution on the HoloLens 2 (that renders lines as 1-2 mm thick when viewed at arm's length and accounting for Fresnel effects), imperfect calibration of the headset for the wearer's eyes leading to incorrectly rendered depth, imprecision in hologram placement relative to the saw blade or tolerance in manual read-outs of physical lengths from a tape measure.

The nature of using virtual models as templates for fabrication and assembly in mixed reality is that the degree to which physical parts match these virtual templates is up to the discretion and skill of whoever is following them. Templates can be followed quickly and approximately to reduce fabrication time, or slowly and carefully to increase precision. In previous research, we have leveraged this ability for fabricators to improvise in response to the contingencies of the given task to work with unwieldly materials and wide tolerances without introducing the risk of dramatic accumulative error. The approach outlined in this research instead focuses on shifting information that is difficult to follow visually (compound cut angles) to notational instructions (angular dimensions) to improve precision necessary for larger scale construction from framing timber.

Visual templates in mixed reality reduce fabrication time by alleviating the need for excessive notation in the form of linear dimensions of part lengths and joint locations. Error in following visual templates in mixed reality is typically introduced from the inability to correctly judge the depth of the template geometry from stereoscopic displays. This error can be reduced by displaying templates on an existing physical surface. The physical surface then provides an exact reference point for the depth of the virtual template, and fabricators only need to accurately follow the shape of the template in two dimensions. By aligning the template with the geometry of cuts in the end stock of a part, fabricators can further improve the reliability and precision of the mixed reality template.

Virtual models in mixed reality can be used as accurate and reliable guides to part placement in mixed reality, as has been demonstrated by several masonry structures completed to construction tolerances [33, 34]. These templates can even reach manufacturing precision when fabricators are able to work in proximity to fiducial markers that correct for drift in the position of virtual models relative to the physical part [22]. In our approach we utilize a hybrid method for assembly in which the geometry of the structure can be inferred entirely from precisely marked joints, with mixed reality models used to deviate from these joints when required without introducing the risk of not being able to fit subsequent parts.

We expected that the precision of these marked joints would be higher than the precision of parts displayed in mixed reality due to incorrectly judging depth in the 3D position of joints combined with drift in the position of the virtual model relative to physical space at the scale of the structure. Instead, small discrepancies in fabricated part geometries (due to human error or material deformation) could be accommodated by slight shifts in joint locations between parts that still adhered to the overall geometry of the structure displayed in mixed reality. Because the design model consisted of many cantilevering elements the structure deformed under self-weight. The mixed reality design model was used to temporarily prop the structure into the correct geometry while bracing parts were installed, thereby avoiding the challenge of tight-fitting parts into an unintended overall geometry.

4.2 Handcraft vs. Automation

Numerous projects have demonstrated the capacity for 6-axis robotic arms to fabricate and assemble similar structures to that demonstrated in this research [15–17]. However, the challenges of fabricating these structures from reclaimed timber outside of robotics lab environments are significant. These include the task of digitizing the geometry of available reclaimed materials prior to generating robotic toolpaths, or the necessity of transporting reclaimed timber to manufacturing facilities rather than being able to process them directly on site. The aim of this research is to demonstrate a method for fabricating and assembling complex timber structures that can be readily implemented in ordinary workshop environments and without sophisticated computer vision, tool pathing or CADCAM requirements.

Mixed reality environments also enable utilization of material that is easy to improvise but comparatively difficult to automate. For instance, saw operators can select offcuts with ends that have already been cut close to the geometry of the desired part, thereby minimizing secondary offcut waste. Saw operators can also orient parts such that knots, breaks or nail holes are not located in the same position as cuts or joint locations. Deformation in parts or deflection in the structure during assembly can also be overcome by improvising small changes in joint locations following common-sense heuristics that minimize deviation from a mixed reality design model while still maximize joint strength. This kind of resilience to unintended structural behavior during design distinguishes the advantages of a mixed-reality carpentry approach to both conventional and automated construction.

4.3 Improvements

Imprecision was partly due to the challenge of exactly determining the relative position of fiducial markers used to locate virtual models in the cutting environment and the saw blade. This could be improved with a reliable 3D model of the saw geometry, or with exact metrology. Providing fabricators with linear dimensions displayed in mixed reality as well as with a ruler fixed to the saw guide would enable rapid checks for length correctness that would improve overall precision by enabling parts fabricated outside of tolerance to be re-cut more carefully (and slowly).

5 Outlook

We demonstrate a low cost and rapid approach to fabricating double-curved timber structures consisting of T-joints between parts cut from reclaimed framing timber. This structure indicates how mixed reality environments can help design and construction teams make informed decisions about how to match a variety of available lengths and defects to geometry required in a design, thereby extending the productive life of material that would ordinarily go to landfill. We anticipate that this method may extend to other similar approaches to designing for circular economies in which new uses are found for non-standard materials by developing new techniques for forming and shaping them. There are, however, several challenges to advancing this approach, particularly at larger scale and when incorporated into building codes and regulations. These include the need to develop design tools for modelling and analyzing structures from irregular materials, and the exploration of how the geometry of these structures can be clad or integrated with building services.

References

1. Durlinger B, Crossin E, Wong J (2013) Life cycle assessment of a cross laminated timber building, RMIT University.
2. Gustavsson, L., Madlener, R., Hoen, H.-F., et al.: The role of wood material for greenhouse gas mitigation. Mitig Adapt Strat Glob Change. **11**, 1097–1127 (2006). https://doi.org/10.1007/s11027-006-9035-8

3. Jahan, I., Zhang, G., Bhuiyan, M., Navaratnam, S.: Circular economy of construction and demolition wood waste—A theoretical framework approach. Sustainability. **14**, 10478 (2022). https://doi.org/10.3390/su141710478
4. Churkina, G., Organschi, A., Reyer, C.P.O., et al.: Buildings as a global carbon sink. Nat Sustain. **3**, 269–276 (2020). https://doi.org/10.1038/s41893-019-0462-4
5. Larsen NM (2020) Digital workflows for natural wood in constructions. In: D. Holzer, W. Nakapan, A. Globa, I. Koh (eds.), RE: Anthropocene, Design in the age of humans—Proceedings of the 25th CAADRIA Conference—Volume 1, Chulalongkorn University, Bangkok, Thailand, 5–6 August 2020, pp. 125–134. CUMINCAD
6. Mitchell, W.J.: Roll over Euclid: How frank Gehry designs and builds. In: Ragheb, J.F. (ed.) Frank Gehry, Architect, pp. 352–363. Guggenheim Museum Publications, New York (2001)
7. Reisach, D., Schütz, S., Willmann, J., Schneider, S.: A Design-to-fabrication workflow for free-form timber structures using offcuts. In: Turrin, M., Andriotis, C., Rafiee, A. (eds.) Computer-Aided Architectural Design. Interconnections: Co-computing Beyond Boundaries, pp. 361–375. Springer Nature, Switzerland, Cham (2023)
8. Milgram, P., Kishino, F.: A taxonomy of mixed reality visual displays. IEICE Trans. Inf. & Syst. **E77-D**(12), 1321–1329 (1994)
9. Kress, B.C., Cummings, W.J.: 11-1: Invited paper: Towards the ultimate mixed reality experience: HoloLens display architecture choices. SID Symp. Dig. Tech. Pap. **48**, 127–131 (2017). https://doi.org/10.1002/sdtp.11586
10. Circular economy introduction. https://ellenmacarthurfoundation.org/topics/circular-economy-introduction/overview (2023). Accessed 5 Oct 2023
11. Re-emerge. https://www.aaschool.ac.uk/publicprogramme/whatson/re-emerge?utm_medium=website&utm_source=archdaily.com (2023). Accessed 5 Oct 2023
12. Self M Wood Chip Barn / AA Design + Make. https://designandmake.aaschool.ac.uk/project/wood-chip-barn/. Accessed 5 Mar 2024
13. The grand stair. In: MARCH.STUDIO. https://march.studio/thegrandstair (2023). Accessed 5 Oct 2023
14. Menges, A., Schwinn, T., Krieg, O.D.: Advancing Wood Architecture: A Computational Approach. Taylor & Francis Group, London, United Kingdom (2016)
15. Apolinarska, A.A.: Performance-driven design of a reciprocal frame canopy—Timber structure of the future tree. In: Stojakovic, V., Tepavcevic, B. (eds.) Towards a new, configurable architecture—Proceedings of the 39th eCAADe Conference—Volume 1, University of Novi Sad, Novi Sad, Serbia, 8–10 September 2021, pp. 497–504 (2021) CUMINCAD
16. Kohlhammer, T., Apolinarska, A.A., Gramazio, F., Kohler, M.: Design and structural analysis of complex timber structures with glued T-joint connections for robotic assembly. Int. J. Space Struct. **32**, 199–215 (2017). https://doi.org/10.1177/0266351117746268
17. Adelzadeh, A.: ReciprocalShell: A hybrid timber system for robotically-fabricated lightweight shell structures. In: Dokonal, W., Hirschberg, U., Wurzer, G. (eds.) Digital Design Reconsidered—Proceedings of the 41st Conference on Education and Research in Computer Aided Architectural Design in Europe (eCAADe 2023) —Volume 1, Graz, 20–22 September 2023, pp. 651–660 (2023) CUMINCAD
18. Tamke, M., Riiber, J., Jungjohann, H., Thomsen, M.: Lamella flock. In: Advances in Architectural Geometry (2010)
19. Hughes, R., Österlund, T., Larsen, N.M.: Integrated design-for-manufacturing and AR-aided-assembly workflows for lightweight reciprocal frame timber structures. Constr. Robot. **5** (2021). https://doi.org/10.1007/s41693-020-00048-3
20. Yang, X., Amtsberg, F., Skoury, L., et al.: VIZOR: Facilitating cyber-physical workflows in prefabrication through augmented reality. In: Proceedings of the 27th International Conference of the Association for Computer-Aided Architectural Design Research in Asia (CAADRIA), vol. 2, pp. 141–150 (2022)

21. Atanasova LM (2020) Prototype as artefact. In: ACADIA 2020: Distributed Proximities / Volume I: Technical Papers [Proceedings of the 40th Annual Conference of the Association of Computer Aided Design in Architecture (ACADIA)]. [ISBN 978-0-578-95213-0]. Online and Global. 24-30 October 2020. edited by B. Slocum, V. Ago, S. Doyle, A. Marcus, M. Yablonina, and M. del Campo. 350–359. CUMINCAD
22. Kyaw, A.H., Xu, A., Jahn, G., et al.: Augmented Reality for high precision fabrication of Glued Laminated Timber beams. Autom. Constr. **152** (2023). https://doi.org/10.1016/j.autcon.2023.104912
23. Blahut S and Marie-Therese H-F (2023) Mixed reality interactive representations for the assembly of a custom timber tower. In: Dokonal, W, Hirschberg, U and Wurzer, G (eds.), Digital Design Reconsidered—Proceedings of the 41st Conference on Education and Research in Computer Aided Architectural Design in Europe (eCAADe 2023) —Volume 2, Graz., 20-22 September 2023, pp. 751–760. CUMINCAD
24. Chi, H.-L., Kang, S.-C., Wang, X.: Research trends and opportunities of augmented reality applications in architecture, engineering, and construction. Autom. Constr. **33**, 116–122 (2013). https://doi.org/10.1016/j.autcon.2012.12.017
25. Jahn, G., Newnham, C., Berg, N., Beanland, M.: Making in mixed reality. In: Proceedings of the 38th Annual Conference of the Association for Computer Aided Design in Architecture (ACADIA), pp. 88–97. Mexico City, Mexico (2018)
26. Song, Y.: BloomShell: Augmented Reality for the assembly and real-time modification of complex curved structure. In: Anthropologic: Architecture and Fabrication in the cognitive age—Proceedings of the 38th eCAADe Conference, pp. 345–354, Berlin (2020)
27. Jahn, G., Wit, A., Pazzi, J.: [BENT] Holographic handcraft in large-scale steam-bent timber structures. In: Proceedings of the 39th Annual Conference of the Association for Computer Aided Design in Architecture (ACADIA), pp. 438–447. The University of Texas at Austin School of Architecture, Austin, Texas (2019)
28. Goepel, G., Crolla, K.: Augmented reality-based collaboration—ARgan, a bamboo art installation case study. In: RE: Anthropocene, Design in the Age of Humans—Proceedings of the 25th CAADRIA Conference, pp. 313–322. University, Bangkok, Thailand (2020)
29. Goepel, G., Crolla, K.: Secret whispers & transmogrifications: A case study in online teaching of Augmented Reality technology for collaborative design production. In: PROJECTIONS—Proceedings of the 26th CAADRIA Conference, pp. 21–30. The Chinese University of Hong Kong and Online, Hong Kong (2021)
30. Wibranek, B.: Carbon Sequestration in Lumber Columns: Facilitating the Reuse of Lumber Waste for Prefabrication through Computational Design and Augmented Fabrication, pp. 90–103. Green Building & Construction Economics (2023). https://doi.org/10.37256/gbce.4120232309
31. Lok L (2022) Timber de-standardized 2.0: Mixed reality visualizations and user interface for processing irregular timber. In: Proceedings of the 27th International Conference of the Association for Computer-Aided Architectural Design Research in Asia. Sydney, Australia
32. RUBI LAB | HoloWall | AAP Labs. https://labs.aap.cornell.edu/rubi-lab/research/holowall (2023). Accessed 16 Oct 2023
33. Jahn, G., Newnham, C., Berg, N., et al.: Holographic construction. In: Impact: Design with All the Senses, pp. 314–324, Berlin (2019)
34. Mitterberger, D., Dörfler, K., Sandy, T., et al.: Augmented bricklaying: Human–machine interaction for in situ assembly of complex brickwork using object-aware augmented reality. Constr. Robot. **4**, 151–161 (2020). https://doi.org/10.1007/s41693-020-00035-8

Mobile Robotic Brickwork Disassembly for Material Reuse

Begüm Saral[1](✉), Hanzhi Chen[2], Stefan Leutenegger[2], and Kathrin Dörfler[1]

[1] TT Professorship of Digital Fabrication, Technical University of Munich, Arcisstr. 21, 80333 Munich, Germany
begum.saral@tum.de

[2] TT Professorship of Machine Learning for Robotics, Technical University of Munich, Arcisstr. 21, 80333 Munich, Germany

Abstract. Within the context of advancing material reuse and circularity in building construction, the careful dismantling of building structures is a crucial step in salvaging materials. Our research is focused on illustrating the potential of Spatial Artificial Intelligence (Spatial AI)-integrated mobile robotic systems to facilitate and streamline such a careful process of autonomous dismantling. In this pursuit, we present a case study that tests and evaluates methods for the autonomous robotic disassembly of a dry-stacked brick structure of unknown geometry carried out by a mobile robot. By utilizing deep learning-based object-level localization using the onboard sensing capabilities of a hand-in-eye RGB-D camera, a mobile robotic system is designed to detect and localize individual objects, in this case, bricks, within a complex brickwork structure. This information is subsequently employed within the robotic disassembly routine, further encompassing several critical components, including path planning, collision avoidance, and robotic control, all contributing to a seamless disassembly process. Furthermore, our exploration extends its scope to the possibility of capturing and documenting the current as well as the evolving state of the disassembled building structure through digital twins, opening new avenues for efficient and transparent deconstruction processes that align with the goals of material reuse and circularity.

Keywords: Robotic Disassembly · Spatial Artificial Intelligence · Mobile Robotics · Object Localization · Deep Learning · Digital Twins · Material Reuse · Circular Construction

1 Introduction and Motivation

The historical construction practices predating the era of globalized supply chains demonstrate the widespread utilization of building material reuse as an integral component of established construction traditions, particularly in contexts characterized by the scarcity of raw materials. Termed *spolia*, this practice involves the repurposing of architectural elements, encompassing not only economic utility but also an architectural sentimentality—a means of preserving historical continuity that resonates through to contemporary times [1] (Fig. 1). This practice finds heightened prominence, notably

M. Yablonina et al. (Eds.): Robotic Fabrication in Architecture, Art and Design 2024, pp. 299–312, 2026.
https://doi.org/10.1007/978-3-032-22607-5_19

during periods following conflicts, as exemplified in the aftermath of World War II. Instances such as the "Trümmerfrauen" of German and Austrian cities, local women who reclaimed materials from the debris of war, and creative restoration projects like Munich's Pinakotheken, where salvaged bricks were utilized to restore and rehabilitate the museums damaged by the war [2], vividly showcase the embrace of reuse practices during periods of scarcity and need.

Fig. 1. Façade of the Teatro Olimpico in Vicenza, northern Italy (left), Trümmerfrauen collecting rubble (middle), and reconstruction of the damaged section of Pinakotheken (right). Image credits: Kathrin Dörfler (left), Wikimedia Commons/Janczikowsky (middle), Begüm Saral (right).

The latter half of the 20th century has brought a decline in reuse practices, not only due to the increased availability of building materials but also due to the growing difficulty in salvaging materials from complex modern building technologies. An illustrative example is the introduction of cement into brick mortar during the 1970s. This alteration led to the bond between bricks becoming stronger than the bricks themselves, resulting in bricks breaking earlier than the connections between them and rendering their reuse nearly impractical for structures built after this date [3]. This challenge is further underscored by the complexity involved in many other deconstruction processes involving various material systems [4].

The typical end-of-life fate for a building structure, therefore, involves demolition, which entails using destructive machinery to break down the structure and discard the remains. This approach is cost-effective and significantly less labor-intensive, considering the complexity of modern built environments and building technologies. Nevertheless, from the perspective of material circularity, this practice essentially undermines the potential for reusing building materials, and limits the possibilities to downcycling, as the materials cannot be retrieved as intact components.

With the growing scarcity of natural resources and a broader commitment to curbing carbon emissions resulting from the production of new materials, contemporary trends within the architecture and construction sector have begun to steer in an alternative direction. Circular economies and urban mining are emerging as strategies that capitalize on enabling the reintroduction of materials through processes like downcycling, upcycling, repurposing, or reusing. Stockhammer and Entner break down the circular economy into five stages: material cadastre, mining and careful dismantling, reconditioning and storage, commerce, and quality assurance [5]. The influence of digitalization can be exemplified across many of these stages; that is, the assignment of material passports to building components [6] for material cadastre, the notion of Buildings as Material Banks (BAMB) for reconditioning and storage, and corresponding online platforms facilitating the resale of collected materials for commerce [7]. Recently, such approaches were also exhibited in the German pavilion at the Venice Architecture Biennale 23′ [8]. At the level of mining and careful dismantling, while there are certain private enterprises that specialize in professional dismantling [9, 10], this undertaking is notably tedious and demanding, necessitating a high degree of adaptability and physical ability to navigate through unstructured environments. It is evident that there is room for exploration and research in the realm of careful dismantling and material mining, especially with the aid of novel digital and robotic technologies. Present advancements in mobile robotics featuring spatial awareness and the capability of executing intricate tasks within existing unstructured built environments appear as promising technology applicable to facilitate autonomous building deconstruction and material salvaging, which is the central focus of the research proposed in this paper.

2 Research Objectives

This research examines the viability of employing mobile robots equipped with Spatial Artificial Intelligence (Spatial AI) capabilities for the purpose of careful dismantling of building structures for material reuse. For this investigation, a key experiment is conducted, in which a mobile robot autonomously disassembles a dry-stacked brick wall of unknown geometry. This brick wall is previously dry-assembled in an arbitrary arrangement by a human. This wall serves as a representative example of an existing building structure, however, with limited complexity to fundamentally explore and validate the methods proposed. By leveraging deep learning (DL)-based object perception using onboard visual sensing, the mobile robot employed can detect and localize individual bricks within the existing brickwork structure. This information is subsequently employed within the robotic disassembly routine, further encompassing several critical components, including path planning, collision avoidance, and robotic control, all contributing to achieving a seamless disassembly process. Furthermore, a real-time digital twin is employed, which not only facilitates the control of the disassembly routine, but also the capture and documentation of the current and progressing state of the disassembled structure within a concurrent building information model. This in-process data collection facilitates future reconstruction or material reuse by providing a comprehensive digital record of the structure and its constituent components.

3 State of the Art

3.1 Robotic Dismantling of Building Structures

Deconstruction is a multifaceted procedure, particularly when considering that current building technologies are not inherently designed for end-of-life treatments [11], noting that future new-building approaches call for the incorporation of Design for Disassembly

(DfD) guidelines [12]. Several existing examples also highlight the applicability of mobile robotics for the realm of autonomous deconstruction processes. In this context, there exist methodologies and frameworks for robotic deconstruction that align with life-cycle-oriented material refurbishment principles, as well as on-site deconstruction practices that are already in operation, particularly in Japan and Germany [13].

Some on-site instances involve robotic applications aimed at the automated disassembly of multi-layered facade constructions, utilizing automated deconstruction strategies that enable effective material recycling [14]. However, these systems often prioritize predefined method-oriented tasks over recognition and adaptation to the environment. In other semi-autonomous mobile robotic applications, specified deconstruction tasks are implemented to enhance process safety by utilizing visual sensors for human monitoring and decision-making [15]. This suggests that there is still significant room for the development of AI-integrated, intelligent robotic systems adaptable to new materials and environments. Other examples that incorporate deep learning methodologies for the classification of construction waste materials directly on-site showcase the potential for such integrated systems [16]. Overall, there remains substantial potential to further investigate and combine various robotic technologies in the pursuit of a circular building methods [17].

3.2 Onboard Mobile Robot Perception in Unstructured Environments

Recent years showcase various efforts focused on mobile robot perception with minimal site setup and onboard sensing, acknowledging its viability particularly in unstructured construction environments. While some projects leverage onboard LiDAR sensing capabilities to localize a mobile robot in relation to the previously positioned work object for subsequent placements [18], others adopt a similar approach by aligning environmental scans with digitally planned models for robotic additive-manufacturing applications [19]. Some additional projects employ such LiDAR technologies on object-level to interact with objects found on-site, such as rocks, in an adaptive and on-the-fly manner [20].

On the other hand, to enhance fine-grained interaction between the robots and the scenes—and considering the affordability and accessibility of camera-based computer vision approaches as opposed to costly laser-based technologies—researchers have begun utilizing AI tools for spatial comprehension from an object-centric perspective. This involves treating each object within the scene as a distinct component and estimating their attributes such as poses, geometry, textures, and more. The 6° of freedom (6-DoF) pose is one of the most important pieces of information to be retrieved, as it not only helps the robot's decision-making procedure for the next goal action but also aids to improve the localization accuracy of the robot [21–23]. pioneered in this research field using an off-the-shelf object detector to obtain objects' foreground masks, then leveraged photometric and geometric cues to solve their 6-DoF poses with classical non-linear optimization approaches. Another line of work aims to fit deep neural networks capable of predicting 6-DoF poses from RGB(-D) observations in an end-to-end fashion [24–28]. In our work, we also adopt such a data-driven paradigm to conduct object-level scene understanding considering the pose ambiguity brought by the texture-less appearance and symmetric geometry of the bricks.

4 Method

4.1 Overview of the Experimental Study

Our case study has been intentionally set up with controlled complexity to validate the concept and showcase the feasibility of autonomously disassembling building structures for material reuse. It centers on the robotic disassembly of a dry-stacked brick wall of unknown geometry by employing a mobile robot. In the reverse scenario of (conventional) robotic assembly, robots typically follow a routine that involves picking up building elements from predetermined locations, such as a brick storage, and placing them in specified locations in a complex structure according to a given digital model. This assembly task is feasible for robots without advanced robotic sensing capabilities, which can be programmed in a feed-forward manner. In contrast, in a robotic disassembly scenario of an existing building structure, there is no prior knowledge about the geometric configuration of the structure or the individual locations of its building elements. Therefore, a robot intended for disassembly must be equipped with advanced sensing and control strategies to both detect and estimate the geometric knowledge about the overall building structure and identify its individual constituent elements, allowing the robot to compute and carry out a step-by-step disassembly process. The mobile robotic system presented in our research is thus enhanced by Spatial AI capabilities, facilitating to detect, estimate, and manipulate objects, i.e., bricks as part of an unknown geometry in the form of an existing brick wall constructed in an arbitrary configuration by a human. Hereby, the objective is to disassemble any given brickwork structure and to relocate the reclaimed bricks to a predefined brick storage (Fig. 2).

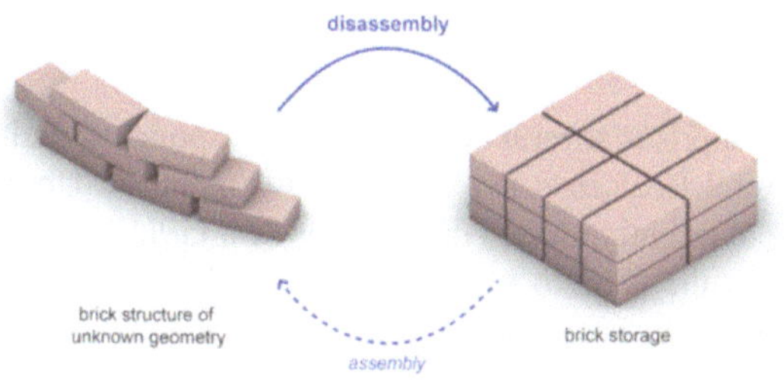

Fig. 2. Disassembly of structures of unknown geometry.

In this direction, we initiate the process by manually assembling a brick wall in an arbitrary configuration without providing any prior information to the robot. The mobile robot is then manually navigated to a location in close proximity to the wall by using the robot's remote-control feature. The robot's subsequent task is to autonomously disassemble a section of the brick wall within its static reach by detecting individual bricks, estimating their poses, identifying the bricks which can be disassembled, picking them up and placing them to a predefined brick storage, in this case, located directly on the mobile robot's back. As depicted in the flowchart in Fig. 3, an automated task management system is employed to achieve this coordinated and iterative workflow. This workflow consists of two steps: first, the robot arm moves to predefined snapshot poses to search for the whole building structure (named workpiece) and estimates its position relative to the robot. Second, it moves to various relative snapshot poses that are calculated by utilizing the workpiece position data, including the workpiece's height and distance from the robot, and searches for individual bricks to pick and place as part of the disassembly routine through these snapshot poses.

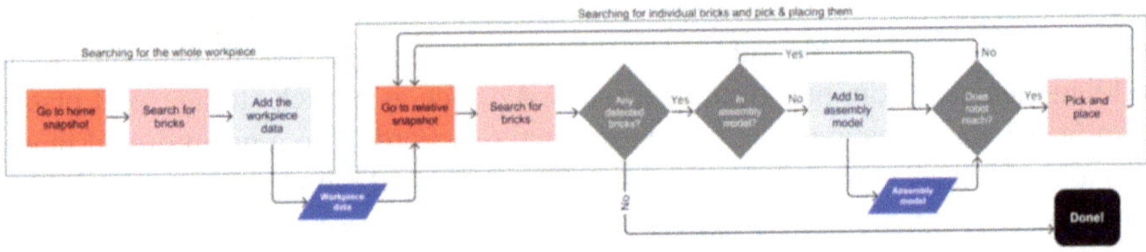

Fig. 3. Automated task workflow.

In the iterative second step, the system continuously scans for bricks using the DL-based brick pose estimator and estimates their location relative to the robot's pose. Upon detection, found bricks are added to a predefined building information model, which serves to track and store the detected and disassembled bricks and their as-found and as-progressed locations. Subsequently, the robot assesses whether it can reach these bricks and, if feasible, plans the pick-and-place routine, and retrieves and stores them one by one based on their estimated poses. This process repeats until no more bricks are detected, ensuring the full disassembly of the given brick wall.

4.2 System Setup

The system setup comprises several key hardware components, including a customized mobile robot, an Intel-Realsense D435i RGB-D camera, an Ubuntu machine, and a Windows machine. Currently, due to the absence of a GPU card on the mobile robot's computing hardware, the Ubuntu machine is required for the DL-based pose estimator. The mobile robot consists of a Robotnik RB-VOGUI+ mobile platform offering omnidirectional movement capabilities, a vertical axis, and a Universal Robots UR10e robot arm with 1.3 m reach and 10 kg payload. Attached to the robot arm is a custom hand-in-eye end-effector featuring a vacuum gripper and the RGB-D camera (Fig. 4).

Regarding the computing setup, the system utilizes a ROS (Robot Operating System) management system to establish seamless low-level communication between various modules, as depicted in Fig. 5.

On the higher and non-time-critical level, the primary user interface is the architectural design and planning environment Grasshopper, which runs on the Windows platform. This environment incorporates both the digital building information model for tracking and storing the disassembled structure, and the robot model. It further serves to manage the automated task executions via a TCP/IP interface to ROS. These tasks include triggering the pose estimator on Ubuntu via ROS services, which communicates to search for bricks and obtain their published poses, requesting and receiving valid

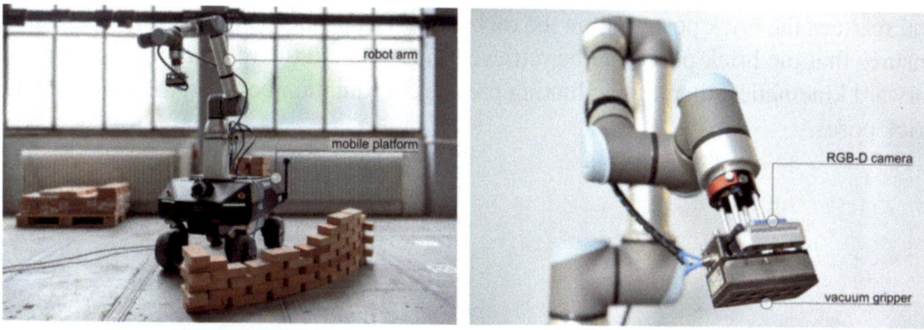

Fig. 4. Mobile robot setup consists of a mobile platform and a robot arm (left) and a customized hand-in-eye end-effector featuring a vacuum gripper and an RGB-D camera (right).

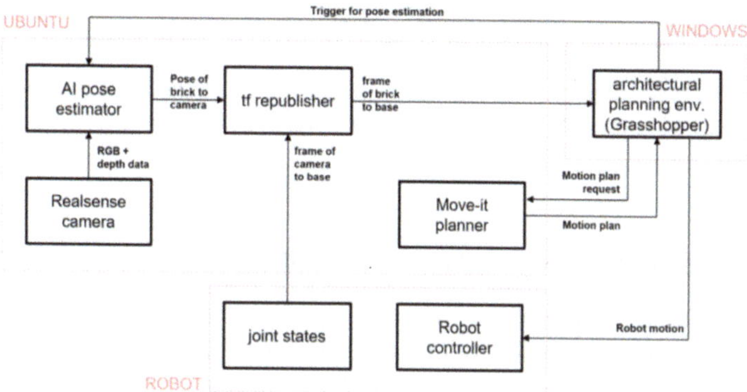

Fig. 5. ROS management system.

motion plans from a MoveIt planner (also running on the Ubuntu machine), and transmitting the planned trajectories to the robot controller for executing the pick-and-place routines.

4.3 Deep Learning-Based Object-Level Localization

To retrieve the 6-DoF brick poses in the workspace, several deep neural networks (DNNs) are used to conduct scene parsing from single-view observation provided by the RGB-D camera. Specifically, the foreground masks of bricks are detected using the PointRend object detector [29]. We obtain the segmented point cloud using the sensed depth and the foreground masks for each brick, followed by a subsampling process to acquire 2,048 points per brick. Further, we pass subsampled points for each object to a point-based DNN [30] to predict the dense object-space coordinates. Finally, by inputting the pixel coordinates in the color image and the predicted object-space coordinates, the 6-DoF transformation between the camera and the brick coordinate system is estimated using the SQPnP algorithm [31]. As a result, the estimated brick poses are initially published relative to the camera frame. To rectify this, a custom pose republisher leverages the established extrinsic calibration of the camera frame relative to the end-effector frame and reissues the brick poses within the robot state's frame tree (Fig. 6). This mechanism ensures that the brick poses can be retrieved in relation to the robot arm's origin using forward kinematics, thereby facilitating precise coordination between the robot and the brick poses.

Fig. 6. The DL-based algorithm detects bricks (left), estimates their poses (middle). Estimated poses are published in the robot state's frame tree (right).

4.4 Relative Snapshot Routines for Searching Bricks

The snapshot strategy for searching bricks and pose estimation exhibited some limitations. Due to the limited capability of the Ubuntu machine, the pose estimation took approximately 8–10 s. Consequently, we opted to employ a strategy where the pose estimator is triggered when the robot arm is positioned at 5 relative snapshot poses around its base ($-60, -30, 0, 30, 60°$) for each layer of the bricks. This approach generally yielded satisfactory data; however, it occasionally led to time loss when the workpiece did not extend to the robot's sides or when some bricks were skipped between the snapshots.

4.5 Process-Level Digital Twin

During the disassembly process, the estimated poses of the previously unknown bricks are instantaneously captured and represented geometrically within a process-level digital twin and stored in a building information model representing the disassembled brick structure and the subsequent storage, as illustrated within the Grasshopper environment in Fig. 7.

When a brick is added to the information model, it is visualized with its original as-found pose within the brick structure, displayed in pink, and after disassembly, this representation is updated to reflect the as-progressed pose within the brick storage, rendered in red. Furthermore, to ensure consistency within the digital twin when the mobile robot is manually relocated, a fiducial marker is utilized. By capturing the robot arm frame relative to the marker frame during each relocation, brick poses collected from every location can be depicted in a unified world coordinate system. This system enables the tracking of the structure's current state as well as the preservation of its original geometry, even when the physical structure itself no longer exists.

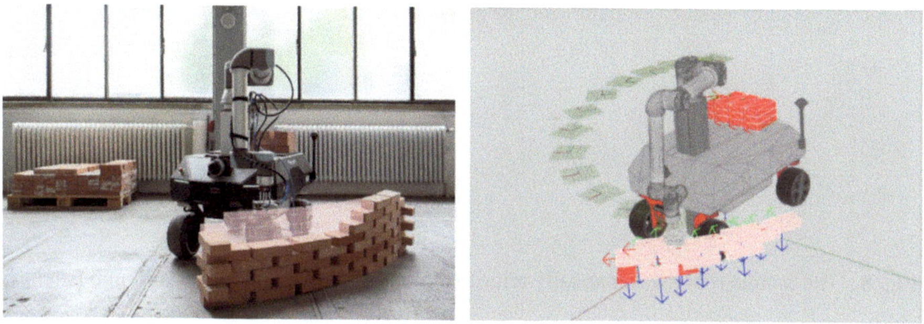

Fig. 7. The disassembled bricks from the brick wall (outlined in the left) are visualized within the digital twin in real-time. A motion plan is computed for the last localized brick (right).

4.6 Planning and Control of Step-by-Step Disassembly Routine

Several preliminary small-scale experiments were conducted with the presented setup to assess the accuracy of brick pose estimation for manipulation purposes. These experiments unveiled minor positional discrepancies that posed a potential risk to the grasping process. Given these discrepancies and the fact that in natural systems, even with humans, tactile sensing typically follows visual sensing, we implemented a tactile force control system within the UR controller to ensure secure manipulation of the individual bricks. In this arrangement, the vacuum gripper initially moves above the estimated pose of the brick element. Subsequently, it activates the force mode, moves along the z-axis towards the picking pose, and upon contacting an element and sensing it through the force sensors, initiates the vacuum suction to firmly grasp the brick. This method has proven to be highly effective in enabling reliable grasping of the bricks.

Once the vacuum gripper securely grasps the brick element, a planning request is initiated to the MoveIt planner, which is responsible for computing an appropriate trajectory from the pick-up location to the designated placement location in the storage on the robot's back. To ensure collision avoidance during this process, the collision meshes of the detected bricks are added into the robot scene and continuously updated as the new bricks are recognized and localized. Consequently, the planned trajectory is obtained with the objective of avoiding any collisions with the existing building structure.

5 Results

The full-scale experiment comprised the disassembly of a brick wall consisting of 58 bricks, in which the mobile robot was manually relocated along the structure five times, each time disassembling a patch of the wall within its limited static reach (Fig. 8).

The mobile robot successfully completed its main tasks, which included localizing bricks within an unknown building structure using the DL-based pose estimation, disassembling, and relocating them to a brick storage, all the while capturing and recording the data in the real-time digital twin (Fig. 9).[1] The cycle times for the camera to transition to a snapshot pose, trigger, and receive poses ranged around 60 s, motion planning

[1] https://vimeo.com/1109961831

Fig. 8. The mobile robot was manually relocated 5 times to disassemble the wall in 5 patches.

averaged 40 s with outliers (see below), and the pick-and-place operation took approximately 100 s. It is worth emphasizing that the robot's speed was maintained at a low level for safety reasons, despite the robot's greater capabilities. On average, the disassembly duration for a single brick amounted to 4:30 min. While the automated process, encompassing estimation, planning, and manipulation, achieved a 100% task success rate for the set of 58 bricks and smaller sets, it is also important to acknowledge that certain limitations affecting the efficiency were observed as outlined below.

Fig. 9. The brick wall was successfully disassembled in a multi-step process.

Pose and Grasping Accuracy. The DL-based algorithm delivered favorable results for both detection and pose estimation when the camera was positioned within snapshots captured over a spherical range of 90° and at distances ranging from 80 to 100 cm, with distances below 70 and above 130 cm giving insufficient poses. Within these snapshots, the detection consistently delivered 1–2 bricks, and the average pose estimation accuracy exhibited deviations of approximately 2.5° in orientation and 4–6 mm in position, with occasional outliers reaching 10–15 mm. These averaged results were obtained by comparing the estimated poses of a set of 15 samples to the ground truth pose data acquired from a measurement tool attached to the robot and calculated using forward kinematics. As a result, the estimated poses demonstrated a level of accuracy that was deemed sufficient for successful grasping tasks with a perfect success rate, considering the tolerances allowed by the vacuum gripper's suction capabilities and the force mode. In terms of lighting conditions, the estimator demonstrated strong performance under various settings, including ambient daylight and indoor lighting. However, it is worth noting that direct sunlight did have an adverse effect on the depth maps and, subsequently, on the detection success. Furthermore, distorted bricks led to detection failures. Nevertheless, it is important to emphasize that even in these conditions, the algorithm maintained adherence to specified thresholds, publishing no foul poses.

Motion Planning. The motion plans generated by the MoveIt planner frequently provided collision-avoiding trajectories between the estimated brick poses and the predefined target poses. Nevertheless, because some trajectories exhibited excessive complexity, an interface was implemented within Grasshopper that allowed a human operator to review the trajectories and either request a new plan or approve them if found satisfactory. While the planning itself required only seconds, the potential replanning iterations occasionally extended the process to over a minute. This system ultimately produced trajectories that met the required criteria but presents opportunities for improvement in the future by implementing higher-level Cartesian and joint constraints.

Data Collection Rate. The pose estimator adequately gathered data regarding the as-found positions of the bricks in the respective robot pose and integrated them into a real-time updated digital twin. Although the positional and orientational deviations mentioned above were noticeable within the digital twin, the data was sufficient for capturing the overall geometry of the building structure.

6 Conclusion and Outlook

This experiment aimed to demonstrate the conceptual viability of using mobile robots for the autonomous, careful dismantling of brickwork structures to retrieve building materials for reuse. Due to the outlined challenges, contemporary brick reclamation methods seldom utilize disassembly strategies; instead, they often resort to demolition and rubble collection techniques, yielding suboptimal reclaimed brick results. We sought to challenge this prevailing consensus by introducing Spatial-AI integrated mobile robotic processes.

Nonetheless, the absence of mortar or any adhesive material between the bricks in the experiment aligns differently from standard brickwork construction practices, thereby reducing the direct applicability of our findings to real-life scenarios. Detecting and estimating the pose of bricks present more significant challenges when the objects are occluded by mortar or when the bricks are not in perfect condition. This is the reason why a dry-stacked brick wall was selected for this experiment, recognizing the need to address this challenge in subsequent steps. Enhancing the DL-based object perception system through the integration of additional simulated material data and further refinement of the training model would better accommodate these real-life conditions. Moreover, such implementation would necessitate the development of robotic or human-assisted mortar removal methods and effective grasping strategies. The structural integrity of the brickwork during disassembly was also not assessed in this study, yet it is crucial for larger-scale disassembly strategies.

While brickwork was chosen for this experiment's primary material system because of its modular nature, other structural systems, such as those designed for disassembly or dry-jointed panel systems, could be promising candidates for future experimentation once the training pipeline is enhanced to generalize across various building materials, shapes, and dimensions.

It is also important to note that the current iteration of the system only employs object-level localization, which restricts the mobile robot's ability to autonomously navigate and position itself between different locations. Integrating scene-level understanding, such as simultaneous localization and mapping (SLAM) systems, can address this limitation. Such additions can significantly enhance the flexibility and automation of the disassembly tasks and enable collaborative systems involving multiple mobile robots working in tandem. Finally, the captured data of robotically disassembled structures has the potential to be employed for reconstructing building structures, if necessary or desired, in the same geometric configuration as the previously existing structure.

References

1. Alessi, A.: Cannibal Architecture: A collection of source material on the embodiment of history. In: Stockhammer, D. (ed.) Upcycling: Reuse and Repurposing as a Design Principle in Architecture, pp. 50–69. Triest Verlag, Zurich (2021)
2. Diener, A.: From waste to architecture: The myth of idealism in 20th century construction. In: Stockhammer, D. (ed.) Upcycling: Reuse and Repurposing as a Design Principle in Architecture, pp. 116–129. Triest Verlag, Zurich (2021)
3. Interreg, F.C.B.R.E.: ROTOR, Bellastock. REUSE TOOLKIT - Solid clay brick, CSTB, Brussels-Environment, Salvo (2021)
4. Entwerfen, L., Nagler Tilmann Jarmer, F., Sack, J.: Badzellen Zerlegen Seminar: Rezyklierbarkeit von Badzellenkonstruktionen. (2023).
5. Entner, P., Stockhammer, D.: From linear to circular construction: Liechtenstein's building stock as the material bank of the future? In: Stockhammer, D. (ed.) Upcycling: Reuse and Repurposing as a Design Principle in Architecture, pp. 132–147. Triest Verlag, Zurich (2021)
6. Byers, B., Cheriyamulla, S., Ewason, J., Hall, D., De Wolf, C.: Using Engraved QR Codes to Connect Building Components to Materials Passports for Circular Construction. (2022).
7. Concular Homepage, https://concular.de/, last accessed 2023/10/12.
8. Femmer, A., et al.: ARCH+ Open for Maintenance - Wegen Umbau Geöffnet. Spector Books OHG (2023)
9. BauKarussell Homepage, https://www.baukarussell.at/, last accessed 2023/10/12.
10. Rotor Homepage, https://rotordb.org/en, last accessed 2023/10/12.
11. Wibranek, B.: Robotic Digital Reassembly, https://tuprints.ulb.tu-darmstadt.de/id/eprint/18578, (2021).
12. Lausselet, C., Dahlstrøm, O.A., Thyholt, M., Eghbali, A., Schneider-Marin, P.: Methods to account for design for disassembly: Status of the building sector. Buildings. **13**(4), 1012 (2023). https://doi.org/10.3390/buildings13041012
13. Lublasser, E., Iturralde, K., Linner, T., Cokcan, S.B., Bock, T.: Automated refurbishment & end-of-life processes-research approaches in German and Japanese construction. In: Proceedings of the CIB*IAARC W119 CIC 2016 Workshop (2016)

14. Lublasser, E., Hildebrand, L., Vollpracht, A., Brell-Cokcan, S.: Robot assisted deconstruction of multi-layered façade constructions on the example of external thermal insulation composite systems. Constr. Robot. **1**, 39–47 (2017). https://doi.org/10.1007/s41693-017-0001-7
15. Lee, H.J., Brell-Cokcan, S.: Towards controlled semi-autonomous deconstruction. Constr. Robot. **7**, 253–263 (2023). https://doi.org/10.1007/s41693-023-00111-9
16. Davis, P., Aziz, F., Newaz, M.T., Sher, W., Simon, L.: The classification of construction waste material using a deep convolutional neural network. Autom. Constr. **122**, 103481 (2021). https://doi.org/10.1016/J.AUTCON.2020.103481
17. Çetin, S., De Wolf, C., Bocken, N.: Circular digital built environment: An emerging framework. Sustainability. **13**(11), 6348 (2021). https://doi.org/10.3390/su13116348
18. Sandy, T., Giftthaler, M., Dorfler, K., Kohler, M., Buchli, J.: Autonomous repositioning and localization of an in situ fabricator. In: Proceedings - IEEE International Conference on Robotics and Automation, pp. 2852–2858. Institute of Electrical and Electronics Engineers Inc. (2016). https://doi.org/10.1109/ICRA.2016.7487449
19. Jenny, S.E., Blum, H., Gawel, A., Siegwart, R., Gramazio, F., Kohler, M.: Online synchronization of building model for on-site mobile robotic construction. In: 37th International Symposium on Automation and Robotics in Construction (2020)
20. Johns, R.L., et al.: Autonomous dry stone. Construction. Robotics. **4**, 127–140 (2020). https://doi.org/10.1007/s41693-020-00037-6
21. Rünz, M., Agapito, L.: Co-Fusion: Real-Time Segmentation, Tracking And Fusion Of Multiple Objects (2017). https://doi.org/10.1109/ICRA.2017.7989518
22. Rünz, M., Buffier, M., Agapito, L.: MaskFusion: Real-Time Recognition, Tracking and Reconstruction of Multiple Moving Objects. (2018).
23. Xu, B., Li, W., Tzoumanikas, D., Bloesch, M., Davison, A., Leutenegger, S.: MID-Fusion: Octree-based Object-Level Multi-Instance Dynamic SLAM. (2018).
24. Kehl, W., Manhardt, F., Tombari, F., Ilic, S., Navab, N.: SSD-6D: Making RGB-based 3D detection and 6D pose estimation great again. In: Proceedings of the IEEE International Conference on Computer Vision (ICCV), pp. 1521–1529 (2017)
25. Zakharov, S., Shugurov, I., Ilic, S.: DPOD: 6D pose object detector and refiner. In: Proceedings of the IEEE/CVF International Conference on Computer Vision (ICCV), pp. 1941–1950 (2019)
26. Park, K., Patten, T., Vincze, M.: Pix2Pose: Pixel-wise coordinate regression of objects for 6D pose estimation. In: Proceedings of the IEEE/CVF International Conference on Computer Vision (ICCV), pp. 7668–7677 (2019)
27. Wang, G., Manhardt, F., Tombari, F., Ji, X.: GDR-Net: Geometry-guided direct regression network for monocular 6D object pose estimation. In: Proceedings of the IEEE/CVF Conference on Computer Vision and Pattern Recognition (CVPR), pp. 16611–16621 (2021)
28. Chen, H., Manhardt, F., Navab, N., Busam, B.: TexPose: Neural texture learning for self-supervised 6D object pose estimation. In: Proceedings of the IEEE/CVF Conference on Computer Vision and Pattern Recognition (CVPR), pp. 4841–4852 (2023)
29. Kirillov, A., Wu, Y., He, K., Girshick, R.: PointRend: Image segmentation as rendering. In: Proceedings of the IEEE/CVF Conference on Computer Vision and Pattern Recognition (CVPR), pp. 9799–9808 (2020)
30. Lin, Z.-H., Huang, S.-Y., Wang, Y.-C.F.: Convolution in the cloud: Learning deformable kernels in 3D graph convolution networks for point cloud analysis. In: Proceedings of the IEEE/CVF Conference on Computer Vision and Pattern Recognition (CVPR), pp. 1800–1809 (2020)
31. George, T., Lourakis, M.: A consistently fast and globally optimal solution to the perspective-n-point problem. In: Andrea, V., Bischof, H., B.T. and F.J.-M. (eds.) Computer Vision – ECCV 2020, pp. 478–494. Springer International Publishing, Cham (2020)

SSWM 20

Lost and Bound: Adaptive Detailing with Robotic Additive Joining for Reclaimed Steel

Inés Ariza[1,2(✉)], Romana Rust[2], Vlad-Alexandru Silvestru[3], Andreas Taras[3], Fabio Gramazio[2], Matthias Kohler[2], and Catherine De Wolf[1]

[1] Circular Engineering for Architecture, ETH, Zurich, Switzerland
arizai@ethz.ch
[2] Gramazio Kohler Research, ETH, Zurich, Switzerland
[3] Chair of Steel and Composite Structures, ETH, Zurich, Switzerland

Abstract. Following circular economy principles, steel components can be re-programmed to be used over multiple life cycles. However, reclaimed building components typically present superficial alterations, often cut away as waste, reducing member sizes and resource efficiency. As an alternative to pre-processing steel parts to bring them back to known standard shapes, this work shows a computational detailing strategy and additive joining approach to adapt to variable, uneven geometries. The method includes digitizing the material stock, integrating the building elements' as-is surface conditions into a node's design, planning robotic procedures with as-built data, and robotic joining with in-place wire and arc additive manufacturing (IPWAAM). The approach profits from the large design space of robotic manufacturing, enabling high-strength material placement with a widespread welding technology in the construction industry. The results of two node prototypes demonstrate a highly malleable connection strategy for reusing imprecise steel elements to meet tolerances that would otherwise require extensive manual labor or pre-processing. With this contribution, we aim to put forward the role that adaptive robotic building strategies can play in supporting the reuse of building components.

Keywords: Element reuse · Connection detailing · Adaptive fabrication · Additive manufacturing · Robotic fabrication · Wire and arc additive manufacturing

1 Introduction and Background

The reuse of building components is a circular economy strategy that has the potential to fundamentally change how the construction and demolition industry manages its waste [1]. However, adopting element reuse presents interdependent economic, legal, and technical challenges [2]. Its wide adoption also requires mindset and skill shifts for architects, engineers, constructors, and decision-makers to get ready to deal with the

M. Yablonina et al. (Eds.): Robotic Fabrication in Architecture, Art and Design 2024, pp. 314–326, 2026.
https://doi.org/10.1007/978-3-032-22607-5_20

inherent variable characteristics of reclaimed building components. After deconstruction, building components suitable for reuse (but not designed for that purpose) often present geometric and surface alterations compared to their standard, unused counterparts. A conventional strategy is to cut non-standard bits out of building components to avoid singularities on their surface [3]. This paper explores how an adaptive detailing approach coupled with an additive robotic joining technique could deal with the inherent variability found in used building components.

Reusing Steel As-is. Of all ubiquitous building systems, steel construction stands out for its broad standardization. However, even structural steel members can become non-standard after their initial use. After each life cycle, typical alterations and surface conditions include residual features such as bolt holes, welds, and cuts due to grinding down existing welds, paint, and corrosion [4]. According to Dunant et al., the costs associated with the speed and high skillset needed for deconstructing and processing steel parts to recreate "like-new" components are too high and become one of the main barriers to their reuse [5]. This research explores a solution to this challenge by minimizing the need for pre-processing steel elements before they can be reused.

Additive Robotic Joining. In the context of design for reuse, conventional welding is often considered a joining technique to be avoided [6, 7]. This is because welds are irreversible, as they modify the base material during fusion. This project examines an alternative welding approach by turning it into an additive joining technique capable of connecting elements at discrete connection points. These connection points run normal to the joined parts' surfaces, making them comparably much easier to take apart than conventional fillet and groove welds. This is achieved through Wire and Arc Additive Manufacturing (WAAM), a robotic 3D printing technique based on Gas Metal Arc Welding (GMAW).

WAAM. This additive manufacturing (AM) method harnesses the principles of GMAW, utilizing an electric arc as the heat source and metal wire as feedstock, to create metal components layer by layer with off-the-shelf welding hardware. The method stands out for its capability to produce parts with very good mechanical properties in a fraction of the time of other metal additive manufacturing methods, making it ideal for industries such as aerospace, maritime, and construction, where the demand for bespoke, large, and complex metal structures is high [8]. In contrast to most studied approaches, where parts are fabricated in a dedicated AM setup and then assembled to other elements manually [9, 10], WAAM is here used *in-place* as a joining technique applied directly onto the building elements or substrate [11]. To achieve this, in-place WAAM, or IPWAAM, integrates adaptive strategies, such as sensing and localization, to handle assembly tolerances and is highly suited for the specific challenges of reusing parts with uneven surface conditions. While most applications in WAAM are applied in continuous layers, bar-like structures can be fabricated with the point-by-point deposition used in this study [12]. Touching the substrate in discrete points, point-by-point WAAM adds enormous versatility to adapt to small surface irregularities.

Digitization. To bridge the gap between physical reality and digital models to design, build, and preserve buildings, methods for digitizing uncertain building material characteristics have been repeatedly studied in architecture and construction. Some recent examples include a variety of techniques for surveying, including photogrammetry [13], 3D laser scanning [14], total stations [15], depth cameras and sensors [16, 17], or any combination of these, constituting fundamental assets of the circular construction toolbox [18].

While the role of robotics, additive manufacturing, and point-by-point WAAM in the circular economy is yet to be quantified, this paper contributes to its methods and workflows with an application of existing digitization and additive manufacturing technologies in a hypothetical circular life cycle through the study of a node of a simply supported beam built with a small stock of reclaimed building components (Fig. 1 and 2a).

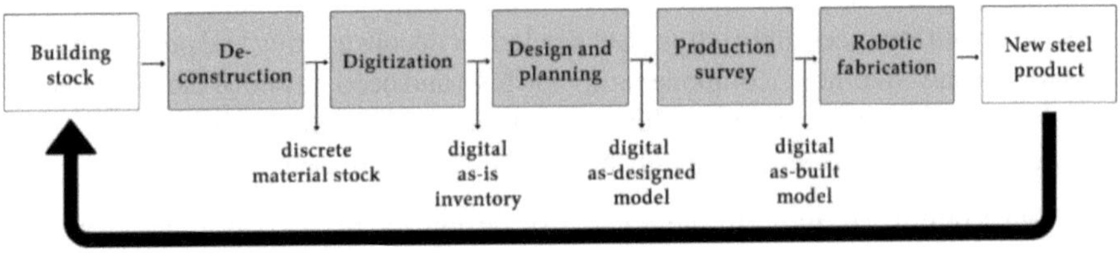

Fig. 1. Hypothetical circular life cycle for reusing steel building components: The work is based on the collaboration between three research groups at ETH Zurich: the Chair of Circular Engineering for Architecture (circular lifecycle and stock digitization), the Chair of Steel and Composite Structures (structural computational design), and the Chair of Architecture and Digital Fabrication, Gramazio Kohler Research (computational design and robotic fabrication) with the support of the deconstruction company Eberhard Bau AG (reused steel stock provider).

2 Methods

2.1 Digitization of Building Stock

To capture the "as-is" characteristics of the elements to reuse, a digitization workflow comparing two scanning techniques was tested: light structured scanning with the professional handheld scanner Creaform Ametek GoScan!50 and photogrammetry [19]. The intended output is a digital model containing geometric and surface information about the building elements. The digital model includes a mesh object with overall dimensions and an associated texture map with information about the geometric features and surface conditions. This paper describes the structured light scanning method consisting of the following steps:

Data Acquisition. First, the profiles are positioned in a standing upright position accessible on all sides in an environment with diffuse light (without direct sunlight). The scanning is done by moving around the object steadily at a constant distance from it. In this process, the scanner projects a known pattern in the object's surface that gets distorted by the surface's geometry [20]. This scene is then captured by the scanner's camera and processed with reconstruction software, i.e., VX Elements [21]. A rough visualization is shown in the software's interface in real-time, allowing the user to track the scanning progress and identify if reconstructed geometries are faulty and need to be re-acquired.

Data Processing. Once the scanning is finalized, the rough reconstruction is recalculated with higher resolution settings, and the result is exported as polygonal meshes in .stl and .obj formats, the latter also including a .mtl texture map (Fig. 2b).

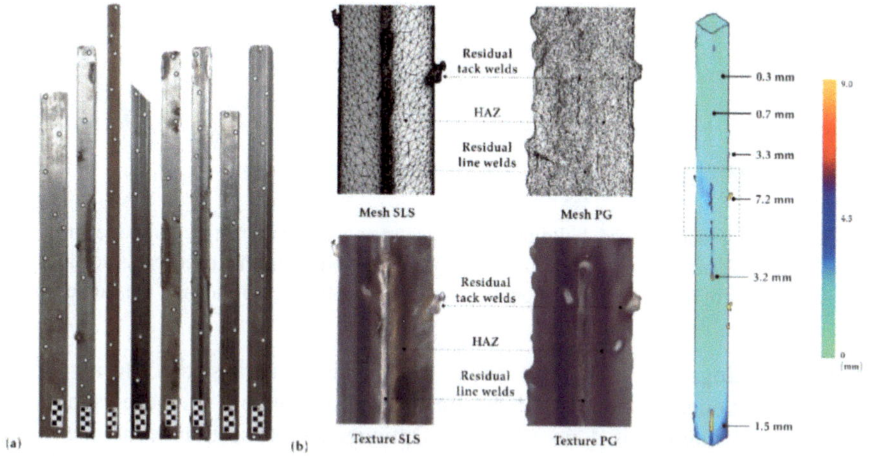

Fig. 2. Digitization: (a) Sample of the material stock showcasing differences in surface geometry and condition, (b) example mesh and texture outputs from structured light scanner (SLS) and photogrammetry (PG) showing residual tack and line welds and heat-affected zones (HAZ) around welds; visualization of features with a deviation analysis from a standardized profile.

The final step includes positioning and aligning the resulting meshes of each profile in a known coordinate system in a 3D modeling software, e.g., *Rhinoceros/Grasshopper 3D* [22]. The structured light scanning technique results are discussed in terms of accuracy and robustness in Sect. 3.

2.2 Design and planning with reused building components and point-by-point WAAM

Planning with reclaimed components not designed for that purpose requires considering their non-parametrized variability. To that end, this project outlines a series of design-and-production steps of a 4 m-long and 50 cm-high simply supported truss-like beam:

Material Distribution of the Beam. The setup to find a feasible truss-like beam geometry is built in *Rhinoceros 3D/Grasshopper* with the *TopOpt* plugin developed by DTU [23]. The design domain is a 4000 x 500 mm rectangle loaded with a vertical load (acting in the 500 mm direction) of 25 kg/m distributed linearly on the upper edge/chord and supported in two points, the upper corners. The result of this step consists of a set of closed curves defining a high-density region, as shown in gray in Fig. 3b.

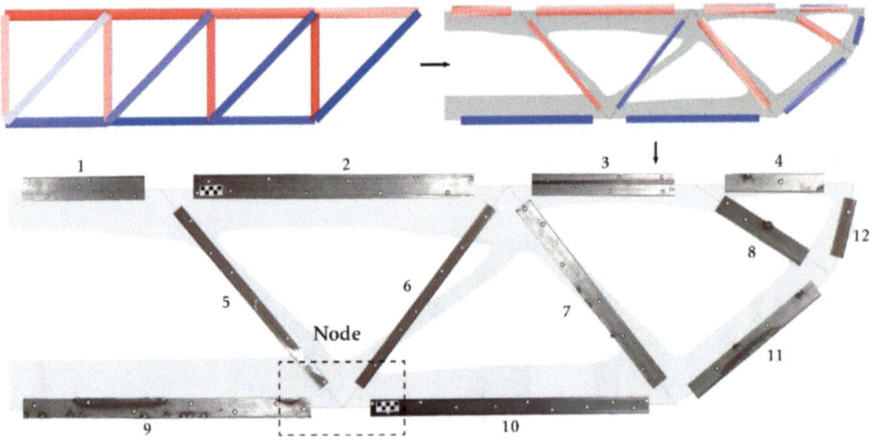

Fig. 3. Planning with reused building components: (a) standard simply supported truss-like beam (half), (b) material distribution result with line model and matched cross-sections, and (c) visualization of available profiles on the resulting truss-like beam showing the studied node.

Stock Matching. The material distribution result is first translated into an approximated line model of chords and diagonals. The assignment of the existing stock to each chord/diagonal is done using the "Optimized cross-section" component from *Karamba3D* [24] with a custom cross-section family built from the inventoried stock data (Fig. 3b and c). From here on, a representative node is investigated in all further steps in two iterative prototypes. For simplicity of the study, bars 9 and 10 are replaced with one continuous bar.

Robot Reachability Space as Design Space. Because the connection is directly produced in place, i.e., in between building components, the design space is highly constrained spatially. Therefore, fabrication feasibility must be considered early on before a connection geometry is generated. This first includes checking for tool accessibility (if the welding torch can access a specific location) and robot reachability (if an accessible location is kinematically reachable by the robot arm). These steps require a robot planning setup that includes the presumed location, position, and geometry of the intervening steel profiles and close-range collision objects in the production scene. In this step, the digitized mesh and texture are used to define no-go zones to avoid, i.e., filtering out unwanted material deposition locations in the design space. The space is then discretized as points and queried for tool accessibility with a collision check in *PyBullet* [25]. For each resulting accessible point, an inverse kinematics (IK) check is calculated using the analytical IK function from the *compas_fab* library [26]. The points with valid IK solutions are then stored as reachable points (Fig. 4a). Finally, a marching cubes algorithm is used to reconstruct the boundary of the reachable space given a minimum threshold of IK solutions found per queried point ([27], Fig. 4b).

Material Distribution of the Node. The reachable space is then used as a design space to run a material distribution script to find locations where material shall be deposited. Prototype 1 uses a *Grasshopper* workflow with the *tOpos* plugin [28] and only considers

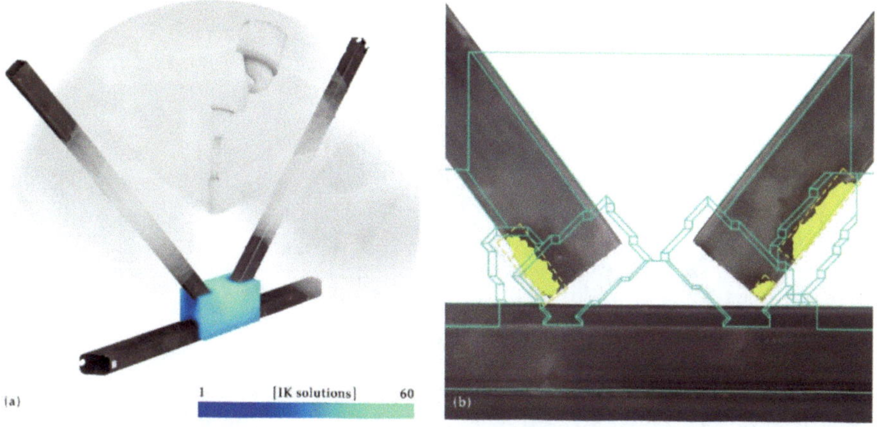

Fig. 4. Planning for point-by-point WAAM: (a) Robot reachability map from less reachable (in blue) to more reachable (in green) points, (b) boundary of reachable design space (in green) showing damaged areas as no-go zones considered for robot path planning (in yellow).

the self-weight of the profiles to be joined as described in Ariza, 2022, resulting in a self-supporting node ([11], Fig. 5a). This first "simplified" method yields plausible material tendency results but lacks the input control variables and metrics to verify it and is not considered structurally correct. The second "verified" method used for Prototype 2 considers the entire beam as a contributor to the intervening forces at the node with full control of the force inputs and parameters of a topology optimization (TO) run in *TOSCA/Abaqus* [29] and results in a load-bearing node (Fig. 5b). For this approach, parts of the bars to be joined in the respective node and the design space previously determined based on the robot reachability were imported to *Abaqus* as three-dimensional geometry. In addition, the inner forces resulting from the distributed line load of 25 kN/m were input for the bars at their ends away from the node. An objective function for minimizing the strain energy and a constraint for reducing the design space volume to 3% of the initial volume were defined for the optimization task. Furthermore, the member size of the connecting bar-like elements was set to be between 7 mm and 11 mm in diameter. The resulting node geometry was subsequently subjected to a smoothing operation.

Path-Finding for WAAM. The results of both material distribution approaches are then translated into WAAM paths. First, a skeletonization algorithm is run inside the TO mesh to get a linear or curvilinear path (Fig. 5). These first skeleton curves are then iteratively modified to meet a series of WAAM-printability checks, e.g., path orientation to gravity is passed if the angle between the tangents along the path curves and the gravity vector is at least 130° to avoid large path deformations. The resulting curves are then fitted between the profiles and sliced, preserving tool orientations suitable for WAAM at each layer (more information on printability checks and WAAM constraints can be found in [11]).

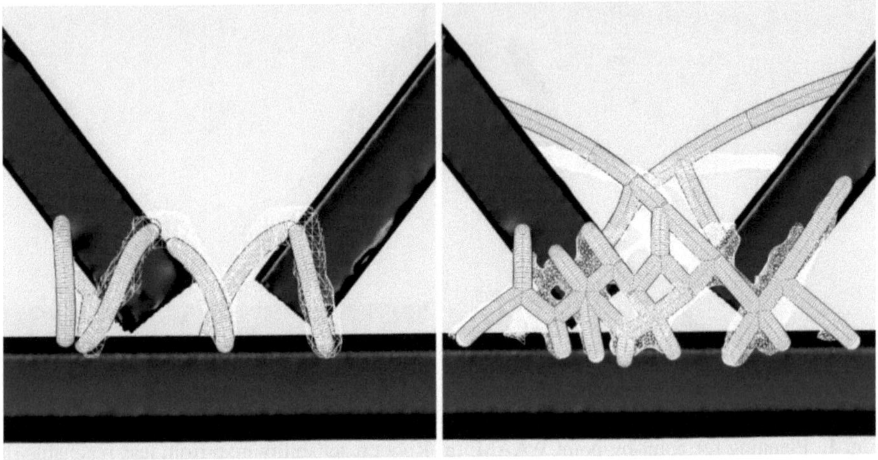

Fig. 5. Path-finding for WAAM of the TO results showing overlap of used inputs: TO mesh result (in white), skeleton curves (in gray), and final sliced path curves (in beige). (a) Prototype 1, "simplified" workflow of a self-supported structure from [11], and (b) Prototype 2, "verified" workflow of a load-bearing structure using *TOSCA/Abaqus*.

2.3 Production survey and adaptive detailing with WAAM

Robotic Setup and Materials. The production setup consists of an ABB IRB 4600/40 robot, a Fronius TPS 500i power source, and a Fronius 60i Robacta Drive Cold Metal Transfer (CMT) torch with a 22° neck. The intervening profiles are placed on a welding table, supported with clamps and flexible third arms. The welding ground is attached to the welding table, and two small welding magnets are placed between the profiles to transfer the electrical current between them. The printed material is Union SG 2-H (see [30, 31] for material properties and complete process parameters).

Pre-Production Survey. The production setup is then digitized to update the planning scene with as-built information, i.e., the actual and exact location of the profiles and all intervening collision objects, particularly the third arms, which are considerably large obstacles near the node location. This step follows the structured light technique presented in Sect. 2.1, plus a localization strategy using markers on the production table. The resulting scanned mesh of the production setup is segmented into types of objects (building elements, clamps, and supporting arms), and the final individual meshes are simplified to be used as collisions in the further path planning step.

Design Adaptation. After surveying the actual position of the elements to be joined, the path curves are fitted to the as-built location of the building elements. This entails minor modifications of the path curves' geometry; therefore, all printability checks need to be passed again.

Robot Planning. To calculate the robot targets for each layer to be printed, the planning setup is updated with the as-built survey, including the actual geometry of the elements to be joined. This setup includes the updated digitized meshes of the profiles and their

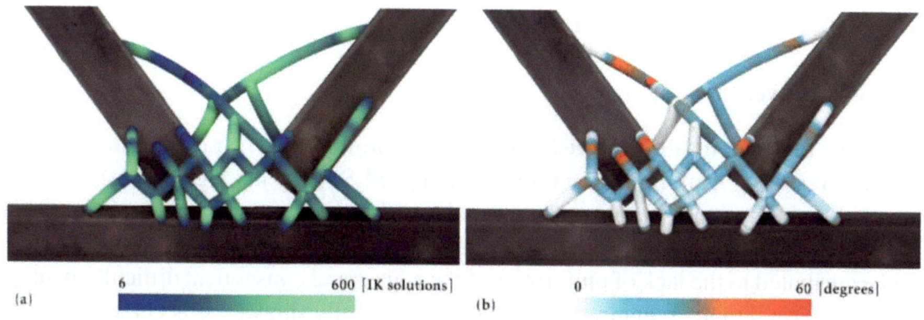

Fig. 6. Path-planning, Prototype 2, left side: (a) Reachability map for the printing path showing a range of reachable configurations from blue (less reachable) to green (more reachable), and (b) layers colored based on the tool deviation to the original path to avoid collisions with the elements from white (less deviation, i.e., less kinematically constrained locations do not require drastic reorientations of the tool) to red (more deviation, i.e., more constrained locations require more significant deviations of the tool). Reduced robot reachability (blue) results in higher tool deviations (red).

damaged/protuberant features as collision objects. The final path curves are then sliced, and a pool of frames representing robot targets is generated at each layer along the path curve. For each frame in the pool of frames, tool accessibility and robot reachability checks are computed in the same fashion as described in "Path-finding for WAAM." This operation results in a reachability map of the path to be printed with multiple possible robot targets per layer (Fig. 6a). To finally select the best robot configurations for each point layer, a "greedy" algorithm that prioritizes WAAM constraints, such as a minimum tool deviation from the path curve, is used as described in [11] (Fig. 6b). These steps rely on data structures and functions of the *compas_fab* and *PyBullet-planning* libraries and a planning algorithm developed by Yijiang Huang [11, 25, 26].

Fabrication Data and Adaptive Printing. With the final robot targets, a sorting step is performed in all branching paths. In all the convergent and divergent curve ends, a checker pattern removes overlapping frames to reduce the amount of material placed in the same spots. Next, a cooling time per layer is pre-calculated based on the distance to the next layer, and with this, the production data is finally exported. The WAAM printing procedure is then executed using the *compas_rrc* library to communicate with the robot controller and the welding machine [32]. The printing is organized in batches of layers between branching nodes. After each batch is completed, a touch-sensing procedure is run to measure all as-built columns and adapt the next layer position as needed: adding or removing material and selecting robot targets from the pre-calculated reachability map. The adaptive printing strategy is explained in detail in [11].

3 Results and Discussion

3.1 Stock Inventory

Accuracy and Robustness of the Scanning Technique. For the profiles' dimensions, the mesh resulting from light structured scanning (SLS) presents highly consistent outputs with maximum deviations to the ground truth/control geometry below 0.5 mm for the width and height and below 2.5 mm for the length. These higher deviations in length can be attributed to the lack of uniformity of the edges (and consequent difficulty in measuring them) and to the scanning method, which generally shows less accuracy on sharp or open edges. For the reconstruction of the surface features, SLS presents acceptable but not highly accurate results on the protuberances presented in Fig. 2, with notably less accurate results for highly detailed features such as spot welds or sharp continuous welds. However, the size and type of features on the reclaimed building elements will determine the accuracy needs. The SLS technique fully meets the needs of the case study in this project, where a 3 mm accuracy is enough to draw a bounding box/collision placeholder around the feature to avoid it.

3.2 Production Survey, Planning, and Adaptive Detailing

Understanding the Accuracy of the As-Built Production Setup. First, the positioning of the scanned scene in the virtual planning setup is verified with TCP measurements on the real robot setup. For these, deviations on both left and right sides range from 0-2 mm. Points measured on the top side of profiles present higher deviations (3-5 mm) and correspond with less accessible surfaces during the scanning process. Second, the overall deviation between the final as-built setup and the original, idealized beam model is measured by comparing the scanned mesh to the mesh of the geometry of the beam. Diagonals show more significant deviations at their upper freestanding edges, up to 15.5 mm. These deviations come from the manual assembly of the profiles and are considerably larger than what a pick-and-place robot arm could achieve. A consistent discrepancy is found on the profile corners, as the exact geometries of the fillet radius were taken from standard datasheets and may differ from the actual ones. These discrepancies are expected when working with reclaimed components with no material passports.

Adaptive Detailing Based on Reconstructed Results. Before the joining process starts, the path curves are adapted to fit the surveyed position of the building elements. Minor adaptations in the order of 1–2 mm were made for 3 out of 20 entry paths touching the base element on the table and for all the exit paths touching the diagonals, which present higher deviations up to 5 mm, as explained above. Considering the complexity of the printing process, Prototype 2 presents a much more constrained reachable space and has less suitable tool orientations than Prototype 1. First, as more elements are in the scene, there is less space to access. Second, the larger printed volume also constrains the orientation of the tool to avoid collision with its own. The final tool deviations to the print path for Prototype 1 are in the order of 0–60° (mean is 12°); for Prototype 2, they vary from 0–60° (mean 17°), with more significant deviations found in locations near the third arms that support the building elements and near convergent and divergent nodes

(Fig. 6). Regarding this aspect, the smaller and less interconnected the print volume, the less complex the planning and printing process is, and consequently, fewer deformed paths are found (Fig. 7).

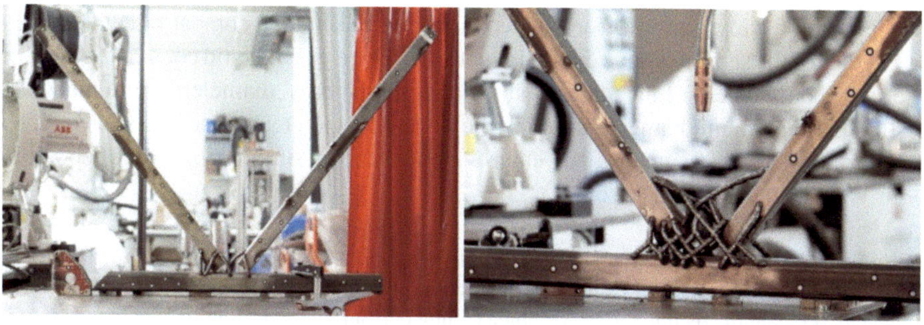

Fig. 7. Printed products: (a) Prototype 1 consists of 10 printed paths, and (b) Prototype 2 consists of a network of printed paths.

4 Conclusion

This paper has illustrated several steps of a circular use of steel over two hypothetical consequent life cycles, including stock digitization, matching stock with a simply supported beam design, design and robotic production of a single node using as-is component data, and final deconstruction. The work contributes concepts and methods for additive joining with adaptive detailing to produce a malleable connector as an alternative pathway for reusing steel with irregular or uneven surface characteristics or when no reversibility has been considered beforehand.

The first digitization step is critical to capture the irregularities that will be considered as no-go zones in the design of the node. With the structured light scanning technique, all critical features found in the stock (residual welds, cuts, and heat-affected zones, all in the order of 1–15 mm deviations from a standard profile) were adequately captured and reconstructed for the purpose at hand. Potential future work should include alternative digitization techniques that ease the scalability of the approach. In addition, the following steps should consider integrating non-destructive analysis methods during digitization of the reclaimed components to ensure they are structurally suitable for reuse or where reinforcement is needed, i.e., using adaptive detailing to upgrade the original parts.

For the design and production of the additive node, two aspects of the as-is conditions of the profiles were considered. First, the areas on the profiles where features exist are categorized as no-go zones and removed from the design space of the node to avoid placing joining material on top of them. Second, following the pre-production survey, the robot path planning considered the as-built state of the steel profiles and their features as obstacles to ensure no collision occurred between their protuberances and the robot during printing. The point-by-point WAAM printing process remains slow; however, enhancing the path-sorting strategy and producing multiple nodes simultaneously could yield more efficient results.

Finally, although only tested here on a small scale, the printed products show promising material-efficient connections that can be scaled up with different WAAM deposition strategies and easily deconstructed. The connection attachment to the building elements in the form of bridges provides a fast deconstruction technique compatible with several life cycles of the same adaptive detailing method. Future work should consider scaling up the size and thickness of the material stock and printed bars' diameters, simultaneously testing the maximum diameter at which the deconstruction technique becomes tedious. Furthermore, an assessment of the environmental impact of point-by-point deposition with WAAM must be carried out.

Acknowledgements. The authors would like to thank Lea Wenger and Giona Peter for their contributions to the stock digitization and production of prototypes; Robert Presl for his technical advice in 3D scanning technologies; Veronica Contucci from Eberhard Bau AG for her support in securing relevant steel stock to study; Yijiang Huang for the development of the greedy algorithm used for toolpath planning; Philippe Fleischmann, Mike Lyrenmann, and Gonzalo Casas for their technical support; Matthew Gordon and Brandon Byers for their suggestions to the research development; and Fronius Schweiz for the welding equipment used in this research.

References

1. Addis, B.: Building with Reclaimed Components and Materials: A Design Handbook for Reuse and Recycling. Routledge, London (2006)
2. Govindan, K., Hasanagic, M.: A systematic review on drivers, barriers, and practices towards circular economy: A supply chain perspective. Int. J. Prod. Res. **56**, 278–311 (2018). https://doi.org/10.1080/00207543.2017.1402141
3. Allwood, J.M.: Chapter 30 - Squaring the circular economy: The role of recycling within a hierarchy of material management strategies. In: Worrell, E., Reuter, M.A. (eds.) Handbook of Recycling, pp. 445–477. Elsevier, Boston (2014)
4. Fujita, M., Iwata, M.: Reuse dismantling and performance evaluation of reusable members. Struct. Eng. Int. **18**, 230–237 (2008). https://doi.org/10.2749/101686608785096531
5. Dunant, C.F., Drewniok, M.P., Sansom, M., et al.: Real and perceived barriers to steel reuse across the UK construction value chain. Resour. Conserv. Recycl. **126**, 118–131 (2017). https://doi.org/10.1016/j.resconrec.2017.07.036
6. Silverstein SA (Scott A (2009) Applying "Design for Disassembly" to Connection Design in Steel Structures. Thesis Massachusetts Institute of Technology
7. Gorgolewski, M.: Designing with reused building components: Some challenges. Build. Res. Inf. **36**, 175–188 (2008). https://doi.org/10.1080/09613210701559499
8. Williams, S., Martina, F., Addison, A.C., et al.: Wire + Arc additive manufacturing. Mater. Sci. Technol. **32**, 641–647 (2016). https://doi.org/10.1179/1743284715Y.0000000073
9. Liu, J., Xu, Y., Ge, Y., et al.: Wire and arc additive manufacturing of metal components: A review of recent research developments. Int. J. Adv. Manuf. Technol. **111**, 149–198 (2020). https://doi.org/10.1007/s00170-020-05966-8
10. Costanzi, C.B., Waldschmitt, B., Knaack, U., Lange, J.: Transforming the construction industry through wire arc additive manufacturing. In: Ruttico, P. (ed.) Coding Architecture: Designing Toolkits, Workflows, Industry, pp. 213–238. Springer Nature Switzerland, Cham (2024)

11. Ariza, I.: Adaptive detailing: Design and fabrication methods for in place wire and arc additive manufacturing connection details. Doctoral Thesis. ETH Zurich (2022)
12. Laghi, V., Palermo, M., Gasparini, G., Trombetti, T.: Computational design and manufacturing of a half-scaled 3D-printed stainless steel diagrid column. Addit. Manuf. **36**, 101505 (2020). https://doi.org/10.1016/j.addma.2020.101505
13. Galantucci, R.A., Fatiguso, F.: Advanced damage detection techniques in historical buildings using digital photogrammetry and 3D surface analysis. J. Cult. Herit. **36**, 51–62 (2019). https://doi.org/10.1016/j.culher.2018.09.014
14. Johns, R.L., Wermelinger, M., Mascaro, R., et al.: A framework for robotic excavation and dry stone construction using on-site materials. Sci. Robot. **8**, eabp9758 (2023). https://doi.org/10.1126/scirobotics.abp9758
15. Yablonina M, Coleman J (2021) Small robots and big projects. In: ACADIA 2021: Realignments: Toward Critical Computation. Online and Global 462–469
16. Wu, K., Kilian, A.: Designing natural wood log structures with stochastic assembly and deep learning. In: Willmann, J., Block, P., Hutter, M., et al. (eds.) Robotic Fabrication in Architecture, Art and Design 2018, pp. 16–30. Springer International Publishing, Cham (2019)
17. Frangez, V., Salido-Monzú, D., Wieser, A.: Surface finish classification using depth camera data. Autom. Constr. **129**, 103799 (2021). https://doi.org/10.1016/j.autcon.2021.103799
18. Gordon, M., von Zimmerman, L., Haradhun, O., et al.: Digitising building materials for reuse with reality capture and scan-to-BIM technologies. In: De Wolf, C., Çetin, S., Bocken, N.M.P. (eds.) A Circular Built Environment in the Digital Age, pp. 41–55. Springer International Publishing, Cham (2024)
19. Creaform Ametek (2016) GoScan!50
20. Wilm, J., Olesen, O.V., Larsen, R.: SLStudio: Open-source framework for real-time structured light. In: 2014 4th International Conference on Image Processing Theory, Tools and Applications (IPTA), pp. 1–4 (2014)
21. Creaform (2023) VX Elements: 3D Scanning Software Platform and Application Suite
22. McNeel R, others (2010) Rhinoceros 3D, Version 6.0
23. Søndergaard A, Maier D, Aage N, Amir O (2013) TopOpt for Grasshopper
24. Preisinger, C.: Linking structure and parametric geometry. Archit. Des. **83**, 110–113 (2013). https://doi.org/10.1002/ad.1564
25. Garrett C (2018) Pybullet-Planning: A Suite of Utility Functions to Facilitate Robotic Planning Related Research on the Pybullet Physics Simulation Engine.
26. Casas, G., Rust, R., Lytle, B., et al.: compas-dev/compas_fab: COMPAS FAB, vol. v0.17.0, (2021)
27. Bernhard M, Clemente R (2019) compas_vol
28. Białkowski, S.: tOpos - GPGPU accelerated structural optimisation utility for architects. In: Proceedings of the 35th eCAADe Conference: ShoCK! - Sharing Computational Knowledge! Rome, pp. 679–688 (2017)
29. (2000) Tosca, Abaqus/CAE
30. Silvestru, V.-A., Ariza, I., Vienne, J., et al.: Performance under tensile loading of point-by-point wire and arc additively manufactured steel bars for structural components. Mater. Des. **205**, 109740 (2021). https://doi.org/10.1016/j.matdes.2021.109740
31. Silvestru, V.-A., Ariza, I., Taras, A.: Structural behaviour of point-by-point wire arc additively manufactured steel bars under compressive loading. J. Constr. Steel Res. **207**, 107982 (2023). https://doi.org/10.1016/j.jcsr.2023.107982
32. Fleischmann P, Casas G, Lyrenmann M (2020) COMPAS RRC: Online Control for ABB Robots Over a Simple-to-Use Python Interface.

Models All the Way Down

Reshaping Wood Waste: Creative Pathways to Enhance Resource Efficiency

Maria-Eleni Papandreou[1,2](✉), Vincent Huyghe[1,2], Joseph Naguib[1], Lucas Gottschild[1], Pit Siebenaler[1], Bora Çobanoğlu[1], Alexandre Dubor[1], and Aldo Sollazzo[1]

[1] Institute of Advanced Architecture of Catalonia, Barcelona 08005, Spain
{marielena.papandreou,alex,aldo.sollazzo}@iaac.net, info@dare-make.com, {joseph.milad.wadie,lucas.gottschild, Pit.Siebenaler,bora.cobanoglu}@students.iaac.net

[2] DARE, Barcelona 08003, Spain

Abstract. In the wood industry, the challenge of managing significant amounts of leftover materials generated from primary and industrial processes persists, often resulting in unsustainable disposal methods. This paper explores an innovative methodology that unlocks new possibilities for resource utilization by harnessing the potential of computer vision, robotic manufacturing, and artificial intelligence (AI) technology, particularly in the realm of image processing. This approach seeks to repurpose residues from primary and secondary timber processing into custom products, spanning applications in both architecture and product design, with the primary objective of waste mitigation. The study focuses not only on the potential reclamation of irregular stock but also on leveraging wood's inherent material properties, such as grain direction, to guide the design process. Through a combination of computer vision and computational tools, a series of glue-up scenarios are generated for crafting unique three-dimensional timber assemblies. Robotic fabrication is employed for precise post-processing of the resultant objects, ensuring high-quality outcomes. The proposed toolchain aims to enable design exploration not just for a single product, but for a multitude of creations. Presented within two case studies, the proposed workflow demonstrates the fusion of technology and creativity, offering a versatile methodology that reimagines wood waste utilization.

Keywords: Computer vision · ROS · Open3D · Diffusion models · ControlNet · Robotic Fabrication · Resource management · Digital Craftsmanship

1 Introduction

In today's industrial landscape, efficiently managing waste has become an increasingly pressing challenge. This concern is especially relevant in the wood industry, where substantial quantities of leftover materials are generated, imposing environmental and economic burdens [1, 2]. A novel approach is imperative—one that transforms these unused materials into valuable products, actively contributing to landfill diversion while simultaneously promoting sustainable design practices.

M. Yablonina et al. (Eds.): Robotic Fabrication in Architecture, Art and Design 2024, pp. 330–342, 2026.
https://doi.org/10.1007/978-3-032-22607-5_21

Through the integration of robotic scanning technologies, wood waste can be systematically cataloged [1], setting the stage for its seamless incorporation into flexible production systems, such as robots. Additionally, image analytics enable an in-depth documentation of the inherent properties of wood [22], such as grain patterns, texture, knots and imperfections. These features can now be revived and used in the digital design processes, reinvigorating the distinctive essence of timber products that was somehow lost during standardization attempts of building materials.

The objective of the proposed workflow (Fig. 1) is to harness the latent potential of slab wood sourced from primary processing, as well as trimmings generated in factory and workshop settings during secondary processing. A toolchain is developed with the capability to effectively manipulate and extract data from a diverse array of wood shapes. Moreover, the establishment of a robust protocol for the conversion of raw scanned data into accurate three-dimensional information is essential for both design and manufacturing. Additionally, Stable Diffusion with ControlNet [18] is leveraged (an open-source machine learning model for image processing and generation tasks) to process image data and effectively transition from unrefined RGB data into precise milling toolpaths. Lastly, the methodology of designing for manufacture and assembly (DFMA) is explained to ensure the practical feasibility of the envisioned products. The goal of DFMA is to evaluate design decisions based on manufacturing process constrains in order to mitigate risks, cut expenses, simplify assembly processes and enhance overall product quality. In summation, this paper presents a comprehensive investigation for unlocking the value within wood residues. By merging advanced technology with the innate beauty of wood, the two case studies aim to showcase the potential of sustainable, creative, and efficient resource utilization.

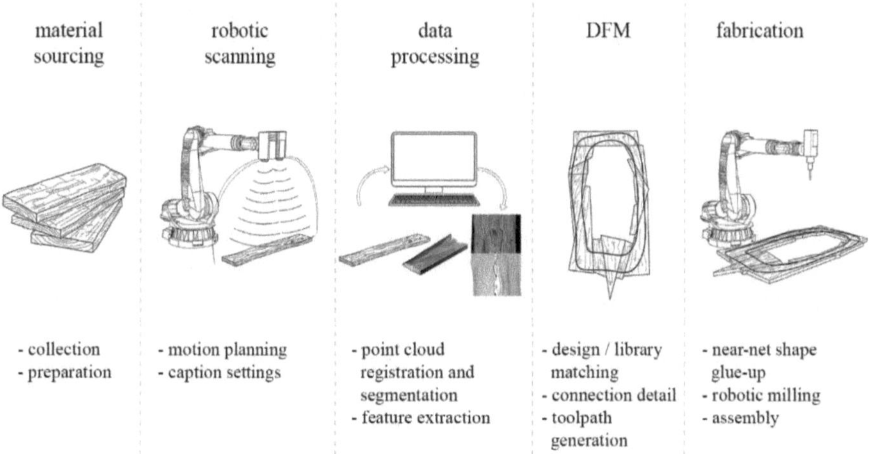

Fig. 1. Workflow diagram

2 Related Works

In alignment with this research focus there is a plethora of academic studies that underscore the value of resource repurposing [1, 2, 5–7] in architecture and design. Sourcing strategies include primary channels such logging sites and sawmills [4]. Secondary sources encompass factory and workshop offcuts, while tertiary sources involve product waste [2, 6]. This commitment to existing resource utilization finds resonance not only in academia but also in the broader market landscape. Circular economy initiatives are embraced by product designers [6] but also practiced by organizations like Herso Circular Houtwerkers [9] and Superuse [10] when it comes to building products. However, the building products are still primarily in 2D, and manual labor remains time-consuming, with the processes far from being automated.

To materialize the vision of sustainable circular practices at a larger scale incorporating scanning technologies plays a pivotal role. One commonly employed method is photogrammetry. While photogrammetry offers a cost-effective solution, its long processing times for 3D model reconstruction make it unsuitable for real-time robotic workflows [23]. Alternatively, other research efforts have delved into the application of Faro technology [4]. Although these scans suffice for design purposes, the precision required for robotic manufacturing and piece localization in front of the robot is achieved through other tracking methods and handheld scanners [4]. A pertinent reference within the realm of robotic scanning is the Scan-N-Plan™ framework developed by ROS Industrial [13]. This framework leverages a depth camera to provide real-time mesh data and enables seamless workpiece registration [16].

In the context of designing with the natural properties of wood, such as their shape irregularities and grain direction, the work by Boleflor is considered relevant [8]. However, their focus is primarily on 2D shapes for floor applications, where the emphasis lies in outlining the shape's perimeter rather than extracting intricate wood features. For similar applications, the work of Carla Saad [3] explores wood pattern recognition using canny edge detection [15] and dilate filters to thicken the detected contours for the purpose of transforming grain data into machine-readable information. While Saad's work offers valuable insights into the geometrical understanding of wood patterns, there is room for further improvement in terms of pattern continuity of the vectorized images.

In conclusion, the primary objective of this research is to contribute to the development of an automated scanning framework for the efficient processing of diverse semi-planar offcuts coming from industrial processing applications. These involve slab wood pieces from sawmills or factory offcuts from secondary processing. The goal is to retrieve high-resolution meshes for design exploration with the precision necessary for numerical manufacturing. In the creative ideation phase, the irregular shapes of the material are leveraged to generate glue up scenarios that result in three-dimensional assemblies. Aligning the assembly logic with the natural contours of the material creates a puzzle-like structure, effectively reducing waste. Furthermore, the extraction of additional layers of data, including grain patterns and knots, is geared towards enhancing the texture of the final 3D object. This integration of wood characteristics into the digital design process opens up new possibilities for material-informed creations, bridging the gap between traditional craftsmanship and digital fabrication.

3 Methodology

A combination of software and hardware tools are employed to effectively address critical challenges of working with wood offcuts. These challenges include the manipulation of irregular stock with high part-to-part variability, streamlining the processing of point cloud and image data and establishing an agile manufacturing infrastructure capable of accommodating diverse geometrical requirements. The primary research objectives are to maximize material utilization while minimizing environmental impact in crafting purposeful objects. In order to achieve the objectives, a methodology has been structured consisting of five interdependent work packages: Material Sourcing, Robotic Scanning, Data Processing, Design for Manufacturing and Fabrication (Fig. 1).

3.1 Material sourcing and pre-processing

The material used in this research is collected from local, small-scale sawmills and workshops. The selection criteria encompass ensuring adequate thickness for nesting 3D geometry, considering grain contours and avoiding any sections with signs of rot or compromised structural integrity. A diverse array of wood pieces was used, ranging from 1500 mm live edge slab softwood to 400 mm short irregular hardwood trimmings.

3.2 Robotic Scanning

The scanning workflow is designed for precise data collection. At the center of this set up is a robotic arm with a structured light depth camera, making use of ROS 1 and MoveIt for motion planning and control. A well-orchestrated sequence of operations

Fig. 2. Scanning process, Projected Patterns of depth camera on a piece of pine

ensures that each offcut is precisely scanned from multiple angles guaranteeing accurate and comprehensive point clouds for further data processing for fabrication.

The robot takes multiple captures of the wood piece from different angles to avoid any occlusions. Larger pieces require more scans for complete coverage (Fig. 2).

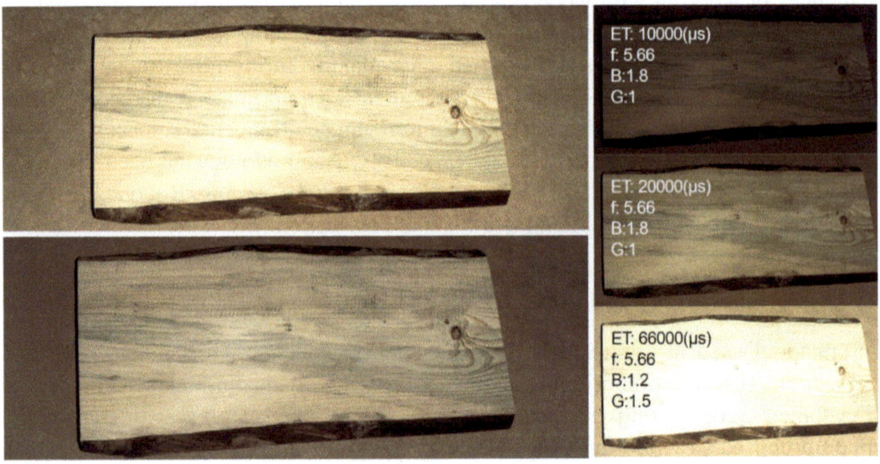

Fig. 3. Capture settings tests for pine offcuts. Automatic scene analysis (top left) with overexposed results (burnt pixels) leading to poor depth information. Custom capture settings (right), resulting in balanced Signal-To-Noise ratio (bottom left) and higher quality point clouds

To achieve high quality point clouds, defining the camera's settings for multi-acquisition HDR (High Dynamic Range) capture is of utmost importance. The goal is to reduce grain, noise and prevent overexposed or underexposed areas. Experimentation with different settings was conducted to determine the optimal combination of exposure time, aperture, projector brightness and gain (Fig. 3). A detailed explanation of the method used to obtain different capture settings is provided in the Zivid knowledge base [19].

3.3 Data Processing

Post-processing of the captured data is the subsequent step. The goal is to create a dimensionally accurate 3D models, but also to effectively extract grain direction and knots through image processing, underpinned by Stable Diffusion and ControlNet.

3.3.1 Point Cloud Processing

A number of captures are taken per piece in order to ensure every surface is clearly visible without occlusions. The captures are subsequently processed using Open3D's [14] Python API. The first step is to transform the point clouds from camera coordinate space to robot coordinate space. The next step is to use multiway [16] point-to-plane registration [17] to further improve the alignment of the different captures (Fig. 4). In the following step RANSAC (RANdom Sampling and Consensus) plane segmentation [24] is used to find the bottom plane. The point cloud is then transformed from this coordinate space into the world coordinate system. Now all points with a z-value lower than or equal to 0 can be removed. In the final step all the point clouds are merged into a single point cloud, which is further resampled (voxel down sampled to original capture resolution) in order to remove superfluous data (Fig. 4). In order to convert the point cloud to mesh the Poisson surface reconstruction [18] method is used.

Fig. 4. Point Cloud post processing for an African cedar offcut

3.3.2 Image Processing

A workflow has been developed to transforms raw RGB data into vector data that can be used for machining purposes, by integrating Stable Diffusion and ControlNet. In the preprocessing phase of the image in ControlNet, four preprocessors were tested: lineart_coarse, lineart_realistic, lineart_standard, and scribble_xdog (Fig. 5). These images were used as inputs to the diffusion model to steer the generated results. Further settings include the control weight that was set at 0.5 to moderate the influence of the control map. Keywords such as 'lineart,' 'line pattern,' 'wood grain pattern,' 'smooth lines,' 'continuous lines,' and 'high quality' were used as prompts to guide the model. The goal is to achieve the closest resemblance to the original wood grain pattern with the least noise and the most continuous lines.

Further settings employed were as follows: DPM++ 2 M Karras was chosen for the sampling method [21] due to its combination of speed and quality, utilizing 20–30 steps. The CFG (Classifier Free Guidance) scale [20] was adjusted within the range of 25–27 to strike a balance between prompt adherence and creative flexibility. Denoising strength was set at 0.5, with the inclusion of VAE (Variational Autoencoder) to conserve VRAM. VAE serves as an add-on that encodes and decodes images into a more compact latent space, thereby accelerating computation.

Fig. 5. Comparison of ControlNet image preprocessors

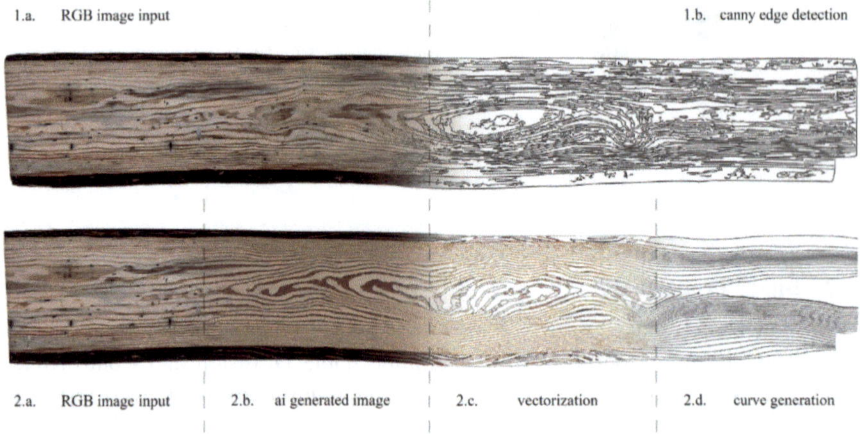

Fig. 6. Curve generation comparison. Process 1. raw RGB image input (1.a.) to vectorized result using open cv canny edge (1.b.) Process 2. raw RGB image input (2.a.) to AI generated enhanced contrast pattern (2.b.) to vectorization (2.c.) and toolpath curve generation (2.d.)

3.4 Design For Manufacturing

Central to our Design for Manufacturing workflow is a dynamic interplay between a material database and a target design. The process commences by matching the selected material to the target design in a digital definition (Fig. 7). This digital definition connects geometry-material-process relationships and can accommodate formal and sizing variability of the object. Parametric software is leveraged to refine detailing connections essential for fabrication and assembly.

3.5 Fabrication

The manufacturing of the design involves a two-step fabrication process. The offcuts are pre-assembled into a near-net-shape [25]. This means the strategic production of sub-assemblies that are closely resembling the final target shape (Fig. 9b). This method is used to produce parts that demand minimal additional finishing. It is a cost-effective and resource-efficient approach, mitigating material waste and machining requirements. Subsequently a final post processing operation is executed to carve the geometry out of this custom stock. The specifics of this fabrication process are further enriched through the lens of each case study.

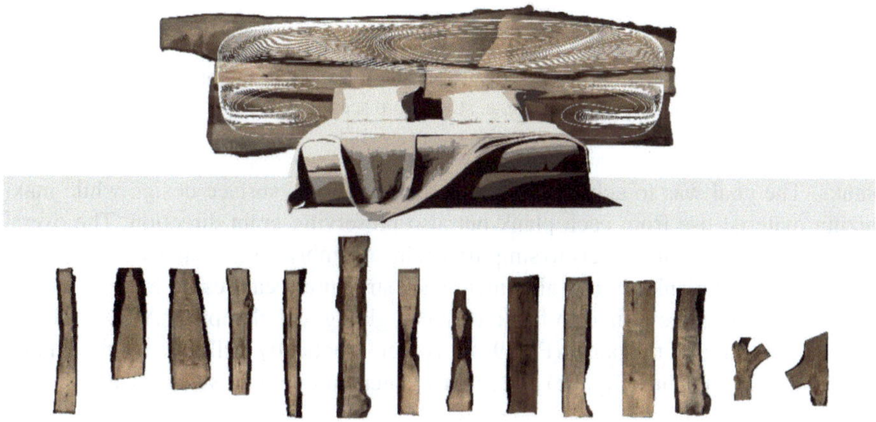

Fig. 7. Material Library (bottom) and target design of Case Study II (top)

4 Case Studies

4.1 re(al) form

The "re(al) form" project is focused on achieving sustainable surface designs through the utilization of marginalized wood offcuts. The utilized pine offcuts were sourced from a sawmill located in Valldaura forest in Barcelona. A digital material library of forty-six wooden planks was created using the scanning method outlined in Section 3.2. Additional feature extraction was carried out using Stable Diffusion providing clear information, such as plank border curves and grain direction (Fig. 6). All extracted data were then consolidated into comprehensive data packages (Fig. 8) for each individual plank, to be employed during the design matching phase.

The digital library was organized based on size, thickness, and grain direction, facilitating more efficient data filtering. A matching process between material library and target design was developed to find the best fit between the long edge curves of the

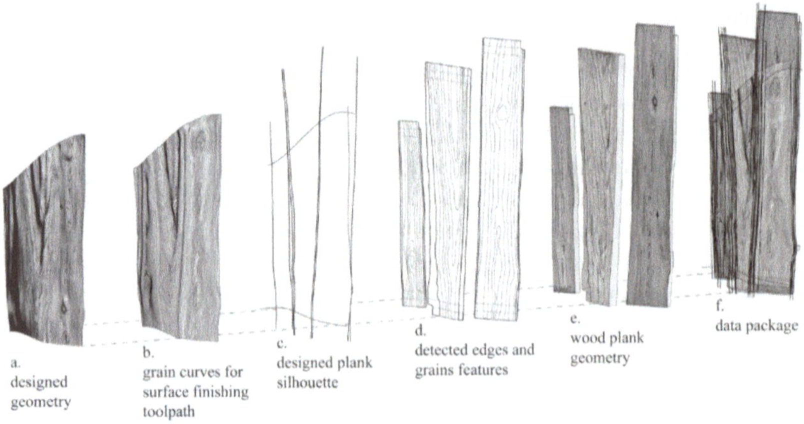

Fig. 8. Data per plank, where a. target shape of a panel, b. finishing operation, c. profiling operation d. grain detection from image processing, e. stock from scanning, f. all data overlaid

planks. The goal was to achieve a curated doubly curved surface design while maximizing material use from each plank but also preserving grain direction. The overall design is organized into panels to simplify onsite assembly. These panels are composed of pre-processed planks and an alignment back structure. Features such as 3D half-lap joints and pockets are milled to facilitate panel gluing and clamping. After producing the near net shape of the panel (Fig. 9b) it is then robotically milled as a single unit to apply the final finishing (Fig. 9c) and remove potential tolerance stack issues.

Fig. 9. (a) Pre-processed plank with half lap joints. (b) Near-net shape of an assembled panel. (c) Finishing operation of the panel. (d) Final prototype with four panels

4.2 GR-AI-N

GR-AI-N is a research project for the production of bespoke furniture products made out of discarded wood, employing the workflow described in Section 3. Two prototypical mirrors (Fig. 10.) were created using chestnut and African cedar offcuts sourced from a local workshop in Barcelona. Additionally, a headboard (Fig. 11.) serves as the final demonstrator of this project, utilizing pine sourced similarly to Case Study 1.

The scanning process provided material information that was used to generate a series of glue up scenarios for the three prototypes. The target designs were developed within a parametric framework, where the outlines controlling the objects' shape could be easily manipulated by the user. Subsequently design data is connected to fabrication data in a single file. By fostering such an agile design environment, design variations can be explored quickly and changes can take place until the first day of final production. Such changes could be driven not only by user preferences but also by the final material availability.

Fig. 10. Manufacturing process of two prototypical mirrors for project GR-AI-N. African cedar pieces and final product (left), Chestnut mirror grain matching detail and sanding (right)

Following the production of the custom stock for each object, milling operations were carried out accordingly. The African cedar prototype focused on optimizing material usage, while the chestnut mirror aimed for smooth transitions between the wood's growth rings. The headboard aimed to demonstrate the application of the workflow in a larger assembly featuring multiple piece connections.

4.3 Case Study Insights

While the current case studies are driven by constraints such as budget and time limitations, they provide valuable insights for the envisioned workflow. An essential element for the deployment of this workflow at scale lies in the conditioning phase of the timber.

Fig. 11. GR-AI-N Headboard, showcasing the application of the workflow in a larger assembly

Prior to scanning, manual planning and cleaning of wood is necessary. Investigating how automation can be leveraged to streamline this particular phase would be beneficial. Additionally, the generation of waste of the new designs persists despite efforts to maximize material utilization. Finally, as this is a craft application, involving the use of glue, further research is imperative for sustainable adhesives, but also in addressing the end-of-lifecycle phase of these new products.

5 Conclusions

The research explores the development of a novel workflow for the effective repurposing of semi-planar timber offcuts into 3D objects for architecture and product design, by seamlessly integrating computer vision, generative AI for image processing and robotic fabrication. Utilizing ControlNet has proven to be a practical approach for vectorizing image data compared to traditional image processing methods. The goal for this approach is to empower digital craftspeople to work with non-standard materials while also harnessing material information for their creative projects.

Additionally, key objectives include waste mitigation and landfill diversion. In the proposed workflow, timber offcuts are evaluated before their disposal, being diverted from decaying, shredding or burning. By repurposing this material into distinctive objects, the retention of captured CO2 is extended.

Looking forward, scalability and the economic feasibility of the workflow are important aspects to explore further. Addressing these aspects will provide a clearer outlook on how this innovative approach can be applied to waste streams coming from larger establishments, such as sawmills and timber factories. This paper lays the foundation for new and exciting possibilities in the integration of image analytics, robotics and creative design, reshaping resource management in timber construction.

Acknowledgments. Project GR-AI-N has received funding from the European Union's Horizon 2020 research and innovation programme under the Grant Agreement no. 952197. Special appreciation goes to VOJEXT for funding support under the residency program "VOJEXT S+T+ARTS Art Residence 2: Robotics in Arts and Crafts." This residency was led by VOJEXT PARTNERS, namely Robotnik and WAAG.

References

1. Yu, B., Luo, J., Shi, Y.: Framework for sustainable building design and construction using off-cut wood. NPJ Mater. Sustainability. **1**(2) (2023)
2. Malé-Alemany, M., Schoen, T., Galli, M., Bors, V., Kozhevnikova, A., van Dijk, L.: "Once my front door–now my coffee table"; Advanced computational design and robotic production with waste wood. In: AMS Institute Conference 'Reinventing the City', pp. 1–5, Amsterdam, Netherlands (2022)
3. Saad, C.: If Only Wood Could Speak...Explorations in Digital Fabrication Processes Based on Timber Grain and Patterns. In: Proceedings of 37 eCAADe and XXIII SIGraDi Joint Conference, pp. 227–234, Blucher, São Paulo (2019)
4. Larsen, N.M., Aagaard, A.K.: Robotic processing of crooked sawlogs for use in architectural construction. Constr. Robot. **4**(1–2), 75–83 (2020)
5. Circular Wood for the Neighborhood–Robotic Building (RB), http://www.roboticbuilding.eu/project/wood-reuse/ (2023). Accessed 09 Sep 2023
6. Product Designer Matthias Gschwendtner's 'New Sources' Project: A Critique of Mass-Market Furniture - Core77, https://www.core77.com/posts/112158/Product-Designer-Matthias-Gschwendtners-New-Sources-Project-A-Critique-of-Mass-Market-Furniture (2023). Accessed 09 Sep 2023
7. Lok, L., Zivkovic, S.: Ashen Cabin. ACADIA 2020: Distributed Proximities 2, 176–181 (2018)
8. In Depth|Bole, https://www.bole.eu/in-depth/ (2023). Accessed 09 Sep 2023
9. Herso circulaire houtbewerkers en de circulaire economie, https://herso.nl/circulaire-economie/ (2023). Accessed 09 Sep 2023
10. Homepage-Superuse Studios, https://www.superuse-studios.com/ (2023). Accessed 17 Aug 2023
11. Zhang, L., Rao, A., Agrawala, M.: Adding conditional control to text-to-image diffusion models. In: Proceedings of the IEEE/CVF International Conference on Computer Vision, pp. 3836–3847 (2023)
12. Scan-N-Plan — ROS-Industrial, https://rosindustrial.org/scan-n-plan (2023). Accessed 21 Sep 2023
13. GitHub-Ros-Industrial/Industrial_reconstruction: Tools for Surface Reconstruction from 2D Depth Images, https://github.com/ros-industrial/industrial_reconstruction (2023). Accessed 21 Sep 2023
14. Zhou, Q. Y., Park, J., Koltun, V.: Open3D: A modern library for 3D data processing. arXiv preprint arXiv:1801.09847, (2018)
15. OpenCV-Open Computer Vision Library, https://opencv.org/ (2023). Accessed 19 Aug 2023
16. Chen, Y., Medioni, G.: Object modeling by registration of multiple range images. In: Proceedings-IEEE International Conference on Robotics and Automation, pp. 145–155 (1992)
17. Choi, S., Zhou, Q.-Y., Koltun, V.: Robust reconstruction of indoor scenes. In: 2015 IEEE Conference on Computer Vision and Pattern Recognition (CVPR), pp. 5556–5565. IEEE, Boston, MA, USA (2015)

18. Kazhdan, M., Bolitho, M., Hoppe, H.: Poisson surface reconstruction. In: Sheffer, A., Polthier, K. (eds.) Symposium on Geometry Processing (2006)
19. Getting the Right Exposure for Good Point Clouds-Zivid Knowledgebase Documentation, https://support.zivid.com/en/latest/academy/camera/capturing-high-quality-point-clouds/getting-the-right-exposure-for-good-point-clouds.html (2023). Accessed 19 Aug 2023
20. Guide: Stable Diffusion's CFG Scale Explained - Once Upon an Algorithm. https://onceuponanalgorithm.org/guide-stable-diffusions-cfg-scale-explained/ (2023). Accessed 19 Aug 2023
21. Stable Diffusion Samplers: A Comprehensive guide-Stable Diffusion Art, https://stable-diffusion-art.com/samplers/ (2023). Accessed 19 Aug 2023
22. Jabo, S.: Machine Vision for Wood Defect Detection and Classification. Thesis, Chalmers university of technology, Göteborg, Sweden (2011)
23. Reljić, I., Seljan, S.: Photogrammetric 3D scanning of physical objects: tools and workflow. TEM J. **8**(2), 383–388 (2019)
24. Cantzler, H.: Random Sample Consensus (RANSAC). Institute for Perception, Action and Behaviour, Division of Informatics, University of Edinburgh, Edinburgh (2004)
25. Near net shape - Wikipedia, https://en.wikipedia.org/wiki/Near_net_shape#:~:text=Near-net-shape%20is%20an,the%20need%20for%20surface%20finishing (2024). Accessed 01 Mar 2024

Artifex Automata Predicting Robotic Chiseling Strokes

Sandro Siefert(✉), Nadja Gaudillière-Jami, Tom Svilans, and Paul Nicholas

Centre for Information Technology and Architecture (CITA), Royal Danish Academy, Copenhagen, Denmark
sandro.siefert@outlook.com

Abstract. *Historic buildings serve as vital links to our past, embodying environmental and cultural values, making their preservation and adaptive reuse a critical issue. However, the diminishing number of skilled craftspeople poses threats to these preservation efforts. This research explores the integration of robotics and machine learning with the traditional craft of stone chiselling, aiming to leverage digital technology for sustainable heritage conservation. A machine learning model trained on skilled craftsperson motion data coupled with a robotic arm demonstrates the ability to predict and execute chiselling strokes on sandstone. While surfaces produced by both the robotic system and the skilled artisan appear closely aligned, the presence of subtle differences challenge existing notions about craft and robotic fabrication. Beyond demonstrating the feasibility of integrating digital technologies into stonemasonry, this research promotes bridging the intricate realm of traditional craft with the integrative complexity of machine learning and the precision of robotics, and explores a paradigm shift in the way craft skills can be perceived and preserved.*

Keywords: Robotic Fabrication · Machine Learning · Digital Craft

1 Introduction

Retrofitting and repurposing existing structures can be more sustainable than demolishing and constructing new buildings. By reusing and adapting historic buildings, the demand for raw materials and energy consumption associated with new construction can be reduced. Preservation and adaptive reuse contribute to the principles of sustainability and mitigate the environmental impact of the built environment. In addition to their sustainability potential, historic buildings can hold significant cultural, historical, aesthetic, and social values [1]. Yet the demolition rate of historic buildings in Denmark, where this research is undertaken, raises concern about the erosion of cultural heritage, social values, and urban character [2]. There is a need to preserve these structures to maintain a link to the past and to promote sustainable architectural practices of maintenance and repair.

The diminishing number of skilled craftspeople in Denmark [3] raises concerns about our ability to tackle such challenges. One reason for the decline in interest of young

M. Yablonina et al. (Eds.): Robotic Fabrication in Architecture, Art and Design 2024, pp. 344–353, 2026.
https://doi.org/10.1007/978-3-032-22607-5_22

people might be that craft, due to the importance of tradition and continuity, has become disconnected from the fast-developing technologies of today. This paper suggests looking beyond tradition and conservatism to new digital technologies and incorporating them within these crafts, without losing the bond to their heritage. The combination of chisel and robot contributes to the challenge of preservation, benefiting from the efficiency and precision that robots offer, automating repetitive or physically demanding tasks and increasing productivity to intervene on a higher number of buildings. Weaving heritage, craft, and digital technology together, this research develops a workflow enabling robotic chiseling, based on the training of a recurrent neural network with the motions of a trained stonemason, to predict and plan chiseling strokes.

2 State of the Art

2.1 Robotic Stone Manufacturing

The stone industry has recently undergone significant change with the increasing introduction of digital tools and workflows. Among these, 3D scanning technology has emerged as an important tool for the creation of high-resolution scans of historic structures, allowing the assessment of damage and the creation of 3D models for stone replacement. The process of stone fabrication with robotic arms and CNC milling machines has become increasingly popular as well. Mark Burry's seminal work at the Sagrada Familia stands as a detailed example of how such workflows have been developed [4]. However, a critical gap in achieving a nuanced surface quality, with chiseled finishes like the manual work of stonemasons, remains in robotic stone carving developments. Current industrial workflows integrate scanning and milling but require manual finishing in the final stage to replicate traditional tooling marks and create a cohesive appearance with the surrounding heritage elements.

Assessing whether this step can be digitized requires tackling two main challenges: understanding the motion of the craftspeople to program the robot accordingly and equipping the robot with traditional tooling and an associated fabrication process. On the understanding of the motion, Shaked et al. describe a protocol for adaptive robotic stone carving [5], beginning with the documentation of manual techniques used by skilled stone artisans through motion capture. This data is then analyzed and used to develop a protocol for stone working deployed by a robotic arm equipped with a chisel-end-effector. Despite an in-depth analysis of key features, the subtle factors behind a successful chisel strike can however remain elusive. This points to the potential of machine learning, which has demonstrated capabilities to identify patterns in complex data sets for various applications.

2.2 Machine Learning for Robotics

Machine Learning (ML) in architectural manufacturing offers potential by streamlining and improving digital design-to-fabrication processes. Prior research demonstrates how ML can simplify the generation of fabrication data by bypassing complicated multi-step processes and adapt fabrication data in response to changes in materials, environment, or design [6]. As an example, Rossi et al. explore the use of collaborative robots in manufacturing processes, combining haptic teaching with ML, a method that showed successful results by enabling the robot to adapt to unexpected changes and obstacles without explicit programming [7]. Research in human-robot collaboration has also investigated the application of recurrent neural network (RNN) ML techniques for predicting human motion trajectories and demonstrated the effectiveness in learning from past human motion trajectories to achieve such predictions [8]. A limitation of this research is the challenge of contextualizing motions such as chisel strikes, where each strike is influenced by previous strikes but no past trajectory is available for future predictions. This highlights the need for additional data sources beyond just motion.

The combination of ML with robotic fabrication processes and its application to specific tooling has also been tackled in prior research. Rossi has explored the implementation of a cyber-physical system that integrates a traditional English wheel with an industrial robotic arm and a Kinect scanner, putting the analogue and digital in context with each other [9]. Brugnaro has developed an approach to this issue that relies on the acquisition of motion data from a skilled human artisan [10], applying ML for data analysis. This enables the training of an autonomous model that interacts with a set of feedback sensors to create a dynamic and material-aware robotic workflow for wood carving. Yet a gap exists in the transfer of such workflows to machine-assisted stone fabrication, as existing predictive models cannot be readily applied to different contexts due to their inherent material specificity. In summary, the study of robotic stone processing is a complex and challenging field that requires in-depth exploration of the intricate techniques and underlying concepts of stonemasonry. The integration of ML methods, which has demonstrated effectiveness in similar endeavors, could prove essential for unravelling the intricacies of this craft. By building upon the references described above, this research tackles the challenge of automating the finishing steps of adding chiseling strokes to a CNC-milled stone.

3 Materials and Methods

Figure 1 introduces the overall workflow developed in the present research. It is segmented into four parts: data acquisition, data processing, prediction, and robotic execution.

3.1 Data Acquisition

For the ML model to be able to predict a chiselling stroke based on an input stone geometry, two distinct sets of data are necessary: the target surface, the stone geometry, and the chisel strokes motion. Furthermore, there are two data gathering steps: first during the ML training phase, and second during the deployment phase. The scenarios for each step are similar but have slightly different setups.

Training To facilitate data acquisition, a motion and 3D-scanning setup as shown in Fig. 2 is used. This setup includes an OptiTrack system, consisting of eight motion-tracking cameras, that monitor the motion of a rigid body attached to the pneumatic chisel at a frame rate of 60 frames per second. The output of this system provides the

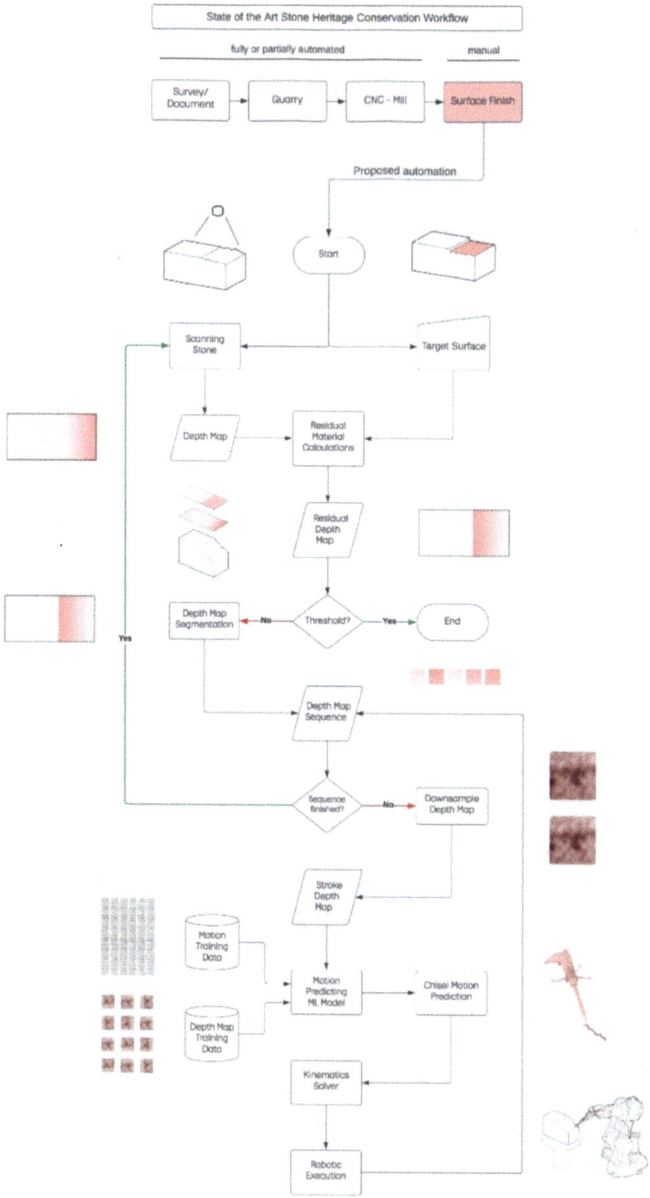

Fig. 1. Overall workflow

contents of the first dataset. The second dataset consists of 3D contextual information of the stone surface before the chiseling stroke occurred, obtained by depth scanning with a RealSense L405 LIDAR scanner positioned above the stone at a calibrated height (Fig. 3).

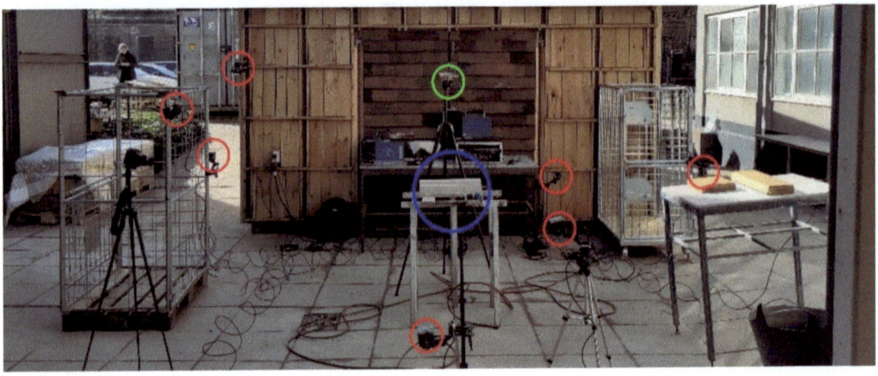

Fig. 2. Data gathering set-up (red: OptiTrack motion tracking cameras; green: RealSense L405 LIDAR scanner; blue: stone)

Fig. 3. Raw motion data (left), virtual representation (right), and raw depth map (left)

Deployment To obtain the stone's surface geometry before every stroke, the same Intel Realsense L405 Lidar scanner is used. The scanner is mounted on the robotic end-effector and accessed over an API provided by the manufacturer. Additionally, to verify the surface measurements of the stone, a calibration is initially required to reference the block's position in relation to the robot.

3.2 Data Processing

Motion Before training, the motion data must be pre-processed. Due to the constant frame rate of the motion capture device and the fact that not every chisel stroke takes the same time, the raw motion data varies in length. This can cause multiple problems in the training process. To ensure a consistent training dataset, motion data must be interpolated and resampled for uniform orientations. Positional data from the motion tracking system poses concerns as well. Hence, the motion is processed in a way so that it is always starting at (0,0,0). This way the positional noise from the motion tracking device is negated.

Stone Geometry The raw depth map contains superfluous details, as the area of interest in which the stroke is executed only occupies a small part of this surface. To refine the data set, the depth map is segmented around the stroke into a 10 × 10 cm square, a size sufficient to capture any reasonable sized stroke while excluding irrelevant data. This square is then down sampled to a 24 × 24 pixel matrix, to further reduce noise and simplify the data (Figs. 4 and 5).

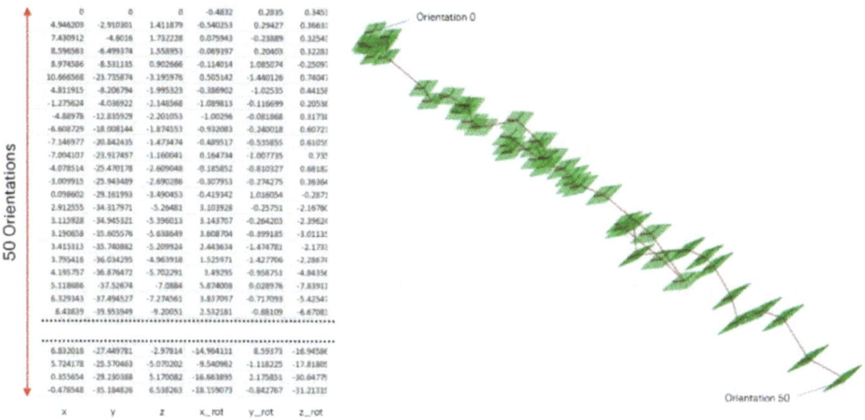

Fig. 4. Pre-processed motion data (left) virtual representation (right)

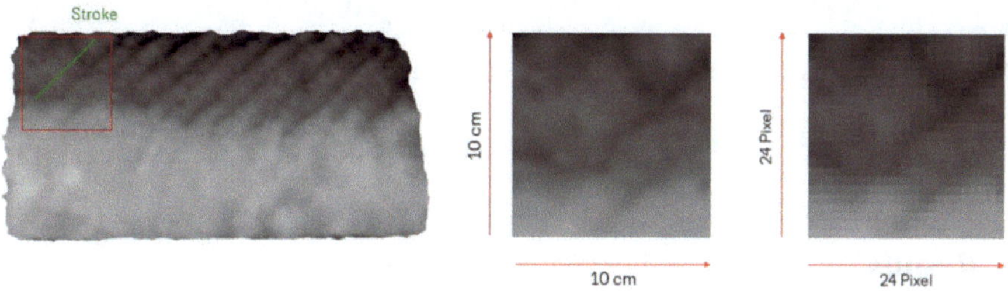

Fig. 5. Raw depth map (left), segmented depth map (middle), resampled depth map (right)

3.3 Prediction

The model architecture chosen for training is a "one to many" RNN, with sequential output prediction of a single depth map as input. The model consists of 5 layers that transform an initial array of input training data of shape (n_samples, 24, 24) to an output training data array of shape (n_samples, 50, 6) (Fig. 6). The input is first encoded with a LSTM layer into an array of shape (n_samples, 100). The array is then repeated with a repeat vector layer (n_samples, 50, 100). After decoding the array with a LSTM layer, a TimeDist (Dense) layer is applied to create the final output array of shape of (n_samples, 50, 6). This translates to each stroke (n_samples) having (50) orientations, each described by (6) values: pos_x, pos_y, pos_z, rot_x, rot_y, rot_z.

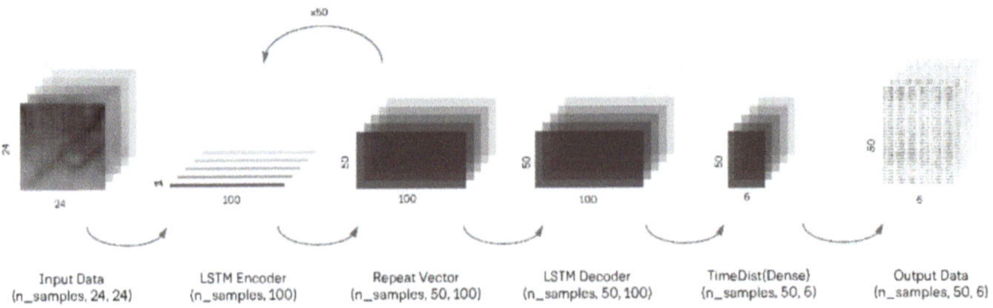

Fig. 6. Machine learning model architecture

3.4 Robotic Execution

The robotic setup consists of an ABB IRB 1600 M2004 equipped with a pneumatically operated chisel end-effector (Fig. 7). The predictions derived from an unseen depth map segment of the stone's depth map are translated into a robotic program and then executed within the area of the stone that the input depth map segment described.

Fig. 7. Robotic chiseling set-up with end-effector and sandstone block (top) and execution sequence (bottom)

4 Results

4.1 Prediction Evaluation

The predicted strokes were compared to the recorded stroke to assess the model performance before the training phase was completed (Fig. 8). Splitting the training data into train and validation groups prior to the training process ensures that the model is not spoiled. The evaluation yields several noteworthy observations. First, the overall trajectory, approach angle, and exit angle of the predicted motion match the actual recorded counterpart within acceptable limits. Second, the prediction accuracy is most pronounced at the beginning of the motion, but gradually deviates as time progresses. Third, distinct sections of rapid counter steering movements happen in the recorded strikes due to the artisan's reaction to unexpected material behaviour. These abrupt movements within the stroke are highly depended on the inner stone structure met by the chisel during the stroke, and therefore not present in a prediction based on the stones current surface.

Fig. 8. Recorded stroke in green (left) and predicted stroke in red (right)

Fig. 9. Chiseling patterns produced by a trained stonemason (left) and the robot (right)

4.2 Chiselling Texture Evaluation

The robotic arm executed chisel strokes on a flat sandstone as guided by the trained model. A trained stonemason then produced the same surface texturing using a conventional chisel technique that mirrored the training data collection process. The resulting stone surfaces (Fig. 9) revealed strong similarities, demonstrating that achieving satisfying textures for the preservation of heritage buildings is possible. However, the resulting surfaces also reveals subtle differences. In comparison to the stone chiselled by the stonemason, the surface chiselled by the robotic arm contains irregularities in the length, placement, and depth of the strokes. This feature reveals that the robot's predictions are working on a stroke-by-stroke basis, without contextual information between consecutive strokes. Experienced stonemasons integrate the strokes into a holistic context, resulting in a surface that appears less disjointed. This links to the lack of counter steering movements in the predicted strikes and raises the question of what sensors to integrate to enable material assessment within the workflow.

5 Discussion

The technical and visual results yielded by the workflow demonstrate the possibility of integrating predictive chiselling into robotically supported industrial stonemasonry. This would allow an extension of the steps of heritage restoration that are robotically managed, expanding into the finishing phase. However, the workflow has only been applied so far to one specific texture, while hundreds of different finishing patterns exist in traditional masonry, leveraging different tool tips. Using the current workflow, training different models for each tool tip and associated chiselling methods would be necessary; as these require individualized datasets, the approach would be very time-consuming. Further research could explore a more holistic approach in the training model. Focusing on potential industrial applications of the workflow has also meant focusing on shapes traditionally chiselled by stonemasons. While helping create a tool that could be outputted directly to industry, it also restricts the applications to more complex shapes. Future developments could focus on the expansion of the formal complexity of target geometries in the workflow.

Expanding the range of textures tackled with the developed process through new datasets also raises the issue of data acquisition. Increasingly fewer stonemasons are being trained in the craft, making future data acquisition difficult. The workflow could leverage human-machine collaboration into knowledge transmission, offering solutions to preserve and share the craft. As the strokes of the robots are predicted based on a set of registered human strokes, a hypothesis to further examine is whether observing the chiselling motions of the robot could help an apprentice learn their craft, much like they observe a trained expert. Additional research comparing learning outcomes in such a set-up with learning outcomes in the traditional stone masonry apprenticeship would provide insights into the degree to which such a preservation and transmission of knowledge would be possible. Beyond its industrial implications, the research presented here therefore address the issue of the role of digital in archiving and preservation.

6 Conclusion

By distilling the nuanced art of stone carving into a series of sequential actions, this research shows that an RNN model can learn and mimic these actions, resulting in the production of stone surfaces that mimic the craft of skilled artisans. Our results challenge preconceived notions about the relationship between craft irregularity and robotic precision by showing that an ML-assisted robotic chisel system can produce manufactured irregularities. Further developments include expanding to other textures, adding sensors and feed-back to the workflow for a more material adaptive framework, exploring a more holistic approach for training, and assessing the potential for knowledge preservation. Additionally, this innovation is not limited to stone carving; it holds the potential to extend its reach to various other crafts, textures, and tools. This research highlights the profound potential of human-machine collaboration. It emphasizes that the goal is not to replace human craft with robotic automation, but to promote, preserve and disseminate it through the fusion of tradition and technology. Moreover, this research expresses the daring assumption that robotics and machine learning can be not only a potent tool of an ever more automated world, but also, given the steady loss of craft knowledge, valuable repositories of knowledge itself.

References

1. Lynch, K.: The Image of the City. MIT Press, Cambridge, Massachusetts (1960)
2. Andersen, R.: Adaptation of circular design strategies based on historical trends and demolition patterns. IOP Conf. Ser.: Earth Environ. Sci. **1085**(1), 012062 (2022)
3. Juul, I.: Challenges for the dual system and occupational self-governance. J. Vocat. Educ. Train. **63**(3), 289–303 (2011)
4. Burry, M.: Robots at the Sagrada Família Basilica: A brief history of robotised stone-cutting. In: Reinhardt, D., Saunders, R., Burry, J. (eds.) Robotic Fabrication in Architecture, Art and Design 2016 (2016)
5. Shaked, T.: Adaptive robotic stone carving: Method, tools, and experiments. Autom. Constr. **129**, 103809 (2021)
6. Thomsen, M.: Towards machine learning for architectural fabrication in the age of industry 4.0. Int. J. Archit. Comput. **18**(4), 335–352 (2020)
7. Rossi, G.: Haptic learning: Towards neural-network-based adaptive cobot path-planning for unstructured spaces. In: Architecture in the Age of the 4th Industrial Revolution: Proceedings of the 37th eCAADe and 23rd SIGraDi Conference, pp. 201–210. ECAADE SIGRADI 2019, Porto (2019)
8. Zhang, J.: Recurrent neural network for motion trajectory prediction in human-robot collaborative assembly. CIRP Ann. **69**(1), 9–12 (2020)
9. Rossi, G.: Re/Learning the wheel: Methods to utilize neural networks as design tools for doubly curved metal surfaces. In: ACADIA//2018: Recalibration. On imprecision and infidelity: Proceedings of the 38th Annual Conference of the Association for Computer Aided Design in Architecture, pp. 146–155. ACADIA 2018, Mexico (2018)
10. Brugnaro, G., Hanna, S.: Adaptive robotic carving. In: Willmann, J., Block, P., Hutter, M., Byrne, K., Schork, T. (eds.) Robotic Fabrication in Architecture, Art and Design 2018 (2019)

Towards Adaptive Additive Manufacturing with Semi-autonomous Robotic Binder Jet 3D Printing of Concrete

Linxiaoyi Wan[1](✉), Jingyang Liu[1], Dana Cupkova[1,2], and Joshua Bard[2]

[1] Carnegie Mellon University, Pittsburgh, PA 15213, USA
linxiaow@andrew.cmu.edu
[2] EPIPHYTE Lab, Pittsburgh, PA 15217, USA

Abstract. This research proposes an integration of image-based feedback monitoring into a process of additive manufacturing with the aim of expanding 3D concrete printing (3DCP) capabilities toward more diverse material feedstock. Building upon our ongoing research focused on current binder-jetting techniques in 3D concrete printing, we address sequential quality control of printing with coarse powders that advance the adaptation of this technology for sustainable construction with a focus on material reuse. By incorporating an image-based monitoring system into binder jetting, we aim to enhance the adaptability and robustness of 3DCP towards using more variable and irregular powder feedstock from waste streams. A comprehensive dataset, collected over a year and comprising approximately 400 samples, was leveraged to formulate an image-based flaw-detection algorithm. Deep learning training was employed to ensure precise segmentation of printed patterns. The adoption of a dual-resolution imaging strategy further accentuated the quality of data collection and interpretation. Central to our flaw-detection mechanism is a feedback loop framework, designed to discern discrepancies between desired and printed binder deposition pathways. This study proposes an integrated approach of binder jetting with flaw-detection feedback and error rectification within 3DCP, holding a transformative potential for the broader construction technology landscape and material reuse.

Keywords: Construction robotics · Image-based monitoring · Additive manufacturing

1 Introduction

Traditional approaches in binder jetting mostly use fine and highly processed powders, paying less regard to the enormous potential offered by coarse powders, waste derivatives, and composite aggregates. This is primarily due to the difficulty inherent in coarse powder characterization, including their uneven distribution and flow during the automation process. Our study proposes a novel approach toward the automation of binder jetting while integrating an image-based monitoring system.

M. Yablonina et al. (Eds.): Robotic Fabrication in Architecture, Art and Design 2024, pp. 355–370, 2026.
https://doi.org/10.1007/978-3-032-22607-5_23

Our recent study on adaptive manufacturing with coarse powders [1] shared a comprehensive study of material compositions, focused on different ratios of cement, sand, and aggregate within the water-based binder jetting process for 3D concrete printing. It highlighted that the mixing ratio has a great influence on parameters such as binder expansion, penetration depth, and the resultant porosity of the mix. From this finding, we were able to determine optimal layer thicknesses corresponding to distinct mix ratios. This in-depth study of the binder-jetted printing patterns helped us to establish a printing layer structure and the binder-to-material saturation rate to validate the digital slicing model to the physical accuracy of printing with coarse aggregate. Optimization of this process helped us reduce the usage of a binder by 6–17%, as well as the wall thickness of the final product by 28%, and reduce its weight by 28.5%. This approach enabled more accurate printing with different mixing ratios of various powders without any change in the printing parameters of different layers. From our previous analysis, which was limited to a single droplet saturation test and short-line printing test for 3 different materials' mixing ratios, we established a more precise understanding of binder behavior.

For higher proportions of aggregates in the concrete composite, there is a higher unpredictability in the horizontal expansion of the binder —a phenomenon supported by large variations in standard deviation, documented from our image analysis of droplet expansion in Table 1. The interaction of penetration depth with specific mixing ratios of concrete composite further explains the complexities in the vertical movement of the binder. The observed pattern was a clear indication that a wider spacing of the binder reduced the bleeding pattern and limitations on penetration depth. For example, in all compositions, the concrete mix with the compact aggregate ratio combined with expansive binder spacing showed less bleeding and superficial penetration. The mix with the increased aggregate ratio and narrowed binder spacing showed pronounced bleeding of the binder (see Table 2 for details). This prior work sets the stage for the current research's trajectory and objectives focused on flaw detection in the printing process of course powders.

Table 1 Single droplet printing test results analysis.

	Material Mixing Ratio (Cement: Sand: Aggregate)		
	1:1:1	1:1:2	1:1:3
Horizontal Expansion Length (cm)	0.663	0.720	0.790
Horizontal Expansion Width (cm)	0.570	0.600	0.670
Horizontal Expansion Area (cm)	0.205	0.225	0.300
Standard Deviation	***0.062***	***0.087***	***0.103***

The flaw detection is addressed by incorporating the image-based monitoring feedback loop. The image-based monitoring in diverse material mixing ratios achieves a dual purpose: it opens up the use of coarse concrete composite powders in 3D concrete binder printing while also introducing a feedback mechanism through computer vision. This

Table 2. Shortline Printing Test Result Analysis.

		Penetration Depth (cm) at Mixing Ratio		
		1:1:1	1:1:2	1:1:3
Saturation Rate	14%	0.3	0.3	0.5
	18%	0.4	0.6	0.8
	22%	0.7	1.0	>1.0

system enhances the flexibility of the printing process which realizes feedback adjustments in binder placements and accentuates the potential for precision in surveilling the expansion of the binder during the printing process.

While the focus of this paper remains on concrete mix applications, these approaches offer insights that show a potential to alleviate the material characterization workload when integrating unconventional powders into the binder-jetting process. Most importantly, our digital image correction pipeline overcomes precision and quality problems that are invariably entailed in heterogeneous mixing materials.

Our study brings out the potential within 3DCP for image-based monitoring, shedding light on an indispensable role in the furtherance of the next generation of adaptive additive manufacturing [1].

2 Situating Research Background

2.1 Robotic Arm in Concrete Digital Fabrication

The technology of binder jetting was first introduced and patented by Emanuel Sachs at MIT in 1993, mixing powders with a water-based binder for printing purposes [2]. Since that time, the application realm of this technology has significantly changed. Many of the advances in binder jetting include contributions from ExOne Co. [3]. Other important advancements are the revolutionary methods like the D-shaped methodology [4] and Voxeljet Company's take on binder jetting [5] that made it possible to print the details of a much more detailed design with increased print fidelity.

Lloret et al. [6] provide a detailed insight into how traditional casting methods integrate with new digital manufacturing processes to improve and develop difficult concrete structures. Another study shows that through robotic and automatic concrete spraying, the technology has been efficiently contributing to the construction of concrete shells, providing versatility and accuracy in building through the adoption of robotic arms [7]. Central to the use of real-time image-driven monitoring in the procedure of binder jetting is the incorporation of sensory apparatus, including optical, acoustic, and thermal sensors. Inclusion among the important steps is digital image correlation (DIC). In addition, several review works were done in connection with in-situ monitoring of Additive Manufacturing (AM) [8]. Moreover, Reiter and Agustí-Juan put forth guidelines and insights regarding the full range of structural and environmental design considerations in digital fabrication with concrete [9].

2.2 Learning Techniques in Flaw Detection Applications

Detection of flaws through Deep Learning promises to equip the construction industry with more adaptive and innovative solutions that enhance its ability to respond to issues of sustainability. Ozegenel et al. offered a comparative analysis of pre-trained convolutional neural network models in the detection of building crack flaws, an insight into the efficacy of various neural networks and their models for flaw detection [10]. Thier work is comparable to Yukang et al.'s deep-learning-based method for photovoltaic defect detection [11]. Kalfarisi et al. dealt with deep learning for crack detection and segmentation to extract accurate positional coordinates from the 3D reality mesh model, providing a quantitative assessment integrated with visualization to better detect the flaws in construction [12].

Further, an externally automated vision-based structure inspection and defect detection system, using the deep learning technique for UAVs, is an overarching approach to defect detection in such structures [13]. In the construction context, the benefits of security can be realized by employing vision and sensing techniques. For example, some authors have developed a hybrid machine-learning framework for the automation of equipment activity identification and earlier-stage fault detection in its occurrence [14]. Another predictive maintenance framework, with a focus on automatic fault detection and diagnostics, is the role of digital twins in predictive maintenance in construction [15]. Zhang also discusses an automatic method of surface crack detection and segmentation in bridges using a computer vision-based deep learning model, in support of a comprehensive solution to the flaw detection problem in bridge construction [16].

3 Methodology

Our workflow is initiated by establishing a predefined geometric pathway for layer-by-layer 3D binder jetting tasks. A pathway provides a location that gives the exact point coordinates for the robot arm to follow, and ensures that binder displacement is laid out. Meanwhile, a built-in image-based assessment is introduced using the RealSense camera after the robot arm has executed one layer of the binder placements. The camera captures a printed layer result and processes an integrated image-based post-print evaluation.

Central to our methodology is the capability to find and quantify differences from the digital model pathway to the printed output. Our framework evaluates the area of the missing portion against the predefined pathway. The ratio of the missing part area to the area of the current layer should not exceed a determined threshold and needs to be corrected before the framework allows the start of the subsequent printing layer (Fig. 1).

When the ratio of a missing pathway exceeds the limit of the determined threshold, the framework turns into a structured correction phase. First, there is an alignment of the printing pattern that visualizes the output and overlays it with the desired outcome, thus constructing a visual comparison. From this comparison, the differences between planned and actual printing could be extracted for the region of actual printing and its internal features. Subsequently, the detection of the missing areas and the computation of their centroids is executed. The robot arm, armed with this data, realigns its position and provides the binder displacement for targeted corrections. This is an iterative quality assurance process, assuring that every single layer that has been laid down in printing meets the desired precision and accuracy. With the help of the Robot Operation System (ROS), we were able to seamlessly incorporate the whole process.

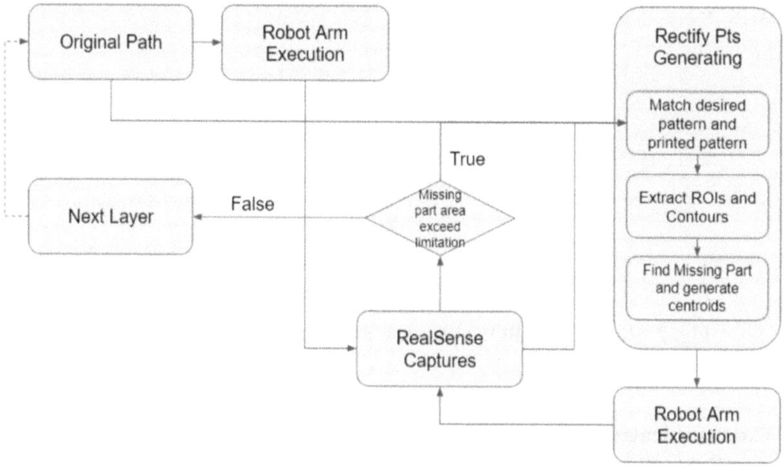

Fig. 1 Pipeline and framework of the whole system.

3.1 Robotic Operation System (ROS) based Simulation Environment Set-Up

ROS Communication System. The Robot Operating System (ROS) [17] has established itself as a powerful middleware for developing complex robotic applications. Our proposed adaptive additive manufacturing system exploits the capability of ROS to lead to seamless integration of the processes of manufacturing, monitoring, decision-making, and flaw rectification (Fig. 2). In this section, we illustrate how the ROS simulator manages communication among disparate nodes.

In this system, the node */amm_node* shall act as the epicenter node, informing the additive manufacturing process by receiving the pathway from geometric files and the rectified pathway from */print_status*. It assures that the result status of the print communicates through the*/print_status* topic, and thus ensures all other associate nodes are informed about the current printing result status.

Simultaneously, the */fdm_node* supports dual service. It orders the camera sensors to take images of the printed paths and transmit them to the */layer_image topic*. Concurrently, the */fdm_node* develops an image analysis process with the corresponding pathway to detect any possible missing parts. The information on detected flaws, such as the attributes of contours and their spatial coordinates, is then relayed through the topic */flaw_data*, subscribed to by the node */flaw_detec*. This information is then further processed and rectified before printing centroids with coordinates are generated.

As the decision-making center, the */decision_node* subscribes to the topic */print_status* and */flaw_detecnode*, along with their data streams, to cull out informed decisions regarding the necessary rectification measures. It then propels these actions, encapsulated as actionable commands, over the /rectification_command topic. Finally, the /rm_node subscribing to the */rectification_command* topic will ensure the rectification measures, whether to execute the re-dispensing binder or jump to the next layer printing task.

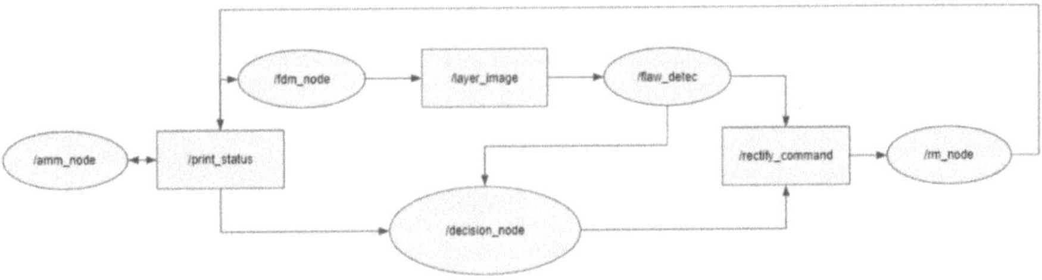

Fig. 2 Rqt graph of ROS nodes and ROS topics workflow integration

ROS Communicates with ABB RobotStudio. Since ABB's RobotStudio software is not available for Linux, the *abb_driver* [18] enabled communication between RobotStudio and ROS.

3.2 Hardware Development and Set-Up

Our primary contribution lies in interfacing the Arduino-based binder extrusion system with signals from an ABB robot controller to dispense the binder meticulously to each predefined point (Fig. 3). A custom print head accommodates the stack for the RealSense camera placement and a rubber tube is used for passing the binding material into the syringe needle. The needle is located within a 3D printed holder attached to the tool plate of the robotic arm. This setup ensures precision and steadiness during printing operations. The system is also supported by a step motor for controlling the position of the plunger in the syringe, which is the repository for the binder.

For careful orchestration of the motor's actions, we have set up a system based on Arduino, which interprets the signals from the robot and transmits digital signals to actuate the rotations of step motors, manipulating the plunger. Special framing was devised to limit the range of movement of the rod connected to the plunger. It was set up inline with the step motor so that the force translation from the step motor to the plunger is directionally focused. The system has an extrusion capacity of not less than 0.027 ml/droplet; hence, not less than 0.027 ml of the binding material is ejected from the nozzle. This forces the step motor to rotate for a duration calibrated to the required amount. Other concepts are also included, such as a leveling bed that includes a control system using Arduino. The bed houses a pair of electric actuators interfacing through the button-controlled signals to form a well-interfaced and friendly front for control.

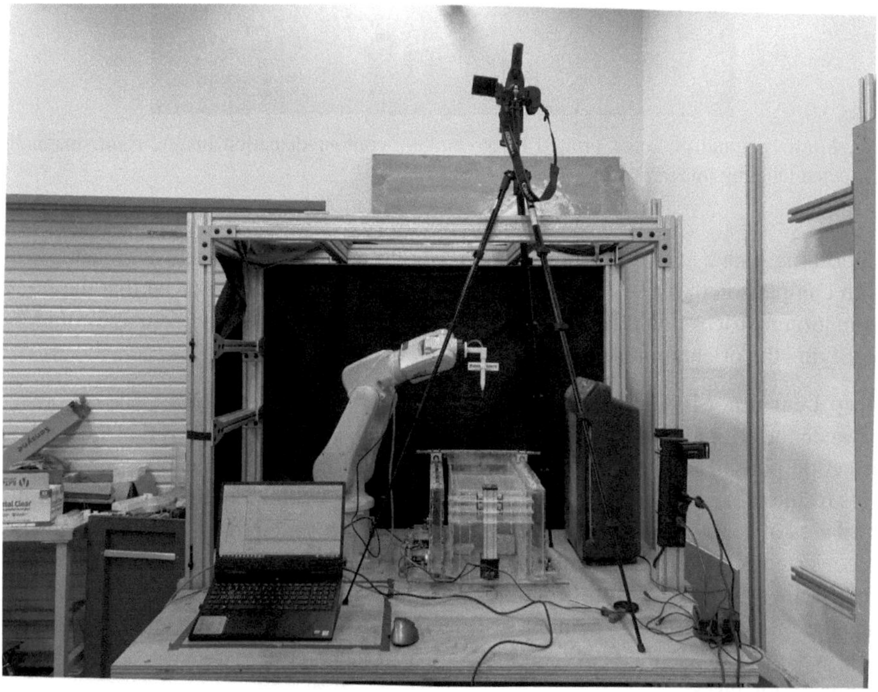

Fig. 3 Lab printing setup [1]

3.3 Flaw Detection Algorithm Development

Data Collection. The concrete mix printing patterns dataset was systematically accrued during a research development that spans one year. During this period, about 30 complex 3D printing tests were carried out so that our water-based binder jetting system could be advanced and further developed. Each of these tests bore a range of layers, minimally 10 to a maximum of 25, aggregating sample size close to 500, representing printed components for every individual layer.

Most of the images were captured using both high and low-resolution output using a dual-resolution technique. In the images, the regions of error were marked to ensure data interpretation exactness. We used custom image masks to identify the visual separation of printed and dry areas. This was the delicate and cumbersome effort of identifying the printed sections manually, where some attributes of the image, such as brightness, contrast, and threshold, were adjusted with great care. Further, we applied morphological operations, particularly dilation and erosion, to improve contour detection with each of the images available in the collection. After this first segmentation, any inaccuracy or abnormality in the masks was corrected by hand to ensure precision (Fig. 4).

During data collection, a slight distortion in photos was also noted. To address this discrepancy, we used low-resolution (1280x720) images, taken through a RealSense camera, as a baseline calibration source for ground truth. This allowed for easy detection of the matching features between the ground truth and the distorted images by

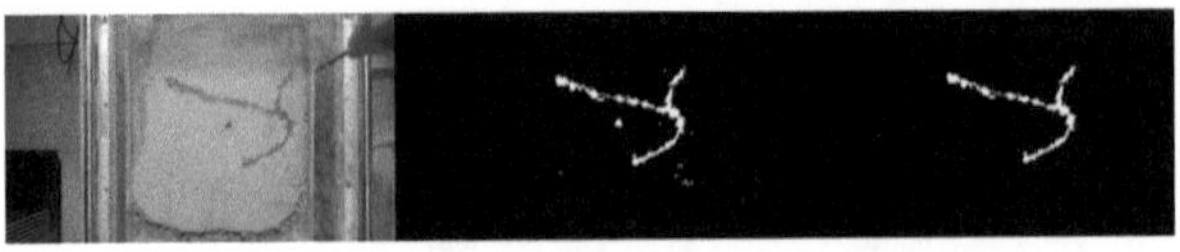

Fig. 4 left: original photo of printed layer; middle: contour detection image; right: manually corrected labeling image.

algorithms such as ORB [19] and SIFT [20]. Eventually, this process was successful, with comparison deriving the transformation matrices. This test showed that the transformation matrices could be used to warp the perspective of distorted images, thereby rectifying the orientation to be congruent with the ground truth.

Deep Learning Model Training for Printed Pattern Segmentation. The difference between the desired and the actual printed path in this research signified that the contour detection phase is crucial to better recognize the different layers of printing patterns. This resulted in an accuracy influenced—for example, by the brightness and contrast—attributes of each captured image. To have this system streamlined towards being able to consistently process such images, we adopted a learning-based approach. The main improvement was made by integrating a deep learning model so that the system was capable of segmenting the contour of the printed part. After some careful consideration, we opted to use the U-Net architecture as it has a proven track record in various segmentation tasks.

The U-Net [21] is a convolutional neural network architecture developed for semantic segmentation with biomedical image analysis. The two major parts of the architecture include the encoder and the decoder. The encoder extracts features of the image from the input. The encoder comprises a convolutional stack followed by max-pooling layers to reduce spatial dimensions. The convolutional encoder, therefore, is designed with the idea of capturing the hierarchical features present in input images at different resolutions. At every resolution of the encoder, the feature maps are upsampled and merged with the corresponding features of the encoder for spatial localization of the segmentation and simultaneously for refining them.

The architecture is built of up-convolutional blocks, which increase the spatial dimensions of the feature maps. The decoder layer is a 1x1 convolutional layer that maps the representational features to the required number of classes and then applies the sigmoid activation to retrieve the segmentation mask. The goal is to train U-Net to minimize the difference between the prediction and ground truth segmentation mask by using, typically, a binary cross-entropy loss (Fig. 5).

Rectify Point Generation. The first stage of flaw detection is related to a careful and detailed assessment of the printed pathway within our image samples. The key focus is a process identification towards successfully printed patterns. The preliminary analysis starts with the comparison of the desired and printed pathways. It is a very important step for which an attempt is made to establish the desired pathway, and results are acquired after detecting and segmenting the pathway in the feedback loop framework so that it helps in finding deviations and anomalies.

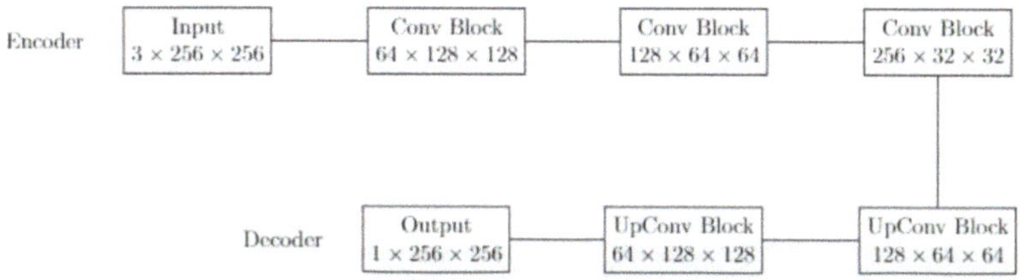

Fig. 5 U-Net Architecture for training.

Fig. 6 left: predefined pathway and printing boundary; middle: original picture; right: overlaid picture of desired printing pathway and capture photo. This image registration process enables matching the locations of the desired pathway and the printed pathway in the pictures. Registering the images allows us to match the locations of the desired pathway in the pictures with the printed pathway.

Alignment of features in images further enables feature matching between the digital model and physical print. This alignment of the desired pathway with the printed pathway in the images provides a comparison analysis of the pathways and commences a correction phase. It is important to align the correct lines of pathways with the actual captured lines of the camera. This process was introduced to simplify the previously used complex procedure using a single droplet spreading radius calculation at every layer. The visual approach is instrumental in offsetting the pathway line in line with the spreading radius of the binder with each mixing ratio of materials. Figure 6 illustrates the process of designation for both print areas inside an inner square form and image dimensions in general inside the outer rectangle form. Building on this foundational step, we extracted the range of prints from both: the images of the desired and printed pathways (Fig. 7). This extraction helps in reducing the environmental noise and fosters accuracy in the later process of contour segmentation. The careful alignment and feature extraction bring a more precise result and strengthen the reliability of the segmentation.

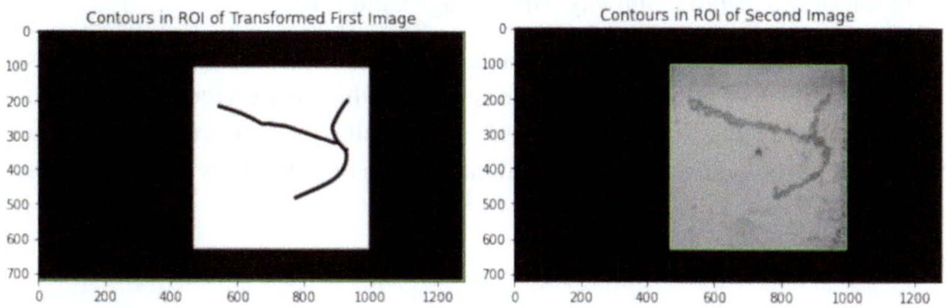

Fig. 7 left: extracted ROI of the predefined pathway; right: extracted ROI of capture photo. This image extraction process helps the system to find contours with less noise.

The trained model is used in printed pathway detections to compare the desired pathway with the one in use. The depiction is translated into a binary form, as the comparison is executed using white color values of the detected pathway pixels (1), while the remaining pixels are represented in black (0) (Fig. 8).

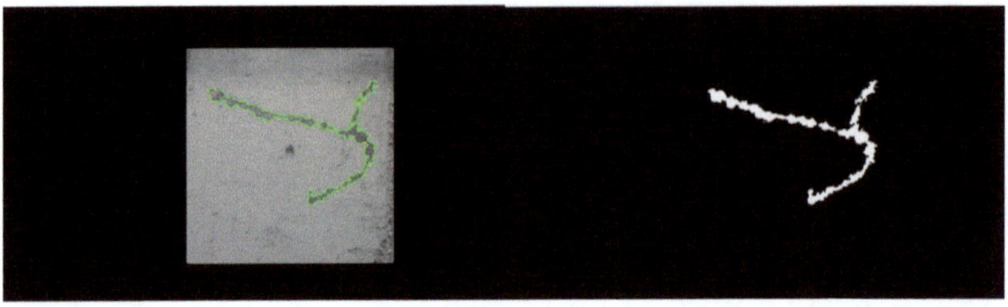

Fig. 8 Detected contours of the printed pathway. Left: Detected contours of the printed pathway. Right: The pixels of the printed part are labeled in white.

The similarity between the expected pattern and the printed pattern is quantitatively determined with the aid of the Structural Similarity Index (SSIM) [22]. This allows rigid and systematic comparison; and allows to location of regions that are deviating from the pathway, while more accurately identifying the regions deviating from the pathway.

The differences obtained by this comparison are those of the deficient pathway contours, while the centroids of these geometric areas are computed directly from the contours. This method produces a precise and detailed representation of the deficient areas that need rectification of the pathway and offers a higher standard for precision than the previous experiment. Analysis of the single droplet binder spreading pattern (Table 1) offers, among others, the radii averages at the different mixing ratios. Thus, by a careful comparison of the differences between the ideal pattern centroids and the pattern used with the radius of the ratio of mixing, judgments can be made as to whether rectification is required. The centroids are obtained relative to the location in the 3D model, using a reference of the outer rectangle to the four corner points (Fig. 9). The relative positions are then passed to a controller for processing by the robot arm through a ROS publisher for a precise correction of the print.

Experiment. The next stage of the flaw detection process necessitated an intensive examination of the printed pathway using image sampling.

This is an essential part required for achieving precision. This phase is significant to ensure a higher level of printing correction accuracy while using a feedback loop between the desired pathway and the actual pathway. The results obtained from the detection and segmentation of the pathway aid in the identification of deviations and anomalies.

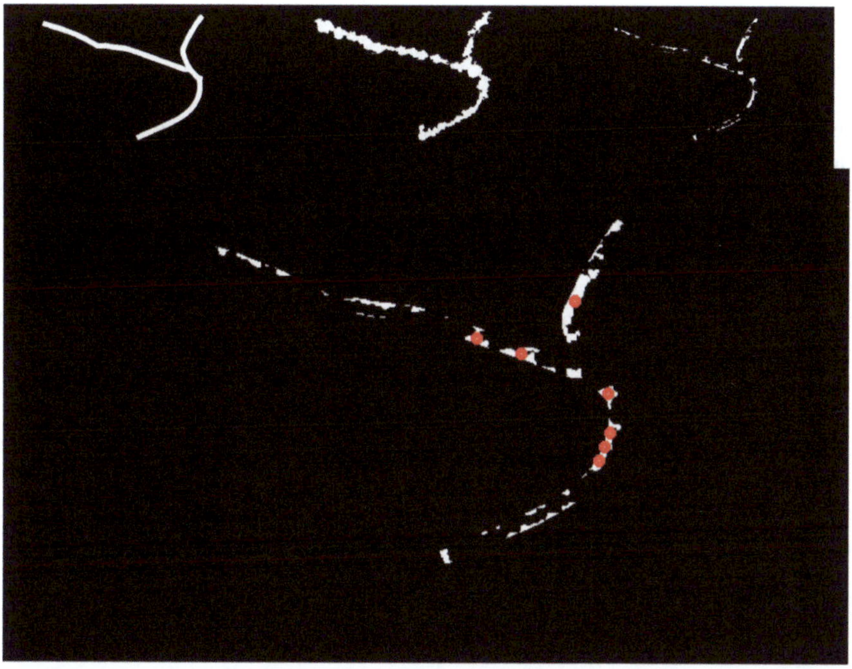

Fig. 9 Generation of rectification points. Up Left: Detected contours of the desired pathway. Up Middle: Detected contours of the printed pathway. Right: Missing part of the printed pathway. Below: Generated points to rectify on printed pathways.

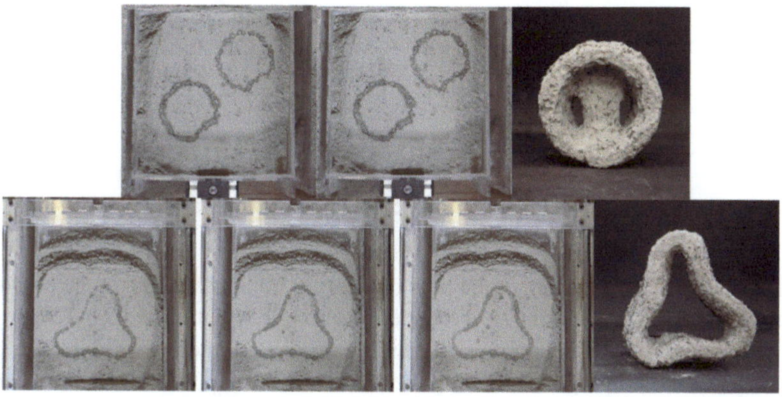

Fig. 10 Comparison of printing pathway and printed object using an image-based feedback loop. Above: One correction feedback loop needed Below: Two correction feedback loops needed.

4 Results

4.1 Test Print

The purpose of this experiment is to strengthen the adaptability of a semi-autonomous binder jetting system over non-uniform tilted surface conditions with coarse powders. Traditional binder-jetting printing processes require precise deposition by using a heterogeneous and highly processed powder and flat substrate. The precision of the print is assured by using evenly processed feedstock and heterogeneous conditions within the print bed. Any powder inconsistency or deviation from the flat print surface results in a print error. This presents a challenge to the usability of coarser powders within printing conditions, as when the print surface deviates it causes uneven integration and absorption of the binder and produces subsequent print errors.

We have designed a new printing bed to test the image feedback monitoring on a tilted print surface to allow for a more adaptive printing framework. The print bed has eight flat sections along its length and a 10° tilt across the shorter axis (Fig. 10). It is designed to simulate the complexities and difficulties involved in printing on uneven terrains and to test the limitations of our semi-autonomous system.

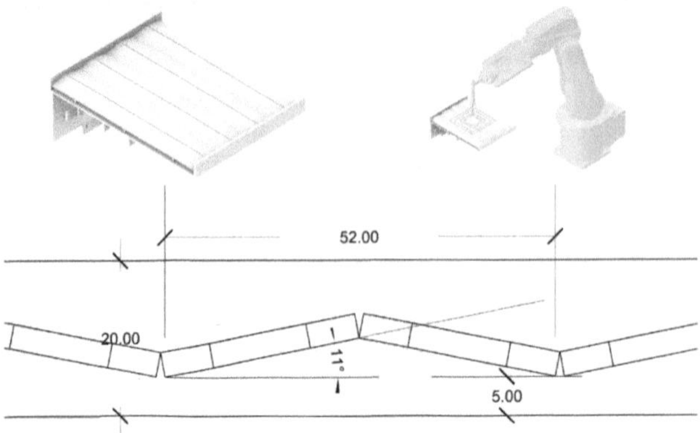

Fig. 10 Tilt printing bed and printing pathway.

The tilted orientation changes the path of printing from orthogonal to the path along the axis of the bed. The test print was produced on an angled printing bed. This experiment resulted in many observations. In particular, the droplets of the binder collect in peaks and valleys of the printing mix. During the print test, the deposition of the binder was broken in the middle sections of each tilted plate, which resulted in gaps in the print. With our image monitoring feedback system method in place, these irregularities have been studied with care. Our system analysis showed areas of discrepancy from the print pathway. Rectifying these discrepancies involved targeting the correction by depositing more binder droplets in the identified areas. Our flaw-correction image-based system is semi-autonomous and thus able to adapt to tilted surface printing (Fig. 11). We were able to increase the shape similarity of the desired pathway and the printed pathway from 93.765 to 96.237.

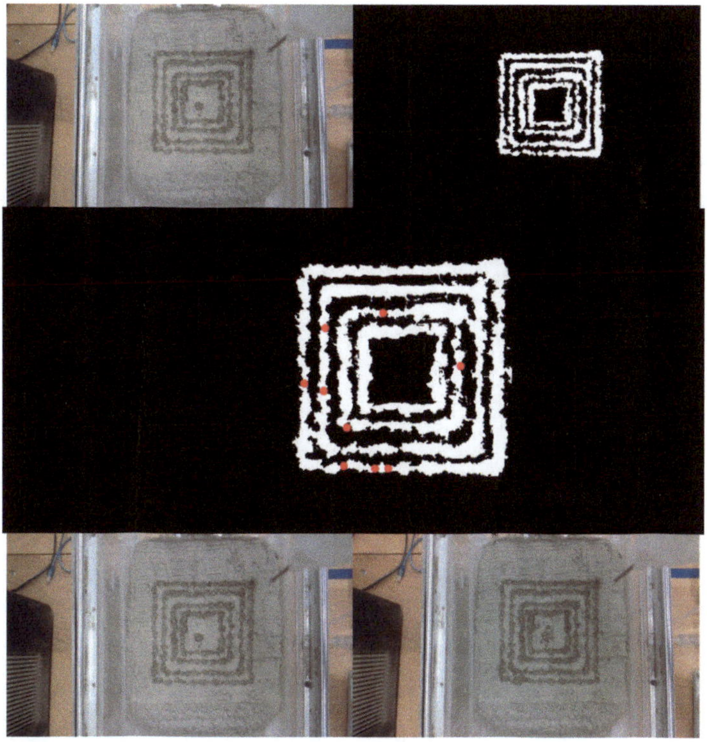

Fig. 11 Up left: actual print; Upright: recognition of printed parts; Middle: Printed pathway on tilt bed and generate rectify points; Bottom left: first round actual print; Bottom right: Rectify points actual print.

4.2 Design Print

Using the above image monitoring integrated printing methodology we designed a proof-of-concept print. It consists of printing a minimal-surface geometry object, thus validating the previously established pipeline, coupled with the lab setup as discussed above. This showcases the system's adaptability for complex shape fabrication across diverse conditions using an image monitoring feedback loop for flaw correction.

The geometry of the object was designed using the SphereCollide and SolidPointCollide components from the Kangaroo plugin and was later fine-tuned through the use of custom shape optimization algorithms (Fig. 12).

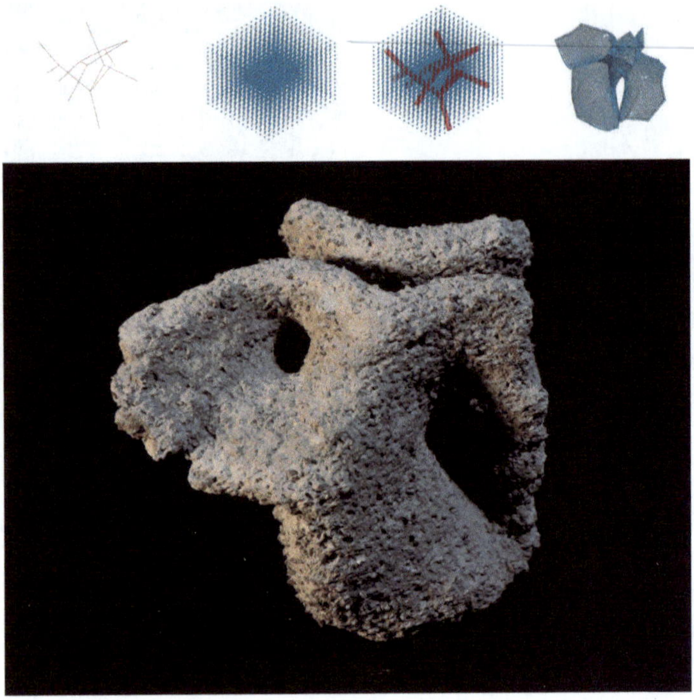

Fig. 12 Minimal surface shape generating process and printing product [1].

5 Conclusion

Our research contributes to current binder jetting techniques in 3D concrete printing while addressing a sequential quality control of printing with coarse powders. This area of focus advances the adaptation of binder jetting for sustainable construction with a focus on possibly using coarse powders from waste streams to support the future of material reuse. By incorporating an image-based monitoring system into binder jetting, we achieved higher adaptability and robustness of binder jetting and opened new pathways towards using more variable and irregular powder feedstock while assuring print cohesiveness. Our main contribution is in integrating additive manufacturing with the image-based monitoring system while focusing on variable material properties and more complex printing setups. Parameters such as the thickness of the print layers, fine-tuning the binder spacing, and absorption were studied to increase the precision of printing patterns and optimal green strength of the green bodies during printing. The flaw detection and correction framework was achieved by developing the image-based monitoring system feedback. The core innovation of this system is in training and integrating the deep learning architecture of U-Net. This process is central for the detection, real-time feedback comparison, and rectification of printing discrepancies, and ensures adaptability of printing with more diverse powder composites, as well as variability of layer parameters.

The primary benefits of the system, such as economization of the consumption of binder, yielding lightweight structures, and refining shape precision, were empirically validated through a series of experimental trials. By advanced binder jetting techniques, combined with image-based monitoring we succeeded in improving the accuracy of printing by 2.6% while decreasing the amount of material usage by 28.5%.

Projecting into the future, this methodology supports two primary directions of AM impact: firstly, more effective integration and automation of flaw detection in manufacturing, and secondly, enabling a wider use of more diverse materials for binder jetting, such as waste streams, in uneven printing conditions.

Acknowledgments. This research was made possible with funding from the Carnegie Mellon University Manufacturing Futures Institute for the research project CRuMBLE: Construction Rubble Manufacturing for Building Life-cycle and Environment (PIs: Dana Cupkova and Josh Bard). To improve the clarity and grammatical accuracy of the textual content, we utilized the capabilities of ChatGPT for copy-editing purposes only. It is important to note that ChatGPT was not involved in the creation of the original text, experimental design, coding, or diagram creation aspects of this study.

References

1. Wan, L., Liu, J., Bard, J., Cupkova, D.: Towards adaptive additive manufacturing: Image-based monitoring for binder jet 3D printing of coarse composite concrete powders. In: Proceedings of the 43rd Annual Conference of the Association for Computer Aided Design in Architecture (ACADIA 2023), Denver, USA, pp. 482–493 (2023).
2. Sachs, E., Haggerty, J., Cima, M., Williams, P.: Three-Dimensional Printing Techniques, US5340656 A; 1993
3. Wohlers, T.: 3D printing and additive manufacturing state of the industry. Annual Worldwide Progress Report. Wohlers Associates (2014).
4. Valente, M., Sibai, A., Sambucci, M.: Extrusion-based additive manufacturing of concrete products: revolutionizing and remodeling the construction industry. J. Compos. Sci. **3**(3), 88 (2019)
5. Cesaretti, G., Dini, E., De Kestelier, X., Colla, V., Pambaguian, L.: Building components for an outpost on the Lunar soil by means of a novel 3D printing technology. Acta Astronaut. **93**(2014), 430–450 (2014)
6. Lloret, E., et al.: Complex concrete structures. Comput. Aided Des. **60**, 40–49 (2015). https://doi.org/10.1016/j.cad.2014.02.011
7. Nuh, M., Oval, R., Orr, J., Shepherd, P.: Digital fabrication of ribbed concrete shells using automated robotic concrete spraying. Addit. Manuf. **59**, 103159 (2022). https://doi.org/10.1016/j.addma.2022.103159
8. Cunha, F.G., Santos, T.G., Xavier, J.: In situ monitoring of additive manufacturing using digital image correlation: A review. Materials. **14**(6), 1511 (2021)
9. Reiter, L., Wangler, T., Roussel, N., Flatt, R. .J.: The role of early age structural build-up in digital fabrication with concrete. Cem. Concr. Res. **112**, 86–95 (2018). https://doi.org/10.1016/j.cemconres.2018.05.011
10. Özgenel, Ç.F., Gönenç Sorguç, A.: Performance comparison of pretrained convolutional neural networks on crack detection in buildings. In: Isarc. proceedings of the international symposium on automation and robotics in construction, vol. 35, pp. 1–8. IAARC Publications, (2018).

11. Cao, Y., Pang, D., Yan, Y., Jiang, Y., Tian, C.: A photovoltaic surface defect detection method for building based on deep learning. J. Build. Eng. **70**, 106375 (2023). https://doi.org/10.1016/j.jobe.2023.106375
12. Kalfarisi, R., Wu, Z.Y., Soh, K.: Crack detection and segmentation using deep learning with 3D reality mesh model for quantitative assessment and integrated visualization. J. Comput. Civ. Eng. **34**(3), 04020010 (2020). https://doi.org/10.1061/(asce)cp.1943-5487.0000890
13. Ko, P., Prieto, S.A., de Soto, B.G.: ABECIS: an automated building exterior crack inspection system using UAVs, open-source deep learning and photogrammetry. In: ISARC. Proceedings of the International Symposium on Automation and Robotics in Construction, vol. 38, pp. 637–644. IAARC Publications, (2021).
14. Harichandran, A., Raphael, B., Mukherjee, A.: Equipment activity recognition and early fault detection in automated construction through a hybrid machine learning framework. Comput. Aided Civ. Infrastruct. Eng. **38**(2), 253–268 (2022). https://doi.org/10.1111/mice.12848
15. Hosamo, H., Svennevig, P.R., Svidt, K., Han, D., Nielsen, H.K.: A Digital Twin predictive maintenance framework of air handling units based on automatic fault detection and diagnostics. Energ. Buildings. **261**, 111988 (2022). https://doi.org/10.1016/j.enbuild.2022.111988
16. Zhang, J., Qian, S., Tan, C.: Automated bridge surface crack detection and segmentation using computer vision-based deep learning model. Eng. Appl. Artif. Intell. **115**, 10522 (2022). https://doi.org/10.1016/j.engappai.2022.105225
17. Quigley, M., et al.: ROS: an open-source Robot Operating System. In: ICRA workshop on open source software, vol. 3, no. 3.2, p. 5 (2009_.
18. https://github.com/ros-industrial/abb_driver
19. Rublee, E., Rabaud, V., Konolige, K., Bradski, G.: ORB: An efficient alternative to SIFT or SURF. In: 2011 International Conference on Computer Vision, pp. 2564–2571. IEEE (2011)
20. Lowe, D.G.: Distinctive image features from scale-invariant keypoints. Int. J. Comput. Vis. **60**, 91–110 (2004)
21. Ronneberger, O., Fischer, P., Brox, T.: . U-net: Convolutional networks for biomedical image segmentation. In Medical Image Computing and Computer-Assisted Intervention – MICCAI 2015: 18th International Conference, Munich, Germany, October 5–9, 2015, Proceedings, Part III (Vol. 18, pp. 234–241). Springer International Publishing (2015).
22. Wang, Z., Bovik, A.C., Sheikh, H.R., Simoncelli, E.P.: Image quality assessment: From error visibility to structural similarity. IEEE Trans. Image Process. **13**(4), 600–612 (2004)

Toolpath Design Calibration for Robotic Spatial Printing with Supervised Learning

Sulaiman Alothman[1(✉)], Hanspeter Pfister[2], and Martin Bechthold[3]

[1] Harvard University, Cambridge, United States
alothman.sulaiman@gmail.com
[2] Graduate School of Design, Harvard University, Cambridge, United States
[3] School of Engineering and Applied Sciences, Harvard University, Cambridge, United States

Abstract. Carbon emissions from construction contribute significantly to global emissions. Advancements are needed to lower the carbon footprint of existing materials, but at the same time we need to find more radical alternative methods and materials. Extrusion 3D printing of natural materials offers a promising sustainable alternative for contemporary construction. However, the nonlinear material behavior inherent in that process presents significant challenges for producing accurate architectural components. The material is wet during and immediately after printing, then deforms, dries, and shrinks at inconsistent rates that make accurate final forms extremely challenging to achieve. In this paper, we introduce a novel method for predicting the toolpath geometries on a layer-by-layer basis using artificial neural networks (ANNs). Tested in the context of clay lattice printing, an unorthodox extrusion scenario characterized by a large feature space and high material uncertainty, our proposed method demonstrates the ability to evaluate and calibrate the toolpath geometry of clay lattices with sufficient accuracy for manufacturing. The work presents an important step toward next-generation solutions for paste-based 3D printing.

Keywords: Paste-Based 3D Printing · Robotic Spatial Printing · Toolpath Design · Neural Networks

1 Introduction

Recent advances in additive manufacturing of earth-based materials indicate promising opportunities for sustainable building construction (Fratello & Rael, 2020; Dritsas et al., 2019). However, the inherent property variations of these materials and their complex behavior during the extrusion and drying process pose significant challenges when attempting to produce accurate building components. Traditionally, designers have sought to understand these fabrication uncertainties through trial-and-error prototyping; they address excessive tolerances through the allocation of safety margins or appropriate detailing strategies. But these common approaches are time-consuming and costly. Advanced simulation methods such as finite-element analysis (FEA) cannot presently capture the highly nonlinear process of 3D printing a variable paste-based material: the

M. Yablonina et al. (Eds.): Robotic Fabrication in Architecture, Art and Design 2024, pp. 372–388, 2026.
https://doi.org/10.1007/978-3-032-22607-5_24

extruded material layer dries and shrinks at inconsistent rates while new layers continue to be deposited, thereby imposing increasing self-weight onto the lower layers in a highly nonlinear process. This affects the structural integrity of the built component and, in the worst-case scenario, causes collapse or major failure during the printing process. Even successful prints might feature large geometrical discrepancies from the intended design, affecting the function, appearance, or assembly process of the component.

Some existing research on paste-based 3D printing uses machine learning models as a means to optimize the process parameters to achieve design solutions with minimal material deformation (Matas et al., 2022; Conev et al., 2020) (Fig. 1). This imposes limits on design space exploration and confines the user within a limited set of solutions. Other approaches use ML to predict a uniform deviation factor based on a global discrepancy measure between the intended design and the printed outcome (Ma et al., 2021) (Fig. 2). This is unnatural for the variable behavior of paste material, as layers keep deforming and shrinking at inconsistent rates after printing subsequent layers. To clarify, current models are often not built to predict geometrical deviations on a layer-by-layer basis, which is important for accurate printing throughout. In addition, geometrical deviations caused by the variable displacement of the printed layers during printing are inevitable, and they are often nonuniform across layers. Thus, prediction models for those deviations, reflecting nonlinear material behavior and other fabrication-related factors (e.g., machine vibration and joint movements, nozzle misalignment, extrusion pressure, and more), become essential for calibrating the toolpath design before printing, ensuring accurate placement for each layer.

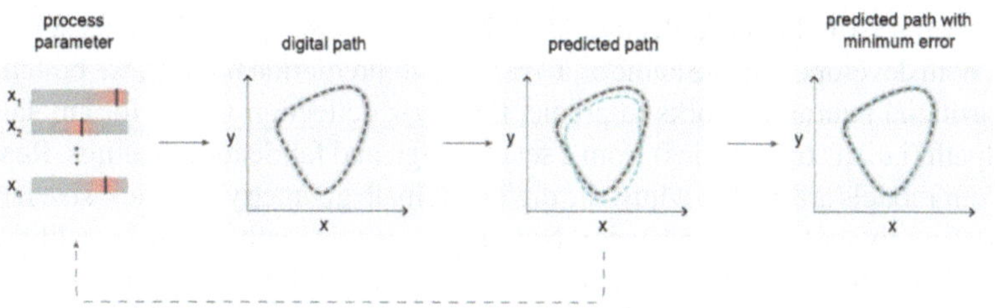

Fig. 1. One strategy for geometric compensation involves adjusting the process parameters until the error between the digital and predicted toolpaths is minimized.

The lack of substantial assessment of geometrical deviation on a layer-by-layer basis is problematic. High-resolution geometrical data regarding material deformation is critical for creating machine learning models for design evaluation and calibration for printing. However, paste-based extrusion processes involve high-dimensional feature spaces, requiring substantial data for accurate and generalizable modeling and, thus, resulting in a time-intensive data acquisition process.

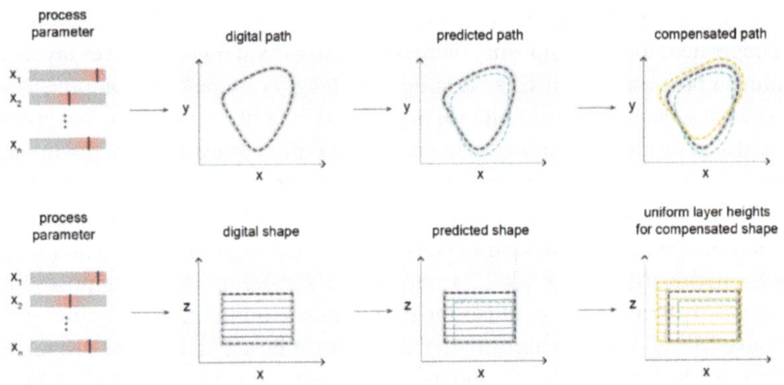

Fig. 2. Another strategy for geometric compensation involves assigning a uniform compensation factor based on a predicted shape.

This paper introduces a method for building prediction models based on neural networks to simulate and calibrate the toolpath design on a layer-by-layer basis to ensure accurate printing throughout. Our goal is to prove the method's effectiveness, robustness, and generalizability by focusing on an unorthodox and challenging case of printing: clay lattices. Lattices minimize material usage compared to solid layer deposition (Fig. 3). They also involve severe deformation and deflections (Fig. 4), thereby serving as a suitable microcosm for validating our method to capture the variability in nonlinear printing processes. The method capitalizes on an optimized dataset collected from a 3D scanning process (AlOthman & Bechthold, 2022) along with a proactive data selection workflow, both developed by the authors. To realize our prediction models, we built three different artificial neural networks to predict the x-y-z deflection values for any lattice unit's toolpath (i.e., its trajectories) from a set of design and fabrication features. Results show that our models are able to adjust the digital toolpath geometry for brick-size lattice structures of up to six layers (or levels). Not only does this method offer a calibration tool for highly nonlinear printing processes that is significant for large-scale sustainable robotic additive manufacturing, but it also provides designers with a rapid visualization tool to navigate their design spaces with real-time feedback about the material outcomes, empowering the design decision process.

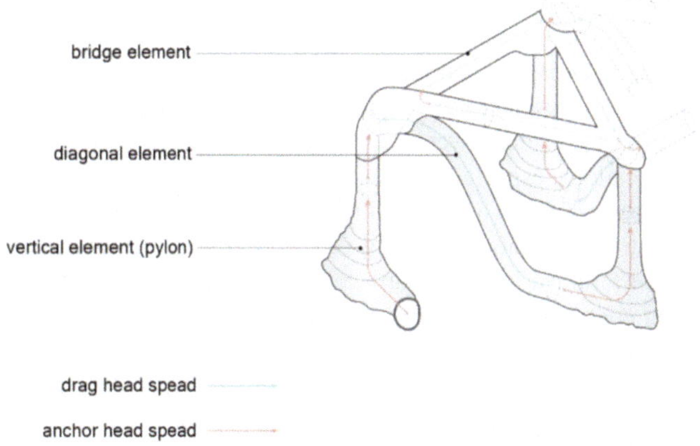

Fig. 3. The lattice unit topology features three key elements.

Fig. 4. Various cases of uncontrolled deformation caused by the material deflection under self-weight.

2 Related Work

Strategies for predicting and compensating for geometrical deviations in paste-based extrusion processes can be divided into two approaches. The first follows a functional form using explicit mathematical formulation for parameter estimation through least square fit or Bayesian analysis. The second uses machine learning models (often referred to as "black box" models) to minimize a loss function (i.e., minimize the error between the true value and the predicted value) without an analytical solution.

Lee et al. (2019) proposed an analytical model to predict the geometric characteristics of cylindrical scaffolds made from polycaprolactone (PCL), specifically the strut diameter, pore size, and porosity. Their results revealed the effectiveness of their model to predict those characteristics. However, their model worked under the assumption that the process parameters for printing were constant. This assumption does not work for paste extrusion applications, especially when the material being printed has properties that change under stress—known as shear thinning behavior—when building multiple layers. In other additive applications, Huang et al. (2014) used a Bayesian approach to estimate parameters for various standard geometries. They presented geometrical deviations in a polar coordinate system for simplification and efficient geometric compensation. Similarly, Hong et al. (2021) utilized a parameter estimation method to predict the cross section of angled struts in truss-based lattice structures. Nevertheless, those methods become challenging to apply on complex geometries.

Zhao et al. (2022) employed a point-wise geometric deviation approach using a convolutional neural network (CNN) for predicting deformation of 3D printed dental crowns. They predicted x-y-z deviations for all sampled points of the 3D model, which were then added to the original sampled points to generate a predicted 3D model. However, their data was simulated and the sampling for their 3D models was low resolution. Ashrafi et al. (2021) used a linear regression model to predict the nozzle distance to the previous layer (representing the vertical deformation) for different numbers of layers and interlayer time intervals. Their study was limited to a few variables and simple rectangular geometry, which might not fully reflect the feature space of a nonlinear application. Most of the aforementioned studies, and others, capture data after the actual print is fully complete. However, during printing, the layers keep deflecting after stacking multiple new layers due to shrinkage, increased self-weight, nozzle maneuvering, extrusion pressure, and other material- and machine-related factors, making calibration difficult to predict for each layer post-printing. Our method focuses on using data about individual layer information directly after printing it. By doing so, the toolpath can be calibrated for each layer before printing for correct material placement.

3 Method

3.1 Spatial Printing for Clay Lattices and Required Calibration

3D printing clay lattice structures involves an unorthodox process called "spatial printing trajectories" developed by the author (AlOthman et al. 2019). The lattice geometry leads to a high degree of fabrication uncertainty due to the overall delicate form that is subject to large deformations throughout the printing and drying process. The novel processes and methods developed and presented here can also be applied to other, more conventional morphologies. Printing clay lattices involves alternating two sets of controlled material behavior: anchoring and dragging to build a lattice unit (Fig. 3). These behaviors are achieved by carefully controlling fabrication parameters, such as the printing head speed and the material extrusion rate. The robotically-guided extrusion system prints a lattice unit by first depositing or compacting material at a low speed and a high extrusion rate in a singular vertical direction (i.e., anchoring) to create a pylon. Then, the material is dragged horizontally, in a curvilinear manner conforming to the material's fluid propensity to preempt shear stress, adjacent to the initial pylon at a high head speed and a low extrusion rate. The pylons are then braced with ties using a horizontally secluded extrusion with low head speed at the pylon peaks to induce anchoring and subsequent high head speed to simulate bridges across the void between the peaks. The toolpath calibration involves adjusting the start position of the horizontal ties, or bridges, for each lattice unit according to the anticipated deflection of the vertical pylons. This ensures that each subsequent lattice level is accurately built up (Fig. 5).

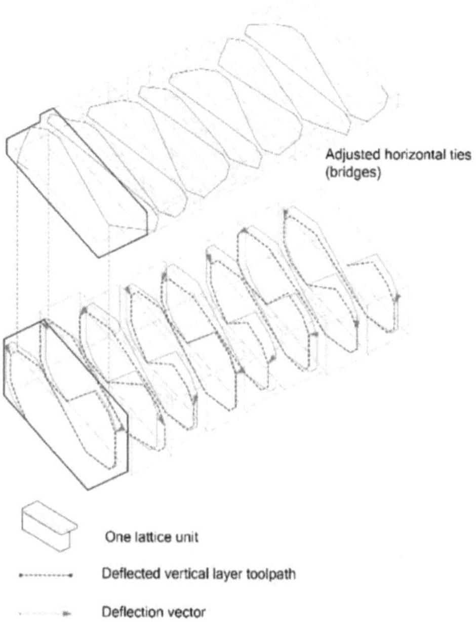

Fig. 5. Toolpath recalibration of horizontal ties (or bridges) based on the vertical layer's deflection vectors.

3.2 Clay Lattice Features and Dataset

The pylons' deflection vectors (or response variables) are recorded for each lattice unit after printing and 3D scanning of each lattice level (Fig. 6, top). These vectors are calculated by selecting the top surface (or interest boundary) around each pylon, where the mesh vertices are averaged to establish an average deflection vector for each pylon (Fig. 6, bottom). The averaged deflection vectors are exported with the input features of that level to a.csv file. The defined input features with their ranges (as shown in Table 1) include: lattice unit dimension of width (LUW) (10–50 mm), length (LUL) (10–100 mm), and height (LUH) (10–40 mm); anchor head speed (AHD) (3–20 mm/s) and drag head speed (DHS) (10–80 mm/s); extrusion rate controlled by the stepper motor (RPM) (20–120 rpm); stepping distance of each level (SD) (−5–5 mm); lattice unit level index (FL) (1–8); and nozzle diameter (ND) (2–10 mm). The parameters' ranges were defined based on knowledge obtained from previous experimental studies on lattice clay printing (Im et al., 2018). We printed different lattices structures, with different numbers of levels, following a design experiment approach that optimized the selection of samples required to approximately cover the nonlinear feature space of lattices that are printable. Printability refers to how closely the actual print matches the intended design (Fu et al., 2021). In our case, we evaluated printability using geometric quantitative and qualitative measures pertaining to lattice structures; this process is covered in different research by the author. Our dataset consists of 244 lattice levels. Each level has 10 lattice units, and each unit has 2 deflection vectors.

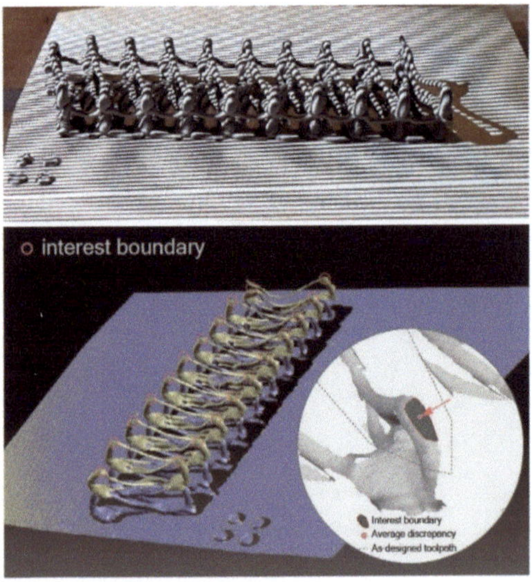

Fig. 6. Top: 3D structured light scanning after printing each pylon layer. Bottom: Interest boundaries around the scanned vertical pylons are selected with a customized processing algorithm; an average discrepancy vector translates the next layer and adjusts the next toolpath.

Within our dataset, there are multiple data points that share identical feature values. These identical points correspond to each lattice level, where each level is constructed with 10 lattice units. However, the deflection vector for each lattice unit within the same level can vary (Fig. 7), while their input features remain the same. These varying deflection vectors reflect the natural variability arising from uncontrolled factors related to the material extrusion process, inherent material variation, and other process-related noise. Although it is common practice to average the responses for identical feature values in most situations, in the paste-based extrusion process, all the data points are considered due to the reasons mentioned above.

Table 1. Data summary for one lattice level.

LUW (mm)	LUL (mm)	LUH (mm)	AHS (mm/s)	DHS (mm/s)	RPM (rpm)	SD (mm)	FL (index)	ND (mm)
14	28	15	8	30	56	0	2	4
data point index	Discrepancy – x (mm)		Discrepancy – y (mm)			Discrepancy – z (mm)		
1	3.35		−0.79			−5.19		
2	−3.95		−1.43			−3.62		
3	4.08		−0.94			−2.78		
⋮	⋮		⋮			⋮		
20	−3.56		−4.17			−3.82		

Fig. 7. Difficulty in controlling process parameters, such as temperature, humidity, machine vibration and jittering, and consistent air pressure in material extrusion, can cause variations in deflection among the lattice units across the same lattice level.

3.3 Data Analysis

We performed data analysis on our dataset to enhance the effectiveness of the modeling process. To identify and quantify the linear relationship between each pair of continuous variables (i.e., input features and response variables), we calculated Pearson correlation coefficients and visualized them on a heatmap (Fig. 8). Among the input lattice features, the lattice unit height (LUH) has the highest correlation (negative) with the Z deflection value. Thus, our training and testing sets for modeling were stratified based on LUH. The other features have no strong or apparent linear relationship with the Z deflection, suggesting that such a relationship in most of the features cannot be modeled with linear modeling techniques. Consequently, we decided to use artificial neural networks (ANNs) for modeling such nonlinear relationships. There is no high correlation among any pairs of input features, indicating their suitability for modeling. When two or more features are highly correlated—called multicollinearity—it can result in unstable estimates of the model parameters that can change erratically in response to small changes in the model or data. There is no strong correlation among the deflection values. Based on this finding, we trained three different models to predict each deflection value, as a single model might need help capturing the variability in each deflection value, thus, reducing the accuracy of the prediction.

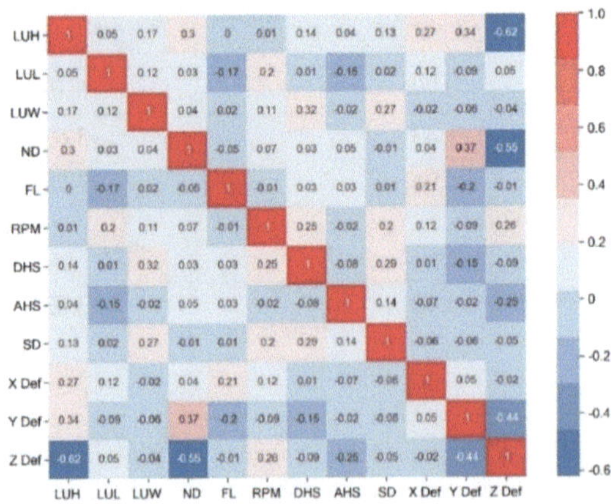

Fig. 8. A heatmap visualizing the Pearson correlation between the input features and the output features (deflection values) of the lattice dataset.

3.4 Predicting Deflection Vectors with ANNs

ANNs are nonlinear data modeling tools used to model and learn the highly nonlinear relationships between inputs and outputs in a dataset (Krogh, 2008). ANNs can have multiple hidden layers, which make them deeper. Studies (Leshno et al., 1993; Hornik et al., 1989) demonstrate that a two-layer-deep neural network with enough width can approximate any target function. Based on these studies, we built an ANN for each deflection value with two hidden layers of dense type, consisting of 20 and 15 nodes, respectively (Fig. 9). We used a rectified linear unit (ReLU) function for all hidden nodes, which outputs the maximum between 0 and the input value (x). The input layer has 9 nodes, corresponding to the number of lattice features, and an output layer with a size of 1 activated by a linear function that results in a continuous value representing one of the deflection values. We used mean absolute error (MAE) as the loss function to minimize, which is computed as the average of the absolute differences between the actual and predicted values. MAE is a more robust measure, as it does not heavily penalize large errors compared to other loss functions. Moreover, it is more interpretable since it utilizes the same scale and units as the original data. In the case of predicting geometrical deflection, MAE directly corresponds to the average absolute deflection error, measured in the same units as the deflection values. We used the Adam optimizer, which is based on gradient descent through backpropagation to update the weight of the hidden nodes towards minimizing the loss function. The implementation of our ANNs was carried out using Keras API, which is part of TensorFlow library in Python. The input features of the samples were standardized to have the same scale to prevent high-range features from biasing the learning process. It is important to note that the training data was standardized first, and then the test data was standardized using the parameters derived from standardizing the training data. This prevents any leakage of information from the training set to the test set, which could lead to overly optimistic (biased) model performance.

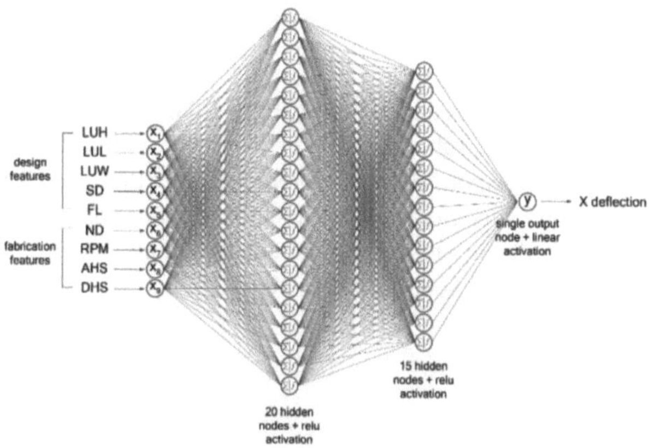

Fig. 9. A diagram showing the proposed neural network architecture for predicting one of the deflection coordinate values.

We trained our ANNs and evaluated them over 100 epochs for MAE training error and recorded the generalization error (test MAE). We compared the ANNs with linear regression models to establish that linear modeling is insufficient for capturing the geometrical changes caused by material deflection in clay lattice printing. Our results showed that the ANN outperformed the linear regression model in both training and test performance for the three deflection models (Fig. 10). For X deflection, the ANN had training and test losses of 1 mm and 1.11 mm, respectively. In comparison, the linear model reported losses of 1.5 mm and 1.51 mm. It achieved losses of approximately 0.8 mm and 0.81 mm respectively for the Y deflection model, compared to just over 1 mm and 1.03 mm for the linear model. The ANN was also superior in predicting Z deflection, with a training and test loss of about 1.12 mm compared to the linear regression model's losses of approximately 2.65 mm and 2.63 mm, respectively.

3.5 Model Deployment and Testing

We integrated our best performance models into Grasshopper3D through a Python development server using Flask. Client requests can be sent from Grasshopper3D. The server receives JSON data containing lattice design and fabrication features and sends it to the neural networks, which generate predictions for the requested lattice features. To account for the inherent variability and minor fluctuations in the printing process, a small random

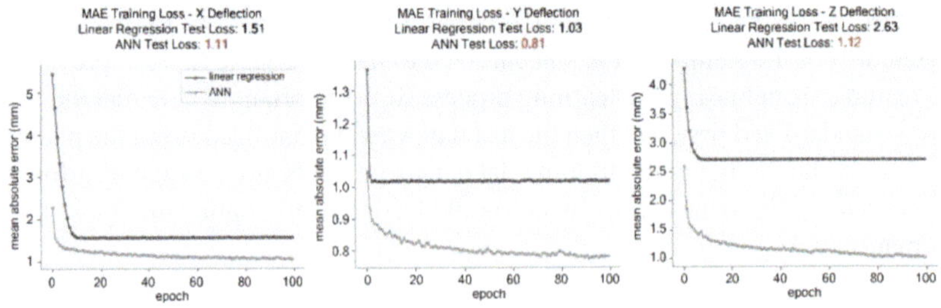

Fig. 10. The training loss of the linear regression and ANN models across 100 epochs.

normal noise (less than 1 mm) is added to predictions. This added noise does not compromise the printability of the subsequent layer, as it falls within the acceptable 1 mm tolerance for lattice printing application. The predictions are then used for toolpath visualization and calibration. This enables designers to fine-tune toolpath geometries and receive instant visual feedback on geometrical changes. This integration enhances the design process in 3D modeling software (Fig. 11).

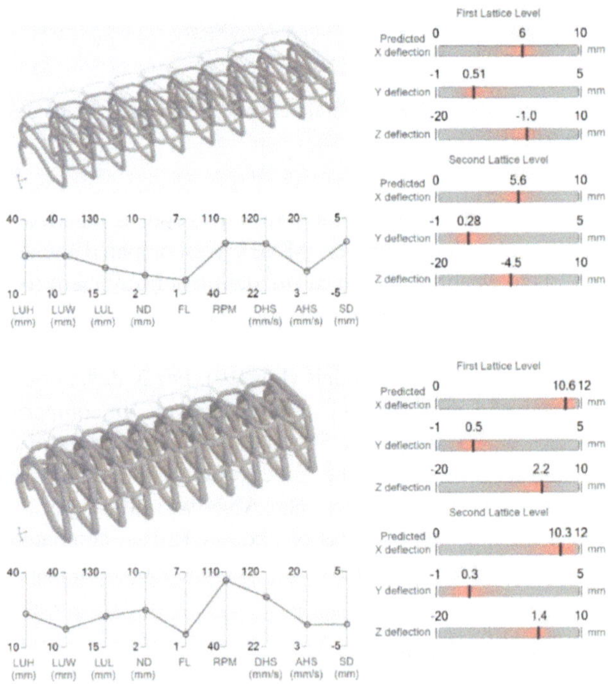

Fig. 11. The trained neural network is deployed in the Grasshopper environment, providing rapid toolpath prediction, calibration, and visualization for different lattice designs.

We evaluated the performance of our trained model by predicting and calibrating the toolpath design of a lattice structure with five levels. We selected a high LUH and a large ND, as we expect them to significantly impact Z deflection—this was based on our correlation analysis (Fig. 8). Though the predictions showed high Z deflection, the lattice structure was printed successfully based on those predictions that were used to calibrate the toolpath geometry for printing (Fig. 12). We understand that these predictions are based on a single average deflection value for a pylon, given specific lattice design and fabrication features. As a result, the calibration of each level may be influenced, as we noticed that the printed structure slanted towards the extrusion direction.

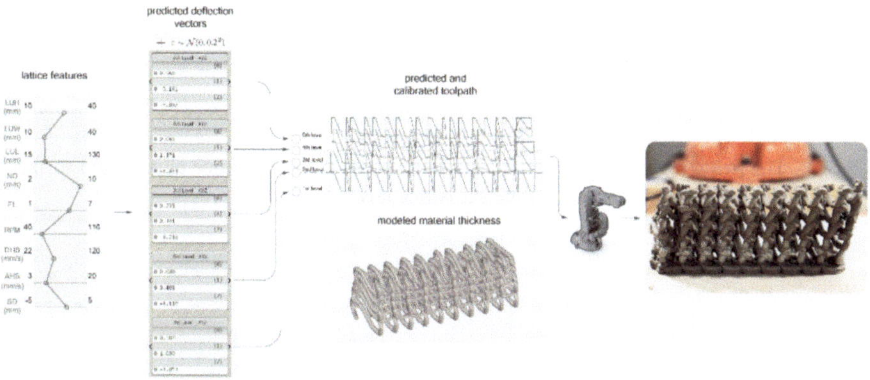

Fig. 12. ANN deployment in Grasshopper3D enables rapid visualization and calibration of toolpath design geometries for printing with sufficient accuracy.

4 Results and Discussion

We have successfully trained neural network models that accurately predict geometrical changes in any lattice unit toolpath, with an error margin of $+-1$ mm on a layer-by-layer (level-by-level) basis across all three coordinate values. These trained models, specific to each deflection coordinate, were seamlessly integrated into 3D modeling software, providing rapid feedback on the precise spatial placement of material by adjusting toolpath trajectories at each lattice level. The modeling approach for training in this study differs from other additive manufacturing processes that deal with geometrical compensation prediction as mentioned earlier. The data analysis on our clay lattice dataset showed that the three deflection coordinates were not highly correlated. This lack of strong correlation suggests that the values for each deflection coordinate exhibit distinct variability patterns, indicating the necessity of training separate models to accurately capture and learn these individual patterns.

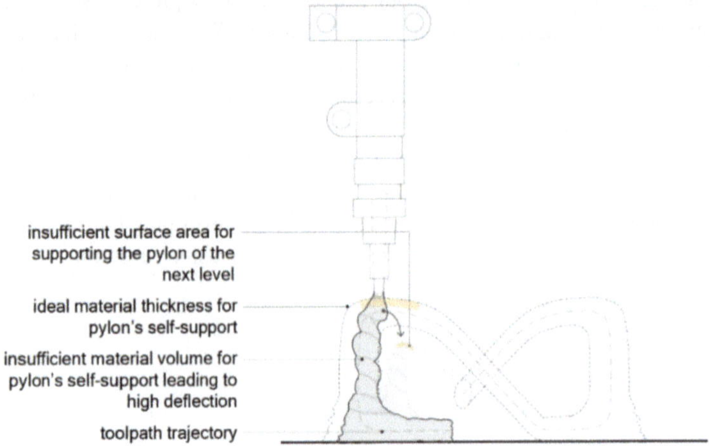

Fig. 13. The pylon serves as the primary structural support for the lattice unit. Its top surface is crucial for the successful printing of the next lattice level.

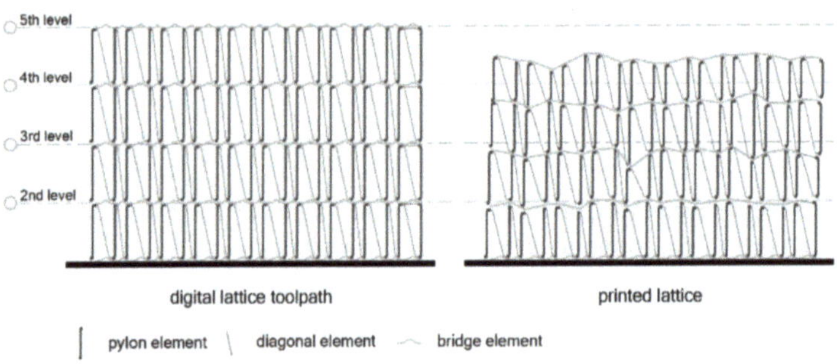

Fig. 14. The pylons of the next level are adjusted to anchor correctly on the ones already printed. The diagonal and bridge elements, meanwhile, serve as lateral supports for the pylons and play no role in the calibration process for layer placement.

The main challenge in clay lattice printing relates to accurately anchoring the pylon element, which plays a crucial role in maintaining the as-designed height of the lattice level and serves as the primary structural support for the subsequent horizontal bracing layer (Fig. 13). Consequently, the pylon element stands as the most significant component in the lattice morphology, directly influencing the success of the lattice printing process. However, the other two lattice unit elements—namely, the diagonal element that braces two opposite pylons and the bridge element that braces two adjacent pylons—also contribute to the stability and structural integrity of the pylon. While these elements are not essential for calibrating the placement of the next layer (since the next layer is consistently deposited on the pylon, as shown in Fig. 14), predicting their deflection is still essential to understand their impact on the pylon's behavior. For instance, if these elements are too thin due to high drag induced by the head speed or low material extrusion rate, they might become susceptible to snapping or breaking, resulting in pylon deflection and bending (Fig. 15). Therefore, having access to geometric information related to the deflection of the diagonal and bridge elements is essential and needs to be addressed in future research to further optimize the clay lattice printing process.

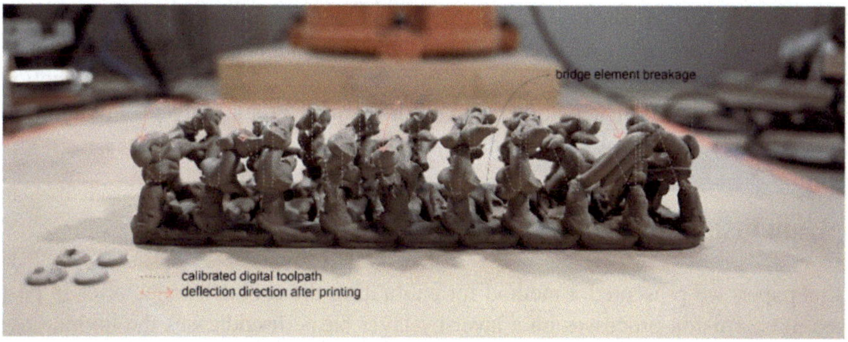

Fig. 15. Pylon deflection occurs due to insufficient lateral stability, often caused by the breakage or discontinuities in the diagonal or bridge elements.

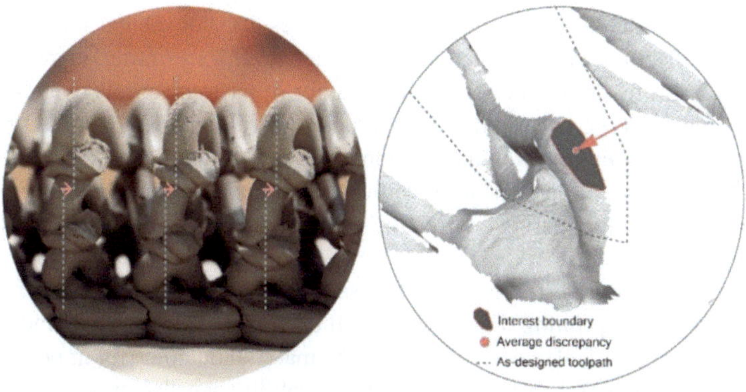

Fig. 16. Because the 3D scan of the built lattice captures only half of the pylon's surface—due to scanning from a single angle—the average discrepancy between the as-designed and as-built models can cause the calibrated trajectory to be off-center.

In our approach, a single 3D point coordinate is predicted based on the input lattice design and fabrication parameters' values. This 3D point represents the average point of the highest peak vertices selected around the interest boundaries, obtained through a customized processing algorithm (as described in another research study by the author). Although this simplification has led to excellent model accuracy and successful printing, it should be noted that the calculated average point might not precisely correspond to the center of the printed pylon. As a result, the placement of the next layer may be off-center (Fig. 16), potentially impacting both the structural integrity (due to the misalignment of load transfer) and design intent of the lattice. To address this limitation and enhance the prediction process, an extension can be made by adopting the 3D point cloud prediction model proposed by Zhao et al. (2022) and Lee et al. (2019) for geometric compensation in additive manufacturing applications. Specifically, instead of predicting a single point, the model can be modified to predict the entire selected interest boundary. This would offer several benefits to the designer or user, allowing for more precise placement of the next anchor and providing additional design flexibility to make adjustments for the next layer's placement. Moreover, this extended approach can also be applied to conventional 3D printing processes that involve stacking horizontal bracing layers. Predicting the shape of the top surface of the deposited layer can have similar advantages.

5 Conclusion and Outlook

In this paper, we presented a method for calibrating the toolpath trajectories in paste-based 3D extrusion processes on a layer-by-layer basis. It addresses the limitations of current methodologies that focus solely on optimizing process parameters or the allocation of uniform compensation measures, without considering the contribution of each layer's deflection to overall geometric changes for precise printing. Our method uses high-resolution geometrical data for each printed layer acquired from a 3D scanning process for clay lattices. We utilized neural networks to predict geometric deviation caused by material deformation (deflection vector), based on a defined set of design and fabrication parameters for lattice printing. Our results showed that such models predicted layer deflection with an acceptable tolerance level for printing, thereby showing their capability in modeling the nonlinear behavior of paste materials in additive manufacturing. Furthermore, we deployed the trained neural network into a 3D modeling environment using a Python web-application library. This integration enables real-time insights and iterative refinement of the lattice printing process, thereby enhancing the design workflow and facilitating the creation of more precise and customized geometries for printing.

Our proposed method in this paper focuses mainly on calibrating material placement for accurate 3D printing. Future work will extend this capability by including the prediction of a geometrical mesh representing the material outcome for improved visualization and evaluation of the final product. This capability can also provide tolerance for toolpath design adjustments. Adjusted toolpath trajectories currently receive a volume reflecting the material thickness based on the nozzle used, yet this does not accurately represent the actual material outcome. Precise information is crucial for both the design and functional analysis of parts and the assessment of aesthetic qualities. These factors have the potential to influence the design decision process significantly. Deep learning techniques offer potential solutions for addressing these challenges and will be explored in future research. In addition, we will focus on testing our method on biomaterials, other shapes featuring different design details, and the more common xy printing technique.

References

AlOthman, S., Bechthold, M.: Non-linear fabrication. In: Proceedings of ACADIA 2022: Hybrids and Haecceities, pp. 68–75. University of Pennsylvania, USA (2022)

AlOthman, S., Im, H.C., Jung, F., Bechthold, M.: Spatial print trajectory: controlling material behavior with print speed, feed rate, and complex print path. In: Robotic Fabrication in Architecture, Art and Design 2018: Foreword by Sigrid Brell-Çokcan and Johannes Braumann, Association for Robots in Architecture, pp. 167–180. Springer International Publishing, Cham (2019)

Ashrafi, N., Nazarian, S., Meisel, N.A., Duarte, J.P.: Experimental prediction of material deformation in large-scale additive manufacturing of concrete. Addit. Manuf. **37**, 101656 (2021). https://doi.org/10.1016/j.addma.2020.101656

Conev, A., Litsa, E.E., Perez, M.R., Diba, M., Mikos, A.G., Kavraki, L.E.: Machine learning-guided three-dimensional printing of tissue engineering scaffolds. Tissue Eng. Part A. **26**(23–24), 1359–1368 (2020). https://doi.org/10.1089/ten.TEA.2020.0191

Dritsas, S., Vijay, Y., Dimopoulou, M., Sanadiya, N., Fernandez, J.G.: An additive and subtractive process for manufacturing with natural composites. In: Willmann, J., Block, P., Hutter, M., Byrne, K., Schork, T. (eds.) Robotic Fabrication in Architecture, Art and Design 2018, pp. 181–191. Springer International Publishing (2019). https://doi.org/10.1007/978-3-319-92294-2_14

Fratello, V.S., Rael, R.: Innovating materials for large scale additive manufacturing: Salt, soil, cement and chardonnay. Cem. Concr. Res. **134**, 106097 (2020). https://doi.org/10.1016/j.cemconres.2020.106097

Fu, Z., Naghieh, S., Xu, C., Wang, C., Sun, W., Chen, X.: Printability in extrusion bioprinting. Biofabrication. **13**(3), 033001 (2021). https://doi.org/10.1088/1758-5090/abe7ab

Hong, R. et al.: Artificial neural network-based geometry compensation to improve the printing accuracy of selective laser melting fabricated sub-millimetre overhang trusses. Addit. Manuf. **37**, 101594 (2021). https://doi.org/10.1016/j.addma.2020.101594

Hornik, K., Stinchcombe, M., White, H.: Multilayer feedforward networks are universal approximators. Neural Netw. **2**(5), 359–366 (1989). https://doi.org/10.1016/0893-6080(89)90020-8

Huang, Q., Nouri, H., Xu, K., Chen, Y., Sosina, S., Dasgupta, T.: Statistical predictive modeling and compensation of geometric deviations of three-dimensional printed products. J. Manuf. Sci. Eng. **136**(6), 061008 (2014). https://doi.org/10.1115/1.4028510

Im, H.C., AlOthman, S., Garcia del Castillo, J.L.: Responsive spatial print. In: Proceedings of ACADIA 2018: On Imprecision and Infidelity, pp. 286–293. Association for Computer Aided Design in Architecture, Mexico City (2018)

Krogh, A.: What are artificial neural networks? Nat. Biotechnol. **26**(2), 195–197 (2008). https://doi.org/10.1038/nbt1386

Lee, J., Walker, J., Natarajan, S., Yi, S.: Prediction of geometric characteristics in polycaprolactone (PCL) scaffolds produced by extrusion-based additive manufacturing technique for tissue engineering. Rapid Prototyp. J. **26**(2), 238–248 (2019). https://doi.org/10.1108/RPJ-08-2018-0219

Leshno, M., Lin, V.Y., Pinkus, A., Schocken, S.: Multilayer feedforward networks with a nonpolynomial activation function can approximate any function. Neural Netw. **6**(6), 861–867 (1993). https://doi.org/10.1016/S0893-6080(05)80131-5

Ma, Y., Schutyser, M.A.I., Boom, R.M., Zhang, L.: Predicting the extrudability of complex food materials during 3D printing based on image analysis and gray-box data-driven modelling. Innovative Food Sci. Emerg. Technol. **73**, 102764 (2021). https://doi.org/10.1016/j.ifset.2021.102764

Matas, A., Molina-Montero, M.d.C., Igual, M., García-Segovia, P., Martínez-Monzó, J.: Printability prediction of three gels for 3D food printing. Biol. Life Sci. Forum. **18**(1), 25 (2022). https://doi.org/10.3390/Foods2022-12986

Vardi, G., Yehudai, G., & Shamir, O.: Width is Less Important than Depth in ReLU Neural Networks. arXiv. (2022). https://doi.org/10.48550/arXiv.2202.03841

Zhao, M. et al.: A point-based neural network for real-scenario deformation prediction in additive manufacturing. In: 2022 IEEE 18th International Conference on Automation Science and Engineering (CASE), pp. 1656–1661 (2022). https://doi.org/10.1109/CASE49997.2022.9926549

Zhu, Z., Ferreira, K., Anwer, N., Mathieu, L., Guo, K., Qiao, L.: Convolutional neural network for geometric deviation prediction in additive manufacturing. Procedia CIRP. **91**, 534–539 (2020). https://doi.org/10.1016/j.procir.2020.03.108s

Correction to: Robotic Fabrication in Architecture, Art and Design 2024

Maria Yablonina, Paul H. Harrison, Brady Peters, Jonathon Anderson, Nicholas Hoban, and Zachary Mollica

Correction to:
M. Yablonina et al. (Eds.): *Robotic Fabrication in Architecture, Art and Design 2024*, https://doi.org/10.1007/978-3-032-22607-5

In the original version of the book, the incorrect title has been updated. In the updated Version correct title "Robotic Fabrication in Architecture, Art and Design 2024" "Foreword by Sigrid Brell-Cokcan and Johannes Braumann, Association for Robotics in Architecture" has been updated.

The updated version of this book can be found at
https://doi.org/10.1007/978-3-032-22607-5

M. Yablonina et al. (Eds.): Robotic Fabrication in Architecture, Art and Design 2024, p. C1, 2026.
https://doi.org/10.1007/978-3-032-22607-5_25

Author Index

A
Al-Ariss, Shadi 231
Alothman, Sulaiman 371
Ariza, Inés 313
Arsenault, Marc 142

B
Baharlou, Ehsan 210
Bard, Joshua 354
Bechthold, Martin 371
Beites, Steven 142
Beyer, Katrin 253
Brell-Cokcan, Sigrid 34

C
Carl, Eppinger 65
Çapunaman, Özgüç Bertuğ 81
Chen, Hanzhi 299
Claßen, Fabian 34
Çobanoğlu, Bora 329
Cupkova, Dana 354

D
De Wolf, Catherine 313
Diamanti, Olga 174
Dillenburger, Benjamin 174
Donnelly, James 231
Dörfler, Kathrin 299
Dörstelmann, Moritz 115
Dubor, Alexandre 329

E
Engholt, Jon Krähling 192

F
Fivet, Corentin 253

G
Garret, Caelan 19
Gaudillière-Jami, Nadja 343
Gottschild, Lucas 329
Gramazio, Fabio 19, 313
Grangeot, Maxence 253
Gürsoy, Benay 81

H
Han, Isla Xi 47
Huang, Yijiang 19
Huyghe, Vincent 329

J
Jahn, Gwyllim 285
Johnson, Matthew 129

K
Kalousdian, Nicolas Kubail 129
Kamp, Carsten 34
King, Nathan 231
Kirner, Lukas 34
Kohler, Matthias 19, 313
Konrad, Sonne 65
Kragl, Philipp 3, 129
Kromoser, Benjamin 266

L
Leder, Samuel 3, 129
Leung, Pok Yin Victor 19
Leutenegger, Stefan 299
Liu, Jingyang 354

M
Maierhofer, Mathias 161
McDonald, Ethan 142
Mehdizadeh, Samim 98
Menges, Achim 3, 129, 161
Mette, Ramsgaard Thomsen 65
Mitropoulou, Ioanna 174
Mohseni, Alale 81

N
Naguib, Joseph 329
Newnham, Cameron 285
Nicholas, Paul 343

M. Yablonina et al. (Eds.): Robotic Fabrication in Architecture, Art and Design 2024, pp. 389–390, 2026.
https://doi.org/10.1007/978-3-032-22607-5

O
Olah, Eszter 115
Oraskari, Jyrki 34

P
Pantscharowitsch, Marc 266
Papandreou, Maria-Eleni 329
Parascho, Stefana 47, 253
Paul, Nicholas 65
Pfister, Hanspeter 371
Pigram, Dave 192

R
Rust, Romana 313
Ruxandra, Chiujdea 65

S
Saral, Begüm 299
Siebenaler, Pit 329
Siefert, Sandro 343
Silvestru, Vlad-Alexandru 313
Siriwardena, Lasath 3
Sollazzo, Aldo 329
Storey, Peter 231
Svilans, Tom 343

T
Taras, Andreas 313
Tessmann, Oliver 98
Tucker, Michael 129

V
van den Berg, Nick 285
Vaxman, Amir 174

W
Walter, Natalie 81
Wan, Linxiaoyi 354
Wang, Qianqing 253

Z
Zamani, Alireza 81
Zareian, Mehrdad 115
Zöcklein, Marit 34

MIX
Papier aus verantwortungsvollen Quellen
Paper from responsible sources
FSC® C105338

If you have any concerns about our products,
you can contact us on
ProductSafety@springernature.com

In case Publisher is established outside the EU,
the EU authorized representative is:
Springer Nature Customer Service Center GmbH
Europaplatz 3, 69115 Heidelberg, Germany

Printed by Libri Plureos GmbH
in Hamburg, Germany